HVACR 401

AIR CONDITIONING CONTRACTORS OF AMERICA
PHCC EDUCATIONAL FOUNDATION
REFRIGERATION SERVICE ENGINEERS SOCIETY

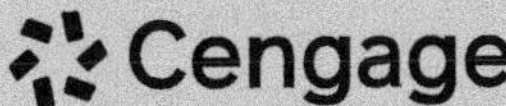

Australia • Brazil • Canada • Mexico • Singapore • United Kingdom • United States

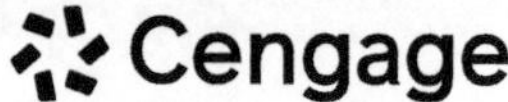

HVACR 401
John Hohman

Vice President, Technology and Trades Professional Business Unit: Gregory L. Clayton

Director of Building Trades: Taryn Zlatin McKenzie

Executive Editor: Robert Person

Product Manager: Vanessa L. Myers

Development: Nobina Preston

Director of Marketing: Beth A. Lutz

Marketing Manager: Marissa Maiella

Senior Production Director: Wendy A. Troeger

Production Manager: Andrew Crouth

Senior Content Project Manager: Kara A. DiCaterino

Senior Art Director: Benj Gleeksman

Library of Congress Control Number: 2011925449

ISBN-13: 978-1-4283-4002-2
ISBN-10: 1-4283-4002-5

Cengage
200 Pier 4 Boulevard
Boston, MA 02210
USA

Cengage is a leading provider of customized learning solutions with employees residing in nearly 40 different countries and sales in more than 125 countries around the world. Find your local representative at: **www.cengage.com.**

To learn more about Cengage platforms and services, register or access your online learning solution, or purchase materials for your course, visit **www.cengage.com.**

Notice to the Reader

Publisher does not warrant or guarantee any of the products described herein or perform any independent analysis in connection with any of the product information contained herein. Publisher does not assume, and expressly disclaims, any obligation to obtain and include information other than that provided to it by the manufacturer. The reader is expressly warned to consider and adopt all safety precautions that might be indicated by the activities described herein and to avoid all potential hazards. By following the instructions contained herein, the reader willingly assumes all risks in connection with such instructions. The publisher makes no representations or warranties of any kind, including but not limited to, the warranties of fitness for particular purpose or merchantability, nor are any such representations implied with respect to the material set forth herein, and the publisher takes no responsibility with respect to such material. The publisher shall not be liable for any special, consequential, or exemplary damages resulting, in whole or part, from the readers' use of, or reliance upon, this material.

Printed at CLDPC, USA, 11-23

Brief Contents

Contents

Preface

This is the fourth book in a 4-year series intended for the training of HVACR technicians, and the first title focused on the in-depth study of a specific HVACR topic. Suitable for use in formal educational or professional settings or for any apprenticeship or training-related program, the series emphasizes a blend of conceptual material and real-world applications.

This book was developed in partnership with the following organizations: Air Conditioning Contractors of America (ACCA), Plumbing-Heating-Cooling-Contractors—National Association Educational Foundation (PHCC Educational Foundation), and Refrigeration Service Engineers Society (RSES). These organizations appointed a team of subject-matter experts, composed of outstanding HVACR professionals and educators, who reviewed the material during development to ensure the curriculum reflects industry needs and takes an approach to which today's technicians can relate.

Throughout the series, topics are revisited in more depth and with additional applications to coincide with students' increasing work experience in the field. The aim of *HVACR 401* is to build upon the learning objectives students will have covered in *HVACR 101*, *201*, and *301*, and to give them an in-depth exploration of heat pumps. Manuscript and proofs were reviewed, improved, and approved by the subject-matter experts. The author has a rich background full of field and teaching experience, and was encouraged to follow a specific set of guidelines in addition to bringing in his own expertise.

Contractors and manufacturers have requested a training program germane to the requirements of the industry. This book and series respond to that need.

Acknowledgments

The publisher wishes to thank the following companies and individuals who provided illustrations and permission to reproduce them:

Air Conditioning Contractors of America
Amprobe® Test Tools
A.O. Smith Electrical Products Company
Carrier Corporation
Climate Control Technologies, Inc.
CoEnergies LLC, Traverse City, MI
Copeland Corporation
Emerson Climate Technologies
General Electric Company
Honeywell International Inc.
Jamie Simpson
NASA
Noranda Metal Industries, Inc.
Refrigeration Service Engineers Society
Shoffner Mechanical Services
Susan Brubaker
Tecumseh Products Company
Trane Inc.
WaterFurnace International

ACCA–PHCC Educational Foundation–RSES Subject Matter Experts

A special thank-you is extended to Merry Beth Hall for her time and dedication as the *HVACR 401* Project and Subject Matter Expert Committee Coordinator. Merry Beth is the Director of Apprentice & Journeyman Training with the PHCC Educational Foundation.

We would also like to recognize and thank the following individuals for their time, effort, and expertise:

Michael Honeycutt (ACCA), Educational Consultant, *HVACR 101* Project Coordinator
John Iwanski (RSES), Director of Publishing, *HVACR 401* Project Coordinator
Scott Balmer (PHCC Educational Foundation), ShoffnerKalthoff Mechanical Services, Knoxville, TN
Greg Goater (ACCA), Isaac Heating & Air Conditioning, Inc., Rochester, NY
Terry Miller (ACCA), Energy Management Specialists, Cleveland, OH
Donald Prather (ACCA), Technical Services Manager, ACCA, Arlington, VA
Dick Shaw (ACCA), Technical Education Consultant & Standards Manager, ACCA, Hesperia, MI
Jamie Simpson (PHCC Educational Foundation), Schaal Heating & Cooling, Des Moines, IA

About the Author

Dr. John E. Hohman

Dr. John E. Hohman has 40 years of experience in various education and industry fields. He has been an instructor and administrator at both the university and community college level. He is currently a consultant and has spent the last 10 years developing national certification, employee enhancement, and apprenticeship programs and assessments.

Dr. Hohman has worked in the industry as a technician, engineer, contractor, wholesaler, manufacturer representative, instructor, and educational administrator. John holds a patent in a refrigeration specialty—cascade refrigeration, part of Environmental Test Conditioning. He has served in the capacity of consultant for his own business and other nationally known companies, professor of HVAC&R at Ferris State University and Mid Michigan Community College, Department Head of Technology at Mid Michigan Community College, Consultant and Trainer for Marshall Institute, Professional Development Coordinator at Central Michigan University and Ferris State University, and Consulting Engineer to Sexton ESPEC. John is the State Director of the Michigan Society for Healthcare Engineers (MiSHE), the National Director of the Mechanic Evaluation and Certification for Healthcare (MECH), and a past Director of Research and Assessment Validation for HVAC Excellence and Director of the National Occupational Competency Testing Institute (NOCTI). John received his associate's degree in Refrigeration, Heating, and Air Conditioning, a bachelor's degree in Teacher Education, a master's degree in Educational Administration from Ferris State University, and a PhD in Education with an emphasis on evaluation systems with Capella University. John Hohman holds many other licenses and certifications in both education and technology.

CHAPTER 1

The One-Way Heat Pump: A Review of the Comfort Conditioning Refrigeration Cycle

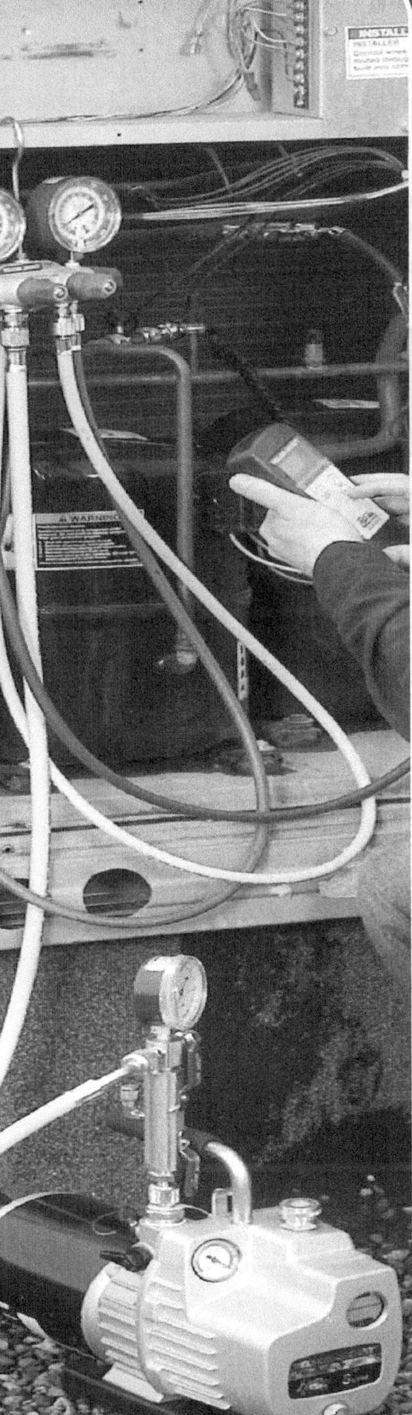

LEARNING OBJECTIVES

The student will:

- Describe the basic refrigeration process
- Identify the four basic components of compression refrigeration systems
- Describe the function of the four basic components of compression refrigeration systems
- Describe saturation as it relates to refrigerants
- Describe superheat as it relates to refrigerants
- Describe subcooling as it relates to refrigerants

INTRODUCTION

The one-way heat pump is a way to describe an air conditioning system. Using the example of a typical split air conditioning system used to cool occupants, the system pumps heat from the occupied space, in one direction, to a place outside the occupied space. The evaporator (indoor coil) absorbs heat from circulating indoor air. The compressor moves the refrigerant from the evaporator to the condenser (outside coil), where the heat is transferred to outside (ambient) air. Pumping heat or moving heat from one place to another has been the function of air conditioning and refrigerating machines from the beginning.

In this chapter the basic refrigeration system, its components, and their function will be reviewed. Three important aspects of moving heat with refrigerants will also be reviewed.

THE BASIC REFRIGERATION CYCLE

All compression refrigeration systems require four components: compressor, condenser, metering device, and evaporator. These four basic components are used in the simplest and most complex refrigeration systems. A split air conditioning

Field Problem

The customer was hot, and so was the outdoor and indoor temperature. His split air conditioning system was not working. Telling the customer that he would do his very best to remedy the problem, the technician listened to the customer and asked a few questions to find out:

- The system had been working fine until a day ago.
- The thermostat had not been changed until the inside temperature went up, but it didn't help.
- There had been no power outage.
- The air filter had been changed a week ago (customer changes filter on the first of each month).
- This was an older system with no service indicator light-emitting diodes (LEDs).

Symptoms: Not enough cooling; system was working.

Possible Causes: Iced evaporator; blower problems; inefficient compressor; low charge; blocked condenser.

Thanking the customer and asking him to take him to the thermostat, the service technician began his work, performing the actions in the following table.

Action	Result
Setting the thermostat for cooling (system switch and temperature selector), he listened for the indoor blower.	The indoor blower started.
He *felt* for air at the first available diffuser.	Air could be felt.
Stepping outside, he *looked* for condenser fan operation.	The outdoor fan was running.
He *listened* for the compressor.	The compressor was running.
Opening the condensing unit service panel, he *felt* the temperature of the suction line.	The suction tube was warm.
He *felt* the liquid line.	It was the same temperature as the suction tube.
He cautiously *felt* the discharge line, first holding his hand close in order to *feel* radiating heat, and then *touching* the line.	There was no radiating heat, so he felt it was safe to touch the discharge line and found it to be only warmer than all of the other tubing, but not hot.

At this point he was fairly certain that the system had lost the charge. He needed to confirm the location of the leak and the extent of repairs. As he finished his checks, he spoke with the customer to deliver the worst news. The condenser had been pierced by a large projectile, likely the cause of a lawn mower. The service technician asked if he could use the customer's phone to call the shop and explain the situation. The phone was then given to the customer to continue the conversation with the shop sales technician regarding replacement of the condenser and the condensing unit, or having a new system installed.

system has these basic components. The compressor sits outside in an enclosure called the condensing unit. In this same enclosure is the condenser. Connected to the inside by tubing is the expansion device, which is mounted on the indoor coil and located within the plenum of a furnace and blower.

Tech Tip

When encountering systems in the field, a mental picture of a basic refrigeration system should come to mind. The four basic system components and their position in the system should form a square. Starting at the right on Figure 1-1 is the compressor. Moving counterclockwise, next is the condenser, the metering device, and the evaporator, with the system ending at the compressor.

As shown in Figure 1–1, when diagramed, the basic refrigeration cycle forms a square where each side is one of the four major components. To make it easier to remember, always start with the compressor and go to the condenser. The compressor always pumps to the condenser. This is going to be an important concept to remember as the lessons on heat pumps continue. Flow continues from the condenser through the metering device to the evaporator. From the evaporator the flow completes the cycle and is ready to continue with another cycle, starting with the compressor. Drawing a line horizontally through this diagram separates the high pressure/temperature from the low pressure/temperature. The condenser is on the high-pressure/temperature side of the system, and the evaporator is on the low-pressure/temperature side. Drawing a line vertically separates those components that handle only liquid from those that can process only vapor. The metering device regulates the flow of liquid refrigerant, while the compressor is designed to compress only vapor. Notice that some of the condenser and evaporator are on the liquid side. That is because the refrigerant is in a liquid state in a portion of both of these heat exchangers. Remember, the refrigerant changes from a vapor to a liquid in the condenser and from a liquid to a vapor in the evaporator. Both of these heat exchangers will have liquid and vapor refrigerant.

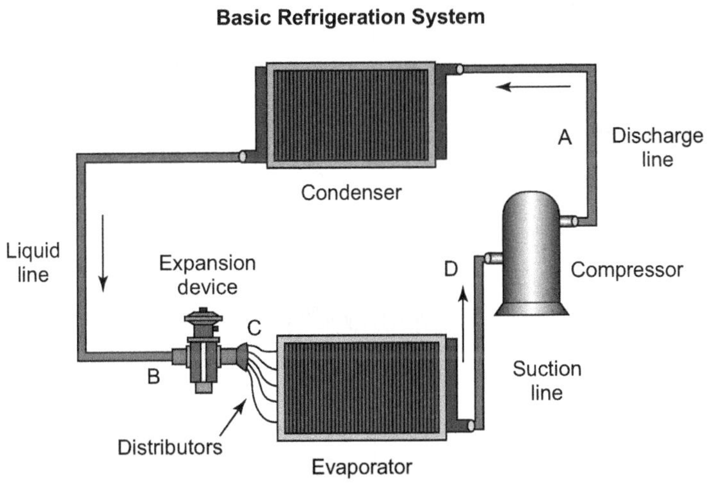

Figure 1–1
The basic refrigeration process can be traced by following the flow from the compressor to the condenser, to the metering device, and to the evaporator before the refrigerant enters the compressor to repeat the process. (Courtesy of Trane)

Tech Tip

There are three important refrigerant lines in a refrigeration system. These lines can provide temperature information that helps to check and diagnose the operation of the system. The discharge line will be the hottest because it conducts highly superheated vapor from the compressor to the condenser. The liquid line may be at outdoor temperature or higher because it conducts liquid refrigerant formed in the condenser to the metering device. The suction line should be slightly warmer than the vaporizing refrigerant as it conducts vapor from the evaporator to the compressor. The fourth line or lines are typically very short and connect the metering device to the evaporator. Sometimes referred to as distribution lines, they are not usually as important as the other lines, but become of interest when some special checks are made.

COMPONENT FUNCTION

Each component must function correctly in order for the refrigeration cycle to be complete. The compressor must increase the pressure, sending the hot vapor to the condenser, where liquid forms and heat is rejected. The metering device regulates the flow of liquid refrigerant to the evaporator, where the refrigerant picks up heat and changes back to a vapor. As shown in Figure 1–2, each of the functions of the components follows in order of refrigerant flow.

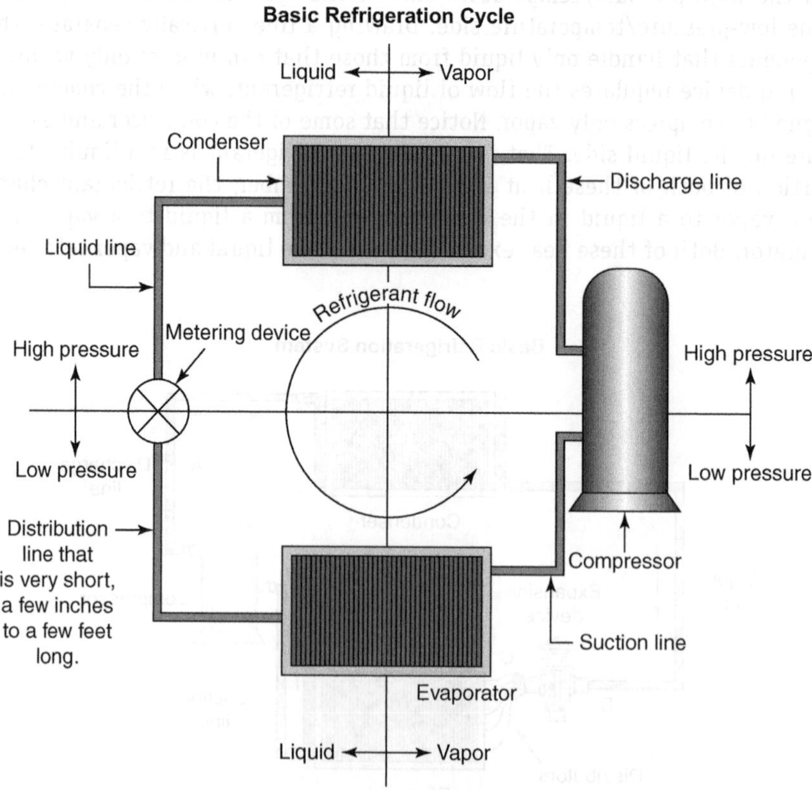

Figure 1–2
Four basic components of every compression refrigeration system are diagramed to show the liquid and vapor split and the high- and low-pressure split. Refrigerant flows from the compressor to the condenser, to the metering device, and to the evaporator before flowing back to the compressor. (Courtesy of Delmar/Cengage Learning)

Compressor

The compressor is used to increase the pressure of suction vapor, moving it from the low-pressure side of the system. The vapor is compressed and forced into the high-pressure side of the system. In this sense the compressor is one point of pressure and temperature change in the system; two are required. The amount of energy needed to move refrigerant vapor from one side of the system to the other increases as the amount of refrigerant increases. The amount of refrigerant moved relates to the amount of cooling needed. As the amount of cooling required increases, the amount of refrigerant moved increases, and the required horsepower of the compressor gets larger.

The compressor might be a centrifugal, reciprocating, scroll, rotary, or screw. All of these types of compressors are used for comfort cooling applications. Some of these different configurations are shown in Figure 1-3. The design of the compressor may be limited to the application for several reasons. One reason might be capacity. Centrifugal compressors might be chosen for an application that requires a large capacity, such as 100 tons. In addition to capacity, considerations could be for: physical size, economy, variable speed, power available, or serviceability.

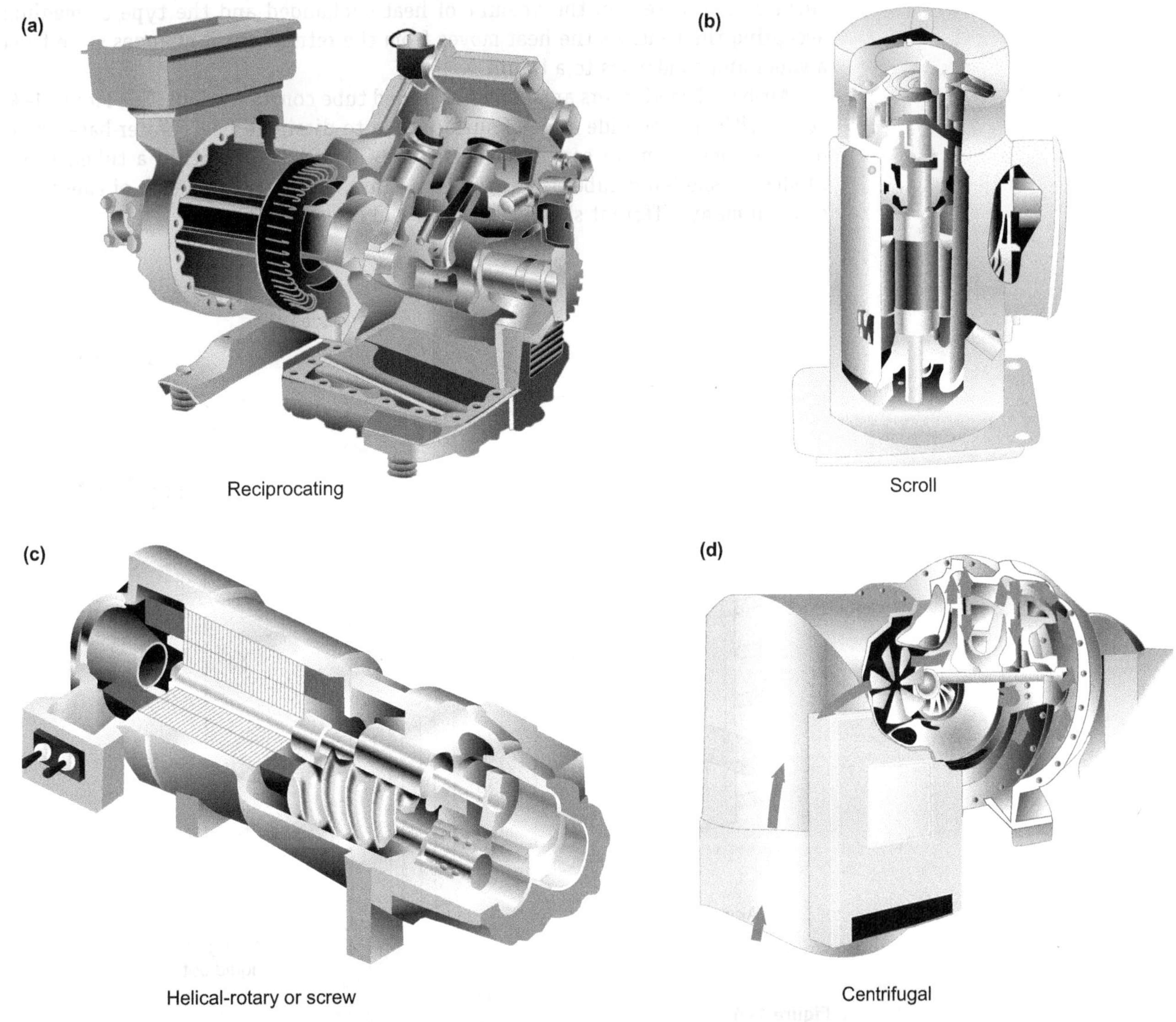

Figure 1-3
Compressors come in many configurations. What they have in common is the ability to compress the refrigerant vapor in volumes great enough to supply vaporous refrigerant to the condenser, where it changes to a liquid. (Courtesy of Trane)

Tech Tip

A ton of refrigeration effect is considered to be 12,000 BTU per hour. If a compressor is selected for a 5-ton air conditioning system, the compressor must be able to pump enough refrigerant to create a refrigeration effect of 60,000 BTU per hour.

Condenser

The condenser is a heat exchanger, designed to move heat from the refrigerant to another medium. The medium is typically air, but depending on the type of air conditioning system, it could be water, earth, or anything else that would easily accept heat energy. The more easily heat moves from the refrigerant, through the walls of the condenser, and into the medium, the more efficient the heat exchange. Engineers will select the size, number of passes (tubes deep), and surface area, based on the amount of heat exchanged and the type of medium accepting the heat. As the heat moves from the refrigerant it changes state from a vapor and condenses to a liquid.

Air-based condensers are generally finned tube configurations. See Figure 1–4. Tubes with fins provide a lot of surface area to dissipate heat. Water-based condensers come in more configurations: tube-in-tube (tube within a tube), tube-in-shell or shell-and-tube (tubes within a tank), and flat plate; each of these may come in many different shapes, bundles, and dimensions.

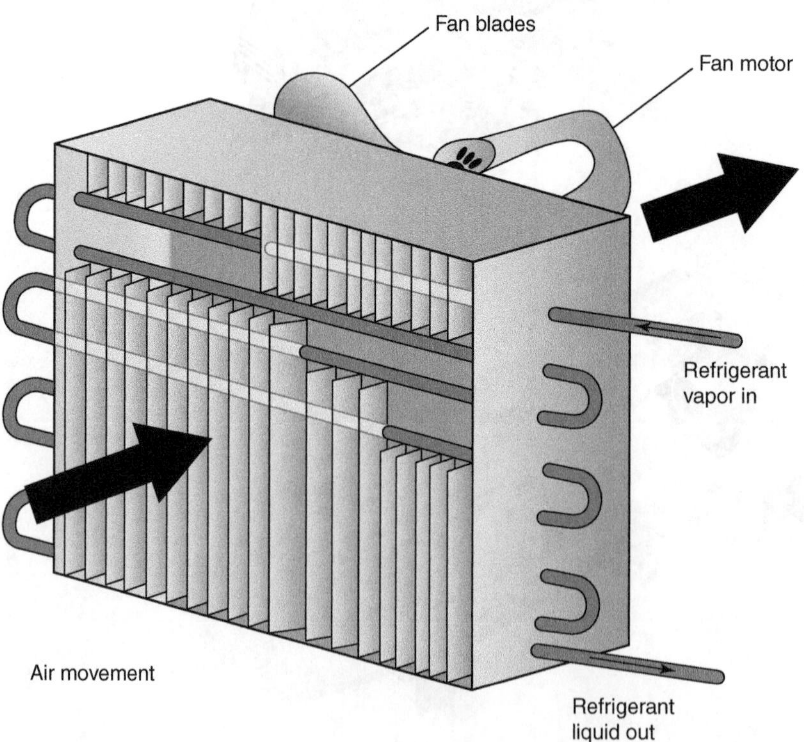

Figure 1–4
The condenser is designed to reject heat and cool the vaporous refrigerant, changing the state of the refrigerant to a liquid. (Courtesy of Delmar/Cengage Learning)

Metering Device

The second point of pressure and temperature change, the metering device, does what the name implies; it meters the flow of liquid refrigerant to the evaporator. See Figure 1–5a and 1–5b for two common metering devices. The metering action may be simple and not require power, or range to the other end of the spectrum and be very sophisticated and require precise electronic control. The simplest type might be a metered orifice, which is a plate with a machined hole. The sophisticated type might be an electronically controlled expansion valve (EXV).

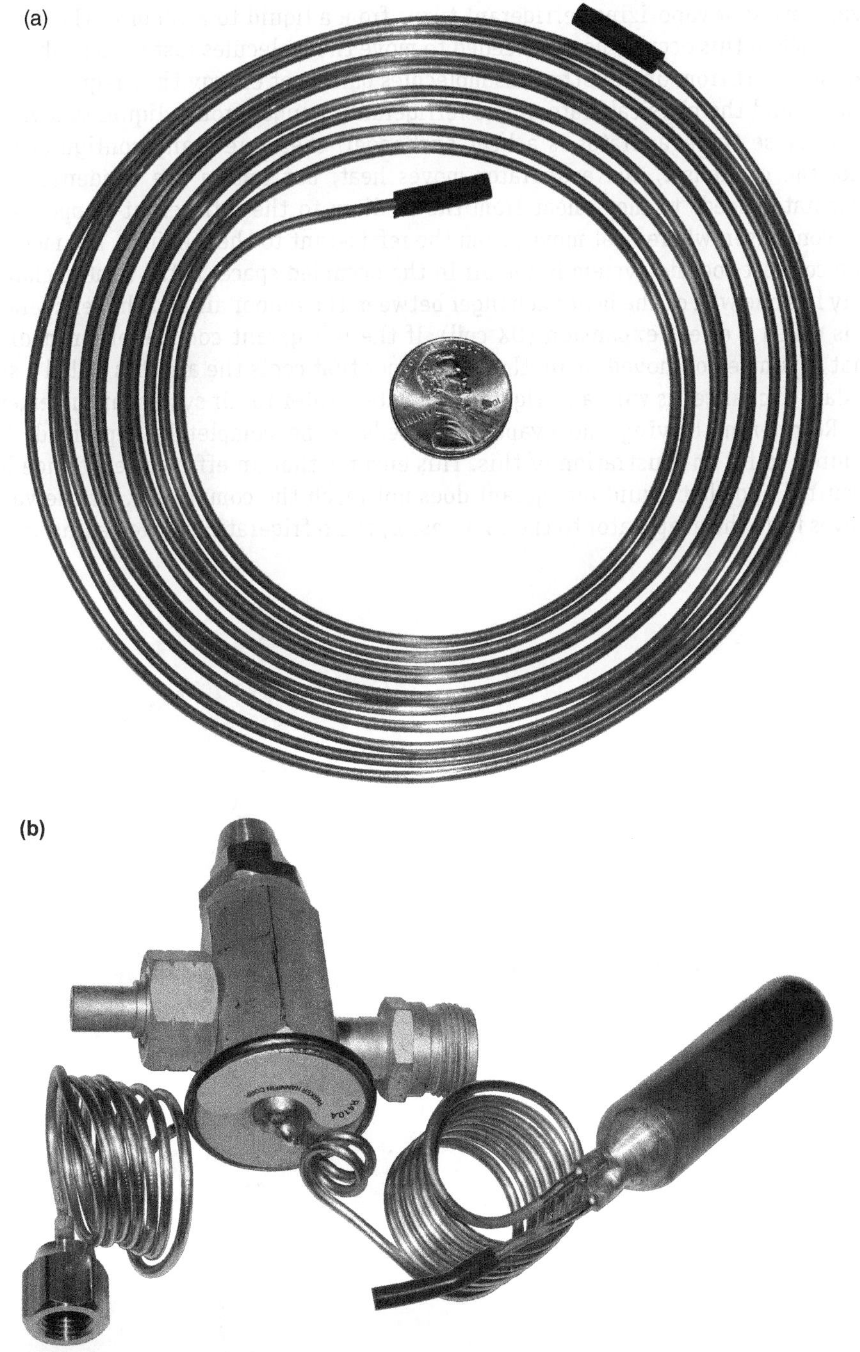

Figure 1–5a,b
Metering devices come in a variety of types. Shown is (a) a capillary tube, and (b) a thermostatic expansion valve. (Courtesy of Delmar/Cengage Learning)

Engineering will select the right valve for the application, but the location is almost always as close to the evaporator as possible. The reason for the location is because refrigerant begins to flash (boil) as it leaves the valve. Depending on the refrigerant, flow, and the pressure, this could happen inside the metering device. If enough heat can be removed before the liquid refrigerant enters the valve, the amount of flashing reduces. In some cases cool suction vapor in the suction line leaving the evaporator is used to cool the liquid line coming to the metering device to prevent such premature flashing.

Evaporator

Evaporating or vaporizing refrigerant turns from a liquid to a vapor in the evaporator. When this occurs, heat is needed to move the molecules faster, to push them farther apart from one another. As molecules gain heat energy they require larger spaces, and the physical state of the refrigerant changes from a liquid to a vapor.

Because the evaporator is a heat exchanger, there are many configurations. Like the condenser, the evaporator moves heat, but unlike the condenser, the evaporator needs to move heat from the medium to the refrigerant—opposite of the condenser, where heat moves from the refrigerant to the medium. The medium in a comfort cooling system is the air in the occupied space. If the heat exchange only has one wall of the heat exchanger between the indoor air and the refrigerant, it is called a direct expansion (DX coil). If the refrigerant cools another medium that is pumped or moved to another exchanger that cools the air, it is called a secondary exchange, as with a refrigerant-to-water, water-to-air system arrangement.

Refrigerant leaving the evaporator needs to be completely vaporized. See Figure 1–6 for an illustration of this. This ensures that an efficient exchange has occurred and that liquid refrigerant does not reach the compressor. As the vapor moves from the evaporator to the compressor, the refrigeration cycle begins again.

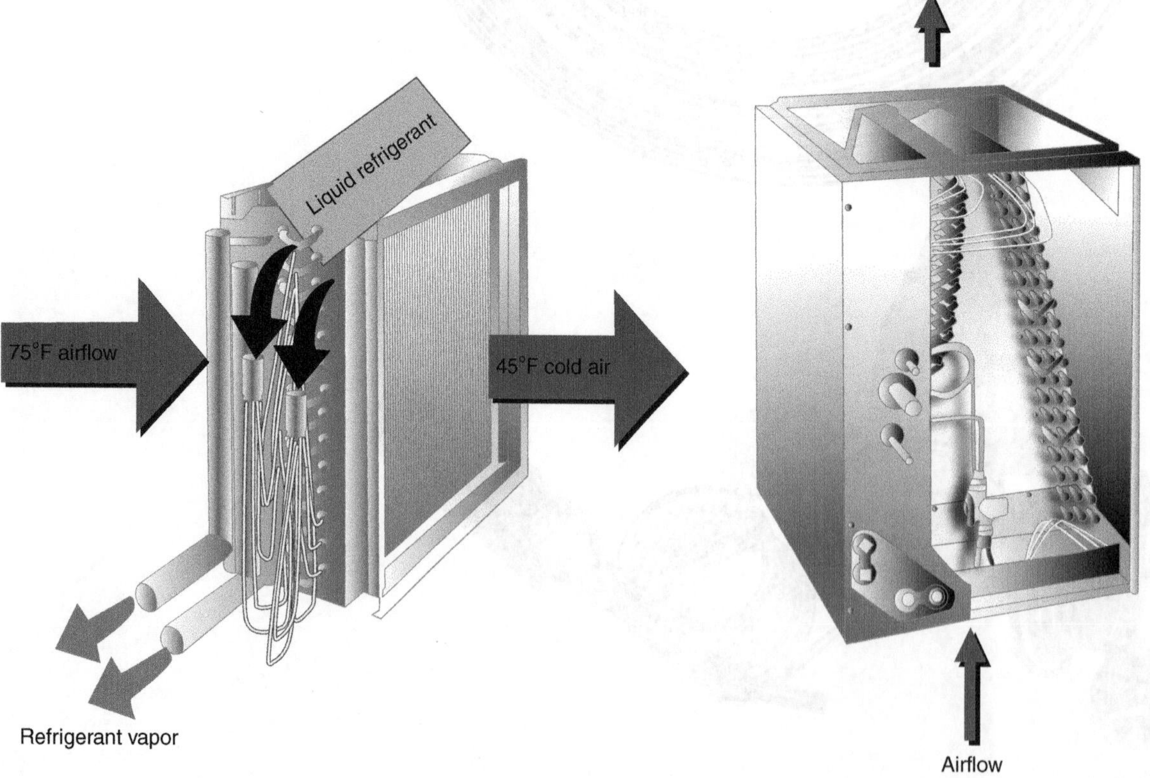

Figure 1–6
The evaporator allows the liquid refrigerant to expand, absorb heat, and change state to a vapor. (Courtesy of Trane)

SUPERHEAT, SUBCOOLING, AND SATURATION

The terms *superheat, subcooling,* and *saturation* need to be completely understood when applied to the refrigerant system. Each term means something independent of the other, but they relate to each other as the refrigerant changes state and moves through the refrigeration cycle. See Figure 1–7 for an example of a pressure/temperature chart and Figure 1-8 for an example of an enthalpy diagram.

Saturation

Before trying to understand superheat and subcooling, it is first necessary to understand the concept of saturation. In the context of refrigerants and refrigeration systems, saturation temperature can be said to be the boiling point of the refrigerant. As with any substance that can take on a liquid or vapor state, any time the temperature of the substance is above its boiling point, it is in the vapor state. Conversely, any time the temperature of the substance is below its boiling point, it must be in a liquid state.

We have also learned that boiling point, or saturation temperature, is a function of pressure. If we raise the pressure on the substance, its boiling point goes up and vice versa. The water in the radiator in your automobile may well be above 212° and yet it is still in its liquid state. This is because the water has no room to expand and turn into a vapor, so it builds pressure in the radiator. If the water is at 220°F and is still in its liquid state, then by definition, its boiling point or saturation temperature has been increased.

This is the whole point of compressing the refrigerant with the compressor. By increasing the pressure to a point where the corresponding boiling point, or

TEMPERATURE °F	REFRIGERANT 12	22	134a	502	404A	410A	TEMPERATURE °F	REFRIGERANT 12	22	134a	502	404A	410A	TEMPERATURE °F	REFRIGERANT 12	22	134a	502	404A	410A
−60	19.0	12.0		7.2	6.6	**0.3**	12	**15.8**	**34.7**	**13.2**	**43.2**	**46.2**	**65.3**	42	**38.8**	**71.4**	**37.0**	**83.8**	**89.7**	**122.9**
−55	17.3	9.2		3.8	3.1	**2.6**	13	**16.4**	**35.7**	**13.8**	**44.3**	**47.4**	**66.8**	43	**39.8**	**73.0**	**38.0**	**85.4**	**91.5**	**125.2**
−50	15.4	6.2		0.2	0.8	**5.0**	14	**17.1**	**36.7**	**14.4**	**45.4**	**48.6**	**68.4**	44	**40.7**	**74.5**	**39.0**	**87.0**	**93.3**	**127.6**
−45	13.3	2.7		**1.9**	**2.5**	**7.8**	15	**17.7**	**37.7**	**15.1**	**46.5**	**49.8**	**70.0**	45	**41.7**	**76.0**	**40.1**	**88.7**	**95.1**	**130.0**
−40	11.0	**0.5**	14.7	**4.1**	**4.8**	**9.8**	16	**18.4**	**38.7**	**15.7**	**47.7**	**51.0**	**71.6**	46	**42.6**	**77.6**	**41.1**	**90.4**	**97.0**	**132.4**
−35	8.4	**2.6**	12.4	**6.5**	**7.4**	**14.2**	17	**19.0**	**39.8**	**16.4**	**48.8**	**52.3**	**73.2**	47	**43.6**	**79.2**	**42.2**	**92.1**	**98.8**	**134.9**
−30	5.5	**4.9**	9.7	**9.2**	**10.2**	**17.9**	18	**19.7**	**40.8**	**17.1**	**50.0**	**53.5**	**75.0**	48	**44.6**	**80.8**	**43.3**	**93.9**	**100.7**	**136.4**
−25	2.3	**7.4**	6.8	**12.1**	**13.3**	**21.9**	19	**20.4**	**41.9**	**17.7**	**51.2**	**54.8**	**76.7**	49	**45.7**	**82.4**	**44.4**	**95.6**	**102.6**	**139.9**
−20	**0.6**	**10.1**	3.6	**15.3**	**16.7**	**26.4**	20	**21.0**	**43.0**	**18.4**	**52.4**	**56.1**	**78.4**	50	**46.7**	**84.0**	**45.5**	**97.4**	**104.5**	**142.5**
−18	**1.3**	**11.3**	2.2	**16.7**	**18.2**	**28.2**	21	**21.7**	**44.1**	**19.2**	**53.7**	**57.4**	**80.1**	55	**52.0**	**92.6**	**51.3**	**106.6**	**114.6**	**156.0**
−16	**2.0**	**12.5**	0.7	**18.1**	**19.6**	**30.2**	22	**22.4**	**45.3**	**19.9**	**54.9**	**58.8**	**81.9**	60	**57.7**	**101.6**	**57.3**	**116.4**	**125.2**	**170.0**
−14	**2.8**	**13.8**	**0.3**	**19.5**	**21.1**	**32.2**	23	**23.2**	**46.4**	**20.6**	**56.2**	**60.1**	**83.7**	65	**63.8**	**111.2**	**64.1**	**126.7**	**136.5**	**185.0**
−12	**3.6**	**15.1**	**1.2**	**21.0**	**22.7**	**34.3**	24	**23.9**	**47.6**	**21.4**	**57.5**	**61.5**	**85.5**	70	**70.2**	**121.4**	**71.2**	**137.6**	**148.5**	**200.8**
−10	**4.5**	**16.5**	**2.0**	**22.6**	**24.3**	**36.4**	25	**24.6**	**48.8**	**22.0**	**58.8**	**62.9**	**87.3**	75	**77.0**	**132.2**	**78.7**	**149.1**	**161.1**	**217.6**
−8	**5.4**	**17.9**	**2.8**	**24.2**	**26.0**	**38.7**	26	**25.4**	**49.9**	**22.9**	**60.1**	**64.3**	**90.2**	80	**84.2**	**143.6**	**86.8**	**161.2**	**174.5**	**235.4**
−6	**6.3**	**19.3**	**3.7**	**25.8**	**27.8**	**40.9**	27	**26.1**	**51.2**	**23.7**	**61.5**	**65.8**	**91.1**	85	**91.8**	**155.7**	**95.3**	**174.0**	**188.6**	**254.2**
−4	**7.2**	**20.8**	**4.6**	**27.5**	**30.0**	**42.3**	28	**26.9**	**52.4**	**24.5**	**62.8**	**67.2**	**93.0**	90	**99.8**	**168.4**	**104.4**	**187.4**	**203.5**	**274.1**
−2	**8.2**	**22.4**	**5.5**	**29.3**	**31.4**	**45.8**	29	**27.7**	**53.6**	**25.3**	**64.2**	**68.7**	**95.0**	95	**108.2**	**181.8**	**114.0**	**201.4**	**219.2**	**295.0**
0	**9.2**	**24.0**	**6.5**	**31.1**	**33.3**	**48.3**	30	**28.4**	**54.9**	**26.1**	**65.6**	**70.2**	**97.0**	100	**117.2**	**195.9**	**124.2**	**216.2**	**235.7**	**317.1**
1	**9.7**	**24.8**	**7.0**	**32.0**	**34.3**	**49.6**	31	**29.2**	**56.2**	**26.9**	**67.0**	**71.7**	**99.0**	105	**126.6**	**210.8**	**135.0**	**231.7**	**253.1**	**340.3**
2	**10.2**	**25.6**	**7.5**	**32.9**	**35.3**	**50.9**	32	**30.1**	**57.5**	**27.8**	**68.4**	**73.2**	**101.0**	110	**136.4**	**226.4**	**146.4**	**247.9**	**271.4**	**364.8**
3	**10.7**	**26.4**	**8.0**	**33.9**	**36.4**	**52.3**	33	**30.9**	**58.8**	**28.7**	**69.9**	**74.8**	**103.1**	115	**146.8**	**242.7**	**158.5**	**264.9**	**290.6**	**390.5**
4	**11.2**	**27.3**	**8.6**	**34.9**	**37.4**	**53.6**	34	**31.7**	**60.1**	**29.5**	**71.3**	**76.4**	**105.1**	120	**157.6**	**259.9**	**171.2**	**282.7**	**310.7**	**417.4**
5	**11.8**	**28.2**	**9.1**	**35.8**	**38.4**	**55.0**	35	**32.6**	**61.5**	**30.4**	**72.8**	**78.0**	**107.3**	125	**169.1**	**277.9**	**184.6**	**301.4**	**331.8**	**445.8**
6	**12.3**	**29.1**	**9.7**	**36.8**	**39.5**	**56.4**	36	**33.4**	**62.8**	**31.3**	**74.3**	**79.6**	**108.4**	130	**181.0**	**296.8**	**198.7**	**320.8**	**354.0**	**475.4**
7	**12.9**	**30.0**	**10.2**	**37.9**	**40.6**	**57.8**	37	**34.3**	**64.2**	**32.2**	**75.8**	**81.2**	**111.6**	135	**193.5**	**316.6**	**213.5**	**341.2**	**377.1**	**506.5**
8	**13.5**	**30.9**	**10.8**	**38.9**	**41.7**	**59.3**	38	**35.2**	**65.6**	**33.2**	**77.4**	**82.9**	**113.8**	140	**206.6**	**337.2**	**229.1**	**362.6**	**401.4**	**539.1**
9	**14.0**	**31.8**	**11.4**	**39.9**	**42.8**	**60.7**	39	**36.1**	**67.1**	**34.1**	**79.0**	**84.6**	**116.0**	145	**220.3**	**358.9**	**245.5**	**385.9**	**426.8**	**573.2**
10	**14.6**	**32.8**	**11.9**	**41.0**	**43.9**	**62.2**	40	**37.0**	**68.5**	**35.1**	**80.5**	**86.3**	**118.3**	150	**234.6**	**381.5**	**262.7**	**408.4**	**453.3**	**608.9**
11	**15.2**	**33.7**	**12.5**	**42.1**	**45.0**	**63.7**	41	**37.9**	**70.0**	**36.0**	**82.1**	**88.0**	**120.5**	155	**249.5**	**405.1**	**280.7**	**432.9**	**479.8**	**616.2**

VACUUM (in. Hg) – RED FIGURES
GAGE PRESSURE (psig) – BOLD FIGURES

Figure 1–7
A typical pressure/temperature chart that can be obtained from a local refrigeration wholesaler. It shows the pressure/temperature relation of saturated refrigerant. (Courtesy of Delmar/Cengage Learning)

saturation temperature, is sufficiently high that the outdoor air looks cool by comparison, we can cool the refrigerant in the condenser coil down below its boiling point or saturation temperature, and cause the vapor refrigerant to return to its liquid state.

Any time we have a refrigerant that contains both liquid and vapor, the temperature of that refrigerant must be at the boiling point. In both the condenser and the evaporator, we have a mix of liquid and vapor refrigerant, so we can determine its saturation temperature by measuring the pressure on the high side or the low side, and using a pressure/temperature chart for that particular refrigerant to convert to saturation temperature.

Superheat

In the low side of the system (the evaporator section), we lower the pressure through the metering device, causing the boiling point to drop. That liquid refrigerant boils off to a vapor as it makes its way through the evaporator. At some point, presumably near the end of the evaporator coil, all of the liquid is boiled off to a vapor. Any heat absorbed in the evaporator is consumed in the change of state, and no temperature change takes place. Remember that as long as we have a liquid and vapor mix, the temperature must be at the boiling point, so on an R-22 system with a low side pressure of 68 psig, the boiling point is 40°F, and the refrigerant (and hence the coil itself) will be at 40°F up to the point where all of the liquid has boiled off to a vapor.

Once we are in a 100% vapor state, the 40°F vapor will still pick up some heat from the remainder of the evaporator coil as well as the suction line passing through a warm space, and that heat absorbed into the now vapor refrigerant will cause a change in the temperature of the vapor (sensible heat). By measuring the suction line temperature and comparing that reading to the boiling point of the refrigerant, that difference in temperature is called superheat. *Superheat* is defined as any heat added to the refrigerant after it has achieved a vapor state. In a very real sense we can use this information to do two things. First, by the virtue of having any superheat at all, we are assured that we have no liquid refrigerant returning to the compressor. Second, the amount of additional heat picked up by the vapor refrigerant will give us an indication of how early or late the refrigerant is boiling off in the evaporator. Given a manufacturer's superheat chart for varying conditions, we may very well be able to make diagnostic judgments as to the operation of the equipment.

Subcooling

In the condenser section, we cool the refrigerant below its boiling point, causing the refrigerant to return to its liquid state. As the hot vapor from the compressor enters the condenser coil, heat is removed. The first effect will be to lower the hot vapor temperature to the boiling point of the refrigerant, or desuperheat the vapor. Once down to the boiling point, we continue to remove heat and this heat is given up in the change of state back to a liquid. Again, if we have a mix of vapor and liquid, the actual temperature of the refrigerant must be at the boiling point, so up until the point of complete condensation in the coil, the temperature of the refrigerant will be at the boiling point. Once all of the vapor is returned to its liquid state, presumably near the end of the condenser coil, it will continue to cool even further, only now that it is in a 100% liquid state, the temperature of the liquid will go down as the heat is removed. This is called subcooling.

By definition, *subcooling* is any heat removed after the refrigerant has completely changed over to a liquid. The presence of subcooling assures us that we have a solid liquid stream of refrigerant heading toward the metering device, and the amount of subcooling that we have is indicative of how early

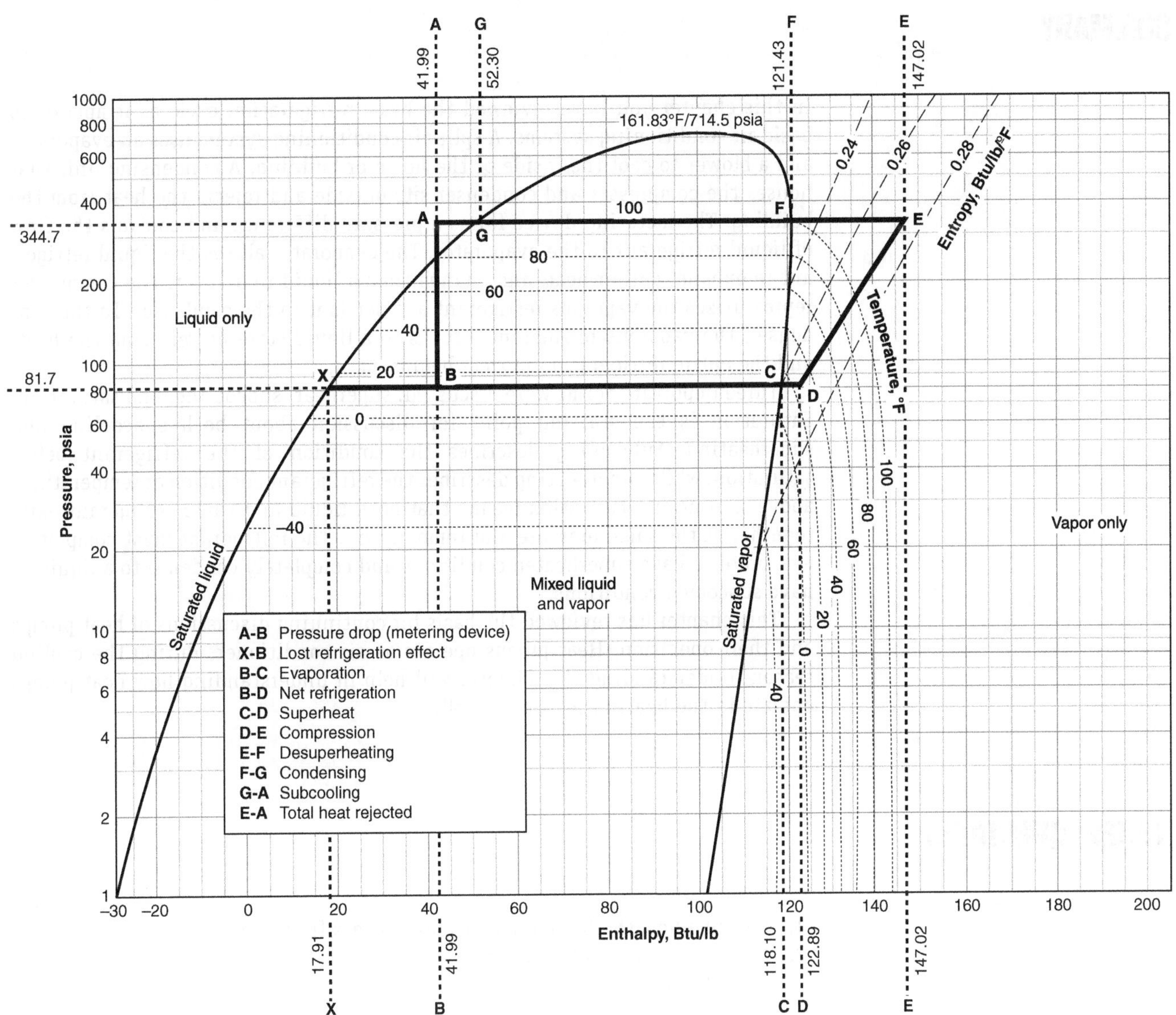

Figure 1–8
Refrigerant 410A pressure/enthalpy diagram showing the condition of refrigerant moving through the system. At point "A," refrigerant that is slightly subcooled to the left of the saturated liquid curve enters the metering device and drops in pressure to "B," the evaporator pressure/temperature (saturation point). From "B" to "C" the refrigerant changes from a liquid to a vapor and acquires some superheat as it leaves the evaporator. From "C" to "D," heat from the compressor adds more superheat before the compressor compresses the vapor from "D" to "E." Superheat is removed from "E" to the saturated vapor curve before vaporous refrigerant changes state to a liquid as it moves toward "A." (Courtesy of RSES)

or late the refrigerant is returning to its liquid state in the condenser coil. Subcooling can be used to make diagnostic judgments regarding the operation of the equipment.

Measuring superheat and subcooling requires an accurate thermometer to measure the suction/liquid line temperatures, pressure gauges to measure the high side and low side pressures, and a pressure/temperature chart for that particular refrigerant to convert the pressures measured to corresponding saturation temperatures. Charts for all types of refrigerants are readily available in a number of forms, and the most common refrigerants have their charts superimposed on the gauges themselves.

Superheat = Suction line temperature − low side boiling point
Subcooling = high side boiling point − liquid line temperature

SUMMARY

In this chapter you have reviewed the basic refrigeration cycle as it applies to basic air conditioning systems. A split air conditioning system uses an evaporator and a blower to cool the inside of the home or building. A condensing unit that houses the compressor and condenser sits outside and rejects the heat from the building. The metering device forms a pressure difference and controls the flow of liquid refrigerant to the evaporator. The evaporator allows the liquid refrigerant to expand, gather heat, and change from a liquid to a vapor. The compressor repressurizes the vaporous refrigerant and sends it to the condenser. In the condenser, the vaporous refrigerant returns to a liquid state by rejecting the heat, and the entire refrigeration process begins again.

Three important terms were discussed: superheat, subcooling, and saturation. Saturation is the starting point for discussion about both subcooling and superheating. Subcooling describes the condition of the refrigerant below saturation, while superheating describes the refrigerant condition at temperatures above saturation. Saturation means that both liquid and vapor refrigerant exist together at the same pressure and temperature. The refrigerant must completely evaporate to have superheated conditions and completely condense to a liquid to have subcooled conditions.

This chapter has reviewed the basis for continuing discussions of heat pumps and their operation. Heat pumps operate as air conditioners during the cooling season. Understanding this chapter will help in understanding how heat pumps operate in the heating season, as well.

REVIEW QUESTIONS

1. Describe how the basic refrigeration process functions.
2. Looking at a real refrigeration system, a picture, or a diagram, identify and name the four basic components of all compression refrigeration systems in the order refrigerant flows through them.
3. Describe the function of the compressor.
4. Describe the function of the condenser.
5. Describe the function of the metering device.
6. Describe the function of the evaporator.
7. Describe saturation as it applies to refrigerants.
8. Describe superheat as it applies to refrigerants.
9. Describe subcooling as it applies to refrigerants.

CHAPTER 2

Reversing the Cycle

LEARNING OBJECTIVES

The student will:

- Describe the function of the indoor coil during the heating mode
- Describe the function of the outdoor coil during the heating mode
- Relate how the indoor and outdoor coils change function
- Name the device that is responsible for changing the function of the indoor and outdoor coils and describe its operation
- Explain how heat is pulled from cold air
- Describe what a thermal balance point is and how it relates to the operation of a heat pump

INTRODUCTION

The heat pump is also known as a reverse-cycle air conditioner. Though this is inexact terminology, we will discuss how reversing of the heat exchangers and refrigerant flow occurs. The indoor and outdoor coils change function and the liquid and vapor lines change function and direction of flow when the heat pump changes its mode of operation from cooling to heating.

We will also discuss how heat is pulled from cold air. Air-source heat pumps (ASHPs) that pull heat from cold outside air are designed with a specific capacity that can be charted as outside air temperature falls or rises. On a thermal balance chart, system capacity and structure heat loss will intersect. The intersection identifies the outdoor condition where the heat pump will maintain indoor air temperature without the need for additional (supplemental) heat.

By the end of this chapter, you should have a good idea of how heat exchangers change function and why. You will also come to understand the device that creates this reversing or function-changing action. How heat is pulled or extracted from cold outside air and how a heat pump can do this will become apparent, along with the limitations imposed by system design considerations.

Field Problem

A new homeowner scheduled a system check. The system was not new to the HVAC company, but the homeowner was new. He had just purchased the home and was unfamiliar with the type of heating system. He had decided to keep the service company that was used by the former owner and had requested the scheduled maintenance ahead of the heating season, just to be prepared. When the technician met with the new homeowner, the owner requested that he accompany the service technician, just to learn more about the system. It became apparent to the technician that this was more of a customer training situation than a system check.

Starting at the customer control, the thermostat, the technician explained its operation and settings: cooling, heating, and emergency heat. Next he checked and recorded the operation of the system in the cooling mode: airflow, temperature difference across the indoor coil, outdoor coil temperature, and so on. At each point, he related the operation of the system to the new homeowner and answered his questions. The technician pointed out the various parts of the system, including the reversing valve. At this point, the customer was confused and asked how the system could run in reverse. Using his best nontechnical communication skills, the technician related the operation of the system to things that the customer might understand. He explained the function change of heat exchangers, the indoor air temperatures delivered, and outdoor air temperatures where heat is extracted. He also included a discussion about the thermal balance point chart in the customer information package hanging on the unit and went over the other material in the package as well.

The technician did the usual types of maintenance tasks: changed the indoor filter, checked the indoor coil for cleanliness, cleaned the outdoor coil, checked the electrical system and connections, and so on. There were no issues with this system, but a grateful new homeowner was completely won over by the technician's knowledge, professionalism, and patience. Before the technician could return to the office, the customer had called, talked to the company owner, and praised the virtues of the technician they had sent to check his system. This system check turned out to be more of a customer relations test than a technical challenge.

THE CHANGING FUNCTIONS OF HEAT EXCHANGERS

Heat pump heat exchangers are referred to by their position, because their function changes depending on the season. The heat exchange coils are called "indoor" and "outdoor" heat exchangers. From this point forward, reference will be made to the indoor and outdoor coils. The indoor coil exchanges heat energy with the indoor air. The outdoor coil exchanges energy with outside air.

The function of the indoor coil during the cooling season is the same as a split air conditioning system. It functions as the evaporator and cools indoor air. The function of the outdoor coil during the cooling season is also the same as a split air conditioning system. But the function of each coil changes in the heating season. During heating, the indoor coil functions as the condenser and the outdoor coil functions as the evaporator. Heat energy is moved from outside to warm occupants on the inside. The same basic effect occurs with a domestic refrigerator. Inside the refrigerator it is cooler than the room air, yet when a hand is placed in the discharge airflow during operation, the air being discharged is warm. Now imagine that the inside of the refrigerator is outside. All of the heat energy available outside is moving into the room and being discharged as warm air.

Figure 2-1 shows heat movement from indoor air to outdoor air during the cooling season. This is the same as a split air conditioning system. During the heating season, heat moves from outside air to indoor air and the functions of the heat exchangers change. The indoor coil becomes the high-pressure, high-temperature condenser, and the outdoor coil becomes the low-pressure, low-temperature evaporator. Outdoor heat energy is absorbed by the outdoor coil, functioning as the evaporator, and the heat is transferred to the indoor coil, functioning as the condenser. The hotter condenser gives up heat energy to the indoor air, warming the indoor space.

Tech Tip

Reversing the functions of the indoor and outdoor coil can only happen if the flow of refrigerant is changed. Discharge vapor must be directed from the compressor to the coil functioning as the condenser. Suction vapor must be directed from the coil functioning as the evaporator to the compressor. To accomplish the change of flow, a special valve is needed called a reversing valve.

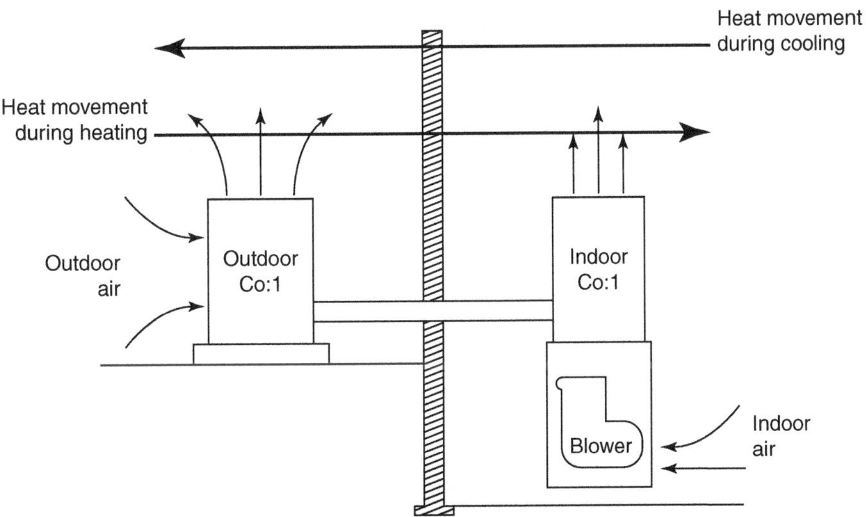

Figure 2–1
During the cooling season, heat moves from inside to outside. During the heating season, heat energy is taken from outside and brought indoors. (Courtesy of Delmar/Cengage Learning)

THE CHANGING FUNCTION AND DIRECTION OF FLOW OF VAPOR AND LIQUID LINES

The suction and discharge lines between the reversing valve and the compressor are the only lines that do not change function; the compressor is always connected to receive suction vapor and to discharge high-temperature/pressure vapor. Between the compressor and the heat exchangers is a reversing valve. The reversing valve directs or redirects the flow of vapor, depending on the mode—cooling or heating. Both the vapor and liquid lines change positions along with the heat exchanger function—condenser or evaporator. Review the cooling and heating modes in Figures 2–2 and 2–3. What the diagrams show is the functional change of the heat exchangers and the direction of flow of the liquid and vapor lines serving each heat exchanger.

In the cooling mode, the indoor coil serves as an evaporator. The evaporator pulls heat from the indoor air. The evaporator inlet is connected to the metering device that is supplying liquid refrigerant. The outlet is connected to the suction side of the compressor. The outside coil is serving as the condenser. The inlet of the condenser is connected to the vapor discharge of the compressor. The outlet is connected to the metering device and is the liquid line. Notice the change that occurs in the heating mode. Both heat exchangers trade functions; the indoor coil serves as the condenser and the outdoor coil functions as the evaporator. Likewise, the vapor and liquid lines for each heat exchanger change direction of flow. High-pressure vapor is sent to the indoor coil, while low-pressure vapor is pulled from the outdoor coil. Condensed liquid refrigerant in the indoor coil is sent through the metering device to the outdoor coil. All of these changes occur because the reversing valve connects the compressor to different heat exchange coils in each mode, heating or cooling.

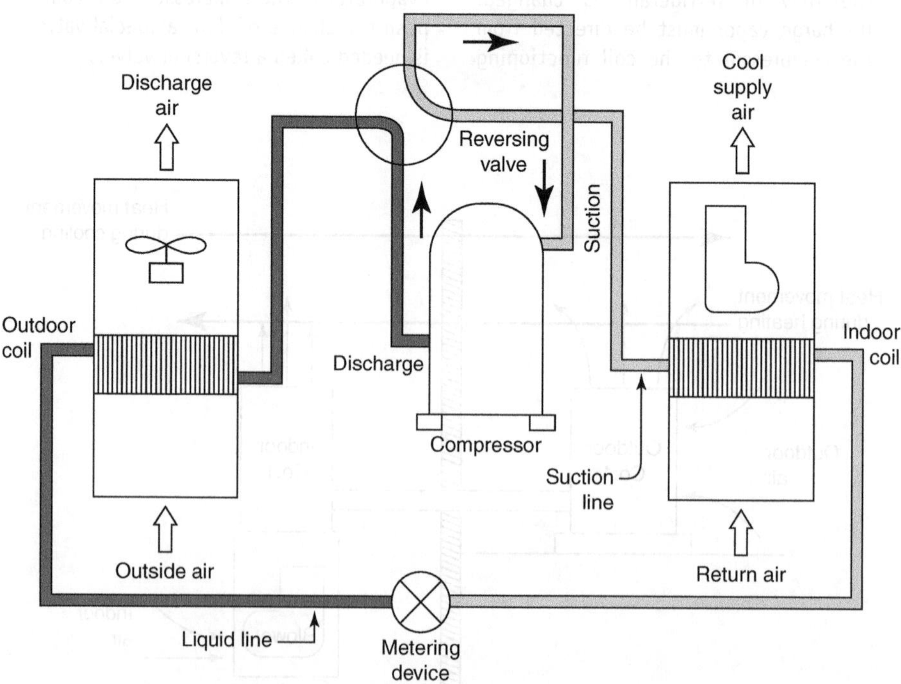

Figure 2–2
The heat pump incorporates a reversing valve that directs the flow of refrigerant from and to the compressor. In cooling mode, the outside and indoor coils work as a split air conditioning system. Notice the suction and liquid line location. (Courtesy of Delmar/Cengage Learning)

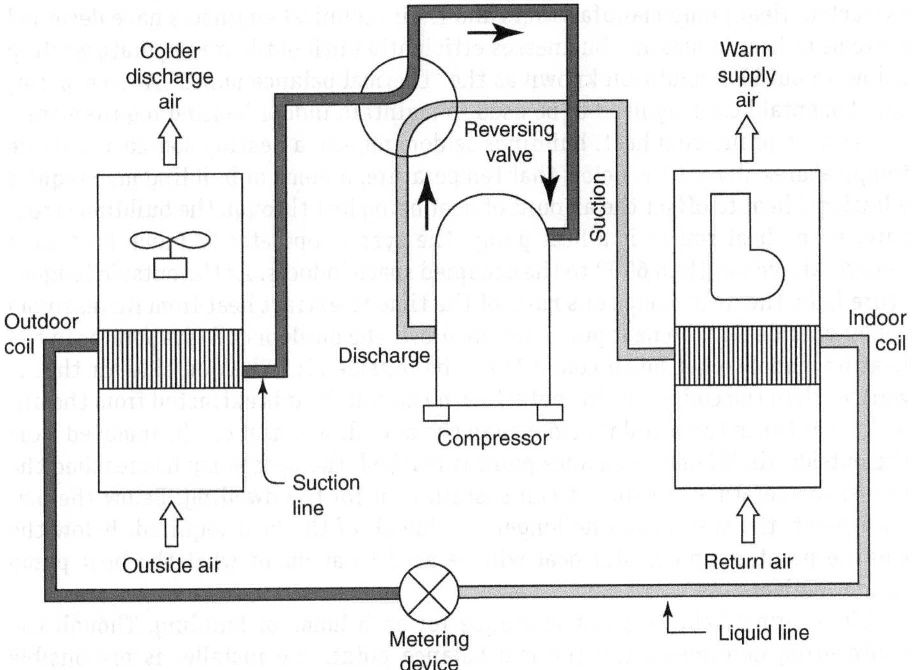

Figure 2–3
In the heating mode, the outside and indoor coil functions have been reversed (or switched). The outdoor coil is colder than the outside air. The coil picks up heat and discharges cooler air. The indoor coil is now the condenser and warms the indoor air. Notice the suction and liquid line location. (Courtesy of Delmar/Cengage Learning)

Pulling Heat from Cold Outside Air

Heat pumps pull heat from cold outside air and send the heat indoors to warm occupied spaces. In a simplistic way, it would be the same thing that would occur if a window air conditioner was mounted in reverse—with the controls and evaporator on the outside and the condenser pointing into the room. In actuality, this arrangement would not work well, but it demonstrates the physical arrangement of the evaporator functioning as the outdoor coil and the condenser as the indoor coil to heat a room. Technically, a window air conditioner is not designed to pull heat from air that is below normal room temperatures of around 70°F.

Tech Tip

There is an abundant amount of heat available almost everywhere. As long as the temperature is above −460°F or −273°C (absolute zero), there is heat. The lower the temperature, the harder the heat is to obtain, and the cost of acquiring the heat gets higher. Air temperature that is 30°F (lower than 32°F; freezing) still has heat in it. To extract the heat, an evaporator is designed to operate below the outdoor air temperature and absorb the heat from the air.

Heat pumps are designed to operate efficiently at temperatures that would be encountered during a normal heating season. Another important point to remember is that there is always heat in outside air. Even as temperatures drop below freezing, the air has enough heat and there is enough volume of air for heat to be extracted. Heat pump manufacturers and their technical engineers have designed systems to heat homes and businesses efficiently until outdoor temperatures drop below an outdoor condition known as the "thermal balance point." At that point, supplemental heat may need to be used to maintain indoor heating requirements.

Because of internal heat, buildings seldom require a heating source at outside temperatures above 65°F. Below that temperature, a home or building may require additional heat to offset the amount of heat being lost through the building structure. If the heat source is a heat pump, the system operates to pump heat from outside air (colder than 65°F) to the occupied space indoors. As the outside temperature falls, the heat pump runs more of the time to extract heat from increasingly colder outside air. The heat pump design allows the outdoor coil (evaporator in the heating mode) to get much colder than the outside air. When outside air that is warmer than the coil comes in contact with the coil, heat is extracted from the air. As long as the outdoor coil is colder than the outside air, heat can be removed from the outside air. When the balance point is reached, the heat pump has reached the lowest temperature at which it can sustain heat for the dwelling. Below the balance point, the system can no longer provide all of the heat required. Below the balance point, supplemental heat will be used to augment what the heat pump cannot deliver.

The thermal balance point is unique for each home or building. Though the engineering determines the thermal balance point, the installer is responsible for setting the thermal balance point and providing the owner with a thermal balance point chart. The installer is also responsible for setting the supplemental heat controls to bring on additional heat below the thermal balance point.

Tech Tip

Supplemental heat should not be confused with emergency heat. All air-source heat pumps are required to have enough electric resistance heat strips to maintain interior space conditions if the heat pump should fail. Those same heat strips can be used off stage 2 on the thermostat to supplement the output from the heat pump at low outside temperatures.

Tech Tip

Every home or building loses heat. The amount of heat loss is determined through a heat loss calculation. Likewise, every heat pump is designed to deliver a certain amount of heat. Both the heat loss calculation and the heat pump capacity are based on outdoor temperature. As the temperature goes down, the amount of heat loss increases and the heat pump capacity diminishes. If these two things were charted, the heat loss line and the system capacity line would cross. The point where the two lines cross is the thermal balance point (see the thermal balance point graph in Figure 2-4).

For this example, refer to the thermal balance point graph in Figure 2-4, where the system capacity at point "A" is 38,000 BTUH when outdoor air is 65°F. Point "C" indicates that the structure requires no additional heat energy at the same outdoor air temperature. As the temperature drops, both the structure loss line and the system capacity line start to merge. At point "E" they cross—the thermal balance

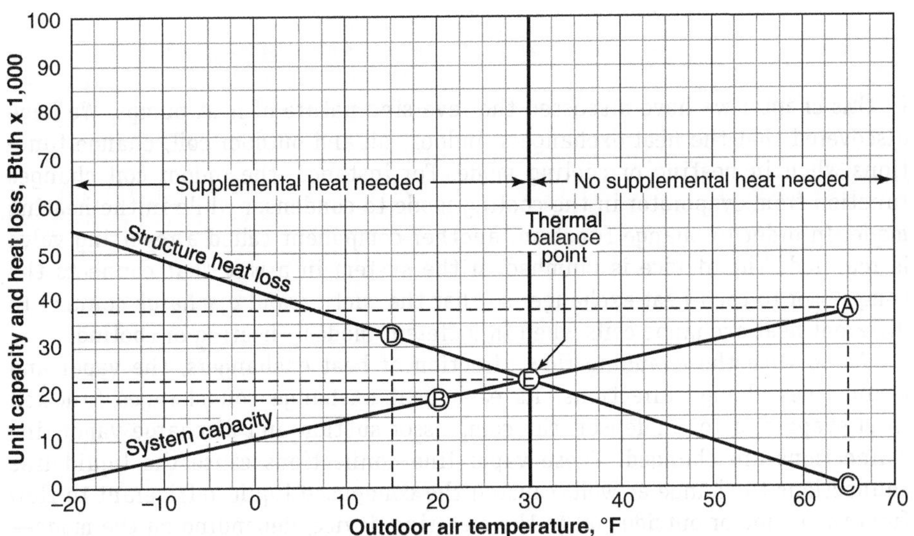

Figure 2–4
The thermal balance point is the point where heat loss and system capacity are in equilibrium. Point "E" is the balance point for this unique system and building example. (Courtesy of RSES)

point. If the outside design is 15°F, the estimated heat loss will be 32,000 BTUH. The system capacity at 20°F will only be 19,000 BTUH. Below 30°F, supplemental heat will be required in addition to the heat output of the heat pump.

Each manufacturer will have recommendations that are specific to the equipment. Some recommend selecting the capacity of the heat pump to be 75–80% of building heat loss. Any remaining supplementary heat is to be made up by supplemental electric strip heaters. The reason for this is because the greatest number of heating hours will be used at conditions that are above design conditions. What this does is reduce the capacity of the equipment to guard against oversizing and reduce initial equipment costs. Oversized systems tend to short cycle, have higher electrical operating costs, increase the need for larger duct systems, and may not provide the correct amount of humidity control during the cooling season.

Tech Tip

If electric resistance heat is used for supplemental heat, 1 kW will provide 3,413 BTU. Using the example in this chapter, the difference between what the heat pump can provide and the structure heat loss is 13,200 BTUH (32,000 BTUH—18,000 BTUH = 13,200 BTUH). Divide 3,413 into 13,200 and you get 3.868 kW of resistance strip heat (13,200/3,413 = 3.868 kW). When outdoor temperatures are 15°F, all 3.686 kW of resistance strips will be operating in conjunction with the heat pump.

SUMMARY

In this chapter we have discussed the reversing nature of heat pumps. We have discovered that the heat exchangers, indoor coil and outdoor coil, change functions when in heating or cooling mode. For instance, the indoor coil changes function from evaporator in the cooling mode to condenser while in the heating mode. In order to change function, another component called a reversing valve is required. This device is plumbed in the system in a way that connects the compressor to each heat exchanger so that the suction and discharge connect to different heat exchange coils when in different modes—heating or cooling.

Along with the changing flow direction of heat exchangers, the vapor and liquid lines change directions. As each heat exchange coil changes function from evaporator to condenser, the compressor suction and discharge vapor line connections are changed. These vapor line connections cause the liquid line connections to change as well, causing the condensed liquid refrigerant to flow from the inside or outside coil to the metering device, depending on the mode—heating or cooling.

Air-source heat pumps pull heat from cold outside air. Even as the air temperature drops, heat pumps can extract enough heat to satisfy interior heat needs. Only when outside temperatures drop below the "balance point" will supplementary heat be necessary. This means that the majority of heating hours can be satisfied by operating the heat pump alone.

REVIEW QUESTIONS

1. Describe the function of the indoor coil during the heating mode.
2. Describe the function of the outdoor coil during the heating mode.
3. Relate how the indoor and outdoor coils change function.
4. Name the device that is responsible for changing the function of the indoor and outdoor coils and describe its operation.
5. Describe the function of the indoor coil during the heating mode.
6. Explain how heat is pulled from cold air.
7. Describe what a thermal balance point is and how it relates to the operation of a heat pump.

CHAPTER 3

The Outdoor Coil: Absorbing Heat

LEARNING OBJECTIVES

The student will:

- Describe how the sun influences the amount of heat available
- Explain how heat is measured in a pound of outdoor air (or in CuFt)
- Relate what "absolute zero" means in relation to the amount of heat available in outdoor air
- Describe the function of and need for supplemental heat
- Explain why a defrost mode is needed for air-source heat pumps (ASHPs)
- Describe the similarity in function of the indoor coil in the cooling mode to the outdoor coil during the heating mode

INTRODUCTION

During the heating mode, the outdoor coil functions as the evaporator. This function is the focus of this chapter. How the outdoor coil functions, how it absorbs heat from colder outdoor air, and how much heat there is in outdoor air will be discussed. The amount of heat versus the temperature of air and how the heat pump system operates have always been curiosities for technicians and homeowners learning for the first time about heat pumps. This chapter will provide the basics of heat in outdoor air and how much might be need to be extracted to heat occupied space.

Field Problem

The heat pump was not keeping up with the heating need and the customer requested a service call because the indoor air temperature was dropping. When the technician arrived, the outdoor air temperature was about 40°F. The thermostat read 68°F and was set for 72°F. The blower was on and delivering air, but the air temperature was 80°F as measured at the closest diffuser.

After speaking with the customer, the technician determined that the system had been functioning well over the past few months, but had started to fail in the past two days. Outside air temperatures had been running between 40°F and 50°F. Air flow at the diffuser seemed to be good (without measuring).

The technician located the outdoor coil and could easily see the problem. The coil was white with frost to the point where the coil looked like a cube of ice. The sensor on the coil was not sensing the frost, so the first step was to find the sensor. When the sensor was found, it was easy to see the problem. The sensor had come undone and was hanging away from the outdoor coil. Manually placing the system in defrost mode, the technician ensured that the coil was frost-free before reattaching the sensor. By allowing the system to run for an extended period and checking the sensor, the technician ensured that the defrost system was functioning correctly before leaving the service call.

THE SUN, HEAT, AND ABSOLUTE ZERO

The sun is responsible for heating the air, causing the wind to blow, and creating the weather patterns. Winter and summer are a result of the position of the sun. As the sun moves more overhead, summer occurs. As the sun moves lower in the southern horizon, winter occurs. During winter, the winds are warmed indirectly by the sun as the wind scrubs warmer objects and land surfaces. Small particles held in suspension also indirectly warm the wind as the sun warms the particles. Sun-warmed wind contains heat that can be extracted to heat occupied space.

The radiant energy from the sun arrives after traveling through space and the atmosphere that surrounds Earth. Some of this energy gets diffused in the atmosphere and some gets absorbed by clouds, trees, buildings, and the ground (Figure 3–1). Very little air is directly heated by the sun, but as the air contacts objects that are warmer, the air picks up heat. The air also loses heat in the same way. As objects re-radiate heat into the night sky, they lose heat and drop below the temperature of the air. Air that comes in contact with these objects conducts heat from the air to the object and the air temperature becomes lower.

THE DEFROST CYCLE AND SUPPLEMENTAL HEAT

Air-source heat pumps will condense water vapor on the outside coil during the heating mode. The outside coil will be operating at temperatures lower than outside air dew point temperatures. The outside air carries various levels of moisture,

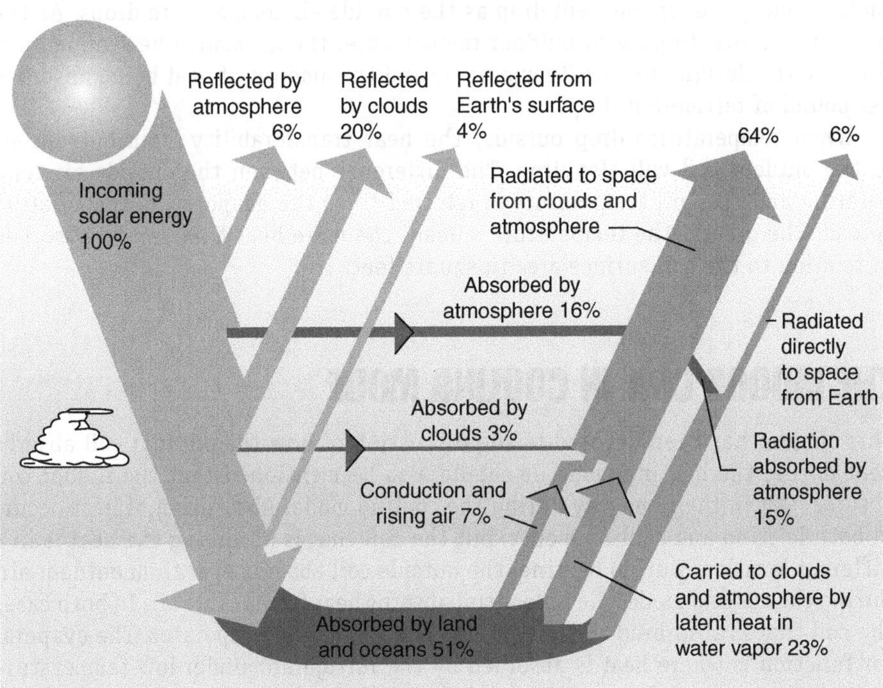

Figure 3–1
Energy from the sun is absorbed largely by the land and oceans. Other objects also absorb energy and are considered part of the land, such as roads, buildings, trees, and so on. Of the energy received, 51% is absorbed by the land, from which 23% is conducted to the air through the latent heat of water vapor and 7% through conduction and convection. Only 15% of the radiation from the sun is absorbed directly by the atmosphere. (Courtesy of NASA)

measured in relative humidity (RH). When dew point temperatures are reached, moisture condenses and frost accumulates on the coil surface. Frost acts as an insulator and reduces airflow. The more frost, the more insulation, and the more inefficient the heat transfer from outside air to refrigerant. A defrost cycle is needed to shed the frost and to restore the efficiency of the coil.

Air-source heat pumps use the ability to reverse modes to defrost the outside coil. A heat pump will switch the function of the indoor and outdoor coils to accomplish the defrost cycle (cooling mode). The reversing valve switches the function of the coils and the outdoor coil becomes the condenser, getting hot enough to melt the frost. The indoor coil picks up enough heat from the air flowing over the coil and moves this heat to the outside coil. For this cycle to work, the outdoor fan is off and is not moving air.

If indoor air needs to be heated during the defrost cycle, supplemental heat is automatically turned on to continue heating the air delivered to occupied spaces. Supplemental heat is often in the form of electric resistance heat strips. The strips are cycled on and off during the defrost cycle in order to eliminate any drop in temperature during the defrost cycle. In this way, the customer is unaware of the defrost cycle and is never uncomfortable during this part of the operation of an air-source heat pump.

OPERATING TEMPERATURES AND PRESSURES

An air-source heat pump is dependent on the air temperatures encountered at the functioning evaporator. In the heating mode, the evaporator is the outdoor coil. Outdoor air temperature will fluctuate, and so will the temperatures and pressures of the refrigerant. As expected, the pressure of the refrigerant on the suction side (outdoor coil) will drop as the outside air temperature drops. As the suction pressure drops with outdoor temperature, the air-source heat pump gets closer to the design pressure/temperature and the amount of heat being absorbed per pound of refrigerant drops.

When temperatures drop outside, the heat transferability from outside air to the outdoor coil will also drop. The difference between the outside air temperature and the coil temperature is referred to as the evaporating temperature spread. The greater the temperature spread, the more heat that can be absorbed in relation to the coil surface area in square feet.

THE INDOOR COIL IN COOLING MODE

This chapter has been devoted to the discussion of how the outdoor coil absorbs heat during the heating mode. It should also be mentioned that the indoor coil absorbs heat in the same way during the cooling mode. Absorption of heat occurs in both heating and cooling modes, but the coil that is absorbing the heat is in a different location. During heating, the outside coil absorbs heat from outdoor air. During the cooling mode, the indoor coil absorbs heat from inside air. In both cases the coil that is absorbing the heat is functioning as the evaporator. The evaporator function is where heat is absorbed by the refrigerant under low-temperature and low-pressure conditions. The condition of the ambient air (indoor or outdoor air) surrounding the evaporator in either heating or cooling mode will dictate the temperature/pressure condition of the refrigerant on the suction side of the compressor.

SUMMARY

In this chapter we have discussed the outdoor coil as a heat absorber. Heat from the sun warms the air and the air is drawn through the outdoor coil to extract the heat. We learned that air at 30°F has heat in it, even though the temperature is below freezing. Air that is warmer is always better, but air-source heat pumps are designed to extract heat energy from air that is below freezing. We also learned that there is an abundance of heat in the air. Until temperatures reach −460°F, there is still heat energy in the air. As the temperature of the outdoor air drops, the amount of heat that a particular heat pump can extract will be reduced. The capacity of the heat pump is dependent on the outdoor temperature. At the same time, the heat pump will be able to extract some heat energy even at low outdoor temperatures.

When outdoor air is cooled below its dew point, it will condense moisture on the outdoor coil. This moisture forms a frost insulator that must be periodically removed to maintain the efficiency of heat transfer. A defrost cycle is needed to remove the frost. This cycle requires the outdoor fan to be off and the system to reverse cycle. Reversing the cycle changes the function of the outdoor coil from evaporator to condenser in order to heat the coil and remove the frost accumulation. Supplemental heaters are used to heat the occupied space during defrost cycles to eliminate any chance that indoor temperature should drop.

The outdoor coil during the heating mode works like the indoor coil during the cooling mode. In both modes the coil acts as the evaporator to extract heat. During heating, the outdoor coil acts as the evaporator to extract heat from outdoor air. During cooling mode, the indoor coil acts as the evaporator to extract heat from the indoor air. In either mode, the temperature and pressure conditions at the compressor suction will depend on the ambient air (indoor or outdoor air) conditions.

REVIEW QUESTIONS

1. Describe how the sun influences the amount of heat available in outdoor air.
2. Explain how heat is measured in a pound of outdoor air (or in CuFt).
3. Relate what "absolute zero" means in relation to the amount of heat available in outdoor air.
4. Describe the function of and need for supplemental heat.
5. Explain why a defrost mode is needed for air-source heat pumps.
6. Describe the similarity in function of the indoor coil in the cooling mode to the outdoor coil during the heating mode.

CHAPTER 4

The Indoor Coil: Rejecting Heat

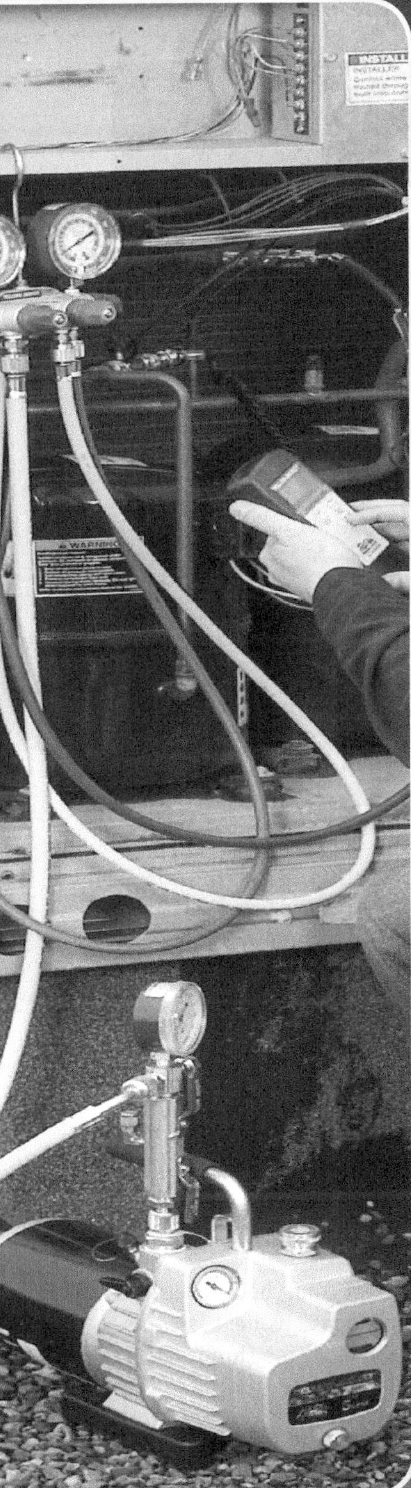

LEARNING OBJECTIVES

The student will:

- Describe how the indoor coil acts as a heat rejector
- Describe the process of desuperheating
- Explain what happens to the liquid/vapor of a refrigerant at saturation
- Explain how a refrigerant can be subcooled
- Describe what happens to the discharge vapor temperature/pressure in the indoor coil as the outside temperature changes

INTRODUCTION

The indoor coil is the point where heat is transferred to the indoor air. The temperature of this coil must be hotter than the indoor air temperature in order for heat to transfer from the coil to the air. If indoor air is 70°F, the indoor coil must be approximately 15°F hotter, or approximately 85°F. How much heat is transferred depends on the amount of heat obtained from outside air. The amount of heat transferred also includes the process of removing superheat, rejecting heat from the refrigerant at saturation to change vapor to liquid, and subcooling the liquid refrigerant. This chapter will explain each of these processes as the heat pump transfers heat to indoor air.

Field Problem

The customer was concerned that the heat pump was not performing well and called to have the system checked. The home had recently been sold to a new owner who was not familiar with the operation of a heat pump. The outdoor temperature that day was approximately 40°F and the indoor temperature was a comfortable 70°F. The homeowner explained that the air temperature from the diffuser didn't seem to be warm; to him, it felt too cool.

The service technician took some time to explain the operation of a heat pump and what the customer should typically expect to see and feel. After the explanation, the service technician proceeded to evaluate the operation of the heat pump. After accurately measuring the outdoor temperature at 41°F, the technician measured the discharge temperature at the compressor. The superheated vapor discharge temperature was 148.6°F. Adding 110°F to the outside ambient temperature, he calculated that the discharge temperature should be 151°F. Subtracting 4°F and adding 4°F, he calculated 147°F to 155°F. The measured discharge temperature of 148.6°F was within this range of temperatures, meaning that the heat pump was operating within the manufacturer's parameters.

The technician completed his system check by checking filters, airflow, air temperature at the closest diffuser, and so on. With all of the checks completed, the technician discussed his findings with the customer and increased the customer's confidence in the heat pump system.

Tech Tip

Some manufacturers require their heat pumps to be checked in a particular way. The 110-degree rule is one example. In this case, the manufacturer's service manual instructs the technician to use the 110-degree method as described in the manual to check the operation of the heat pump.

HOT VAPOR

The indoor coil is acting as a heat rejector. The heat pump is in the heating mode and the indoor coil is functioning as the condenser. Hot vapor is being sent from the compressor to the indoor coil through the vapor line, where the refrigerant vapor changes state from a vapor to a liquid. In order for this change of state to occur, heat needs to be removed from the hot vapor.

Room air is pushed across the indoor coil and it picks up heat from the coil because the coil is hotter than the indoor air. Hot vapor starts to condense as heat is transferred from the refrigerant to the indoor air. During this process, three things occur:

1. Refrigerant vapor is desuperheated.
2. The refrigerant cools to its saturation point, turning vapor to liquid.
3. Subcooling of the liquid occurs before it reaches the metering device.

Desuperheating

Superheat is the amount of heat in a vapor above the saturation point. Desuperheating is the process of removing this heat until the temperature of the vapor reaches saturation. Superheat is the extra heat that is picked up during the evaporation phase of the refrigerant cycle, plus the heat of compression and the amount of heat the compressor motor adds to the refrigerant. Before the refrigerant can condense to a liquid, superheat needs to be removed, desuperheating the hot vapor. Referring to the pressure-enthalpy diagram for R-410A in Figure 4–1, desuperheating is occurring between point E and point F. Because the horizontal line between points E and F relates to pressure, desuperheating is removing the heat above saturation pressure, as well as temperature. On the bottom of the chart, the amount of heat per pound of refrigerant drops during desuperheating, from 147.02 to 122.89.

Saturation

Again referring to the pressure-enthalpy diagram for R-410A in Figure 4–1, saturation begins at point F and continues until point G. From point F to point G, there is an increasing amount of liquid refrigerant forming and a decrease in the

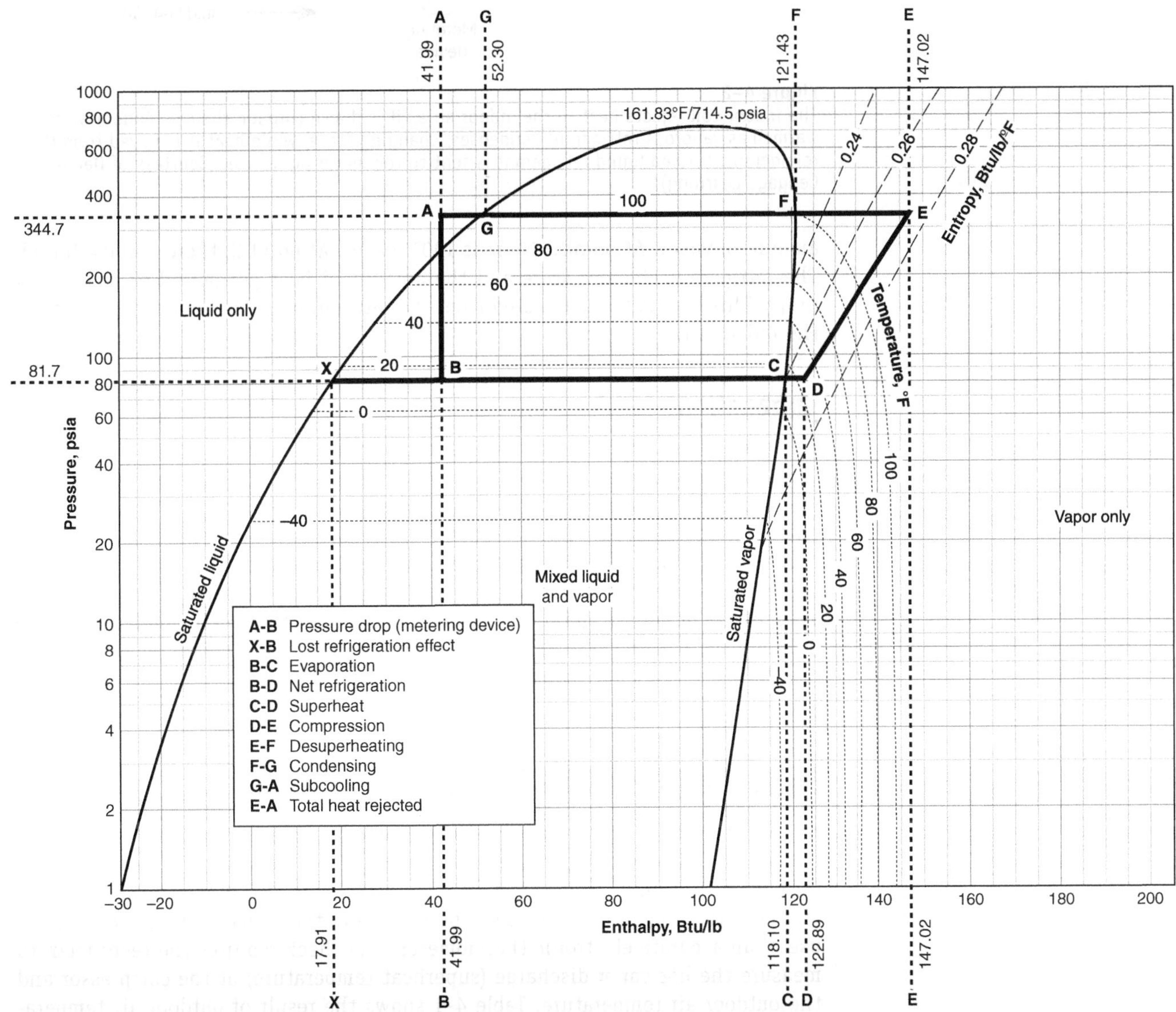

Figure 4–1
Pressure-enthalpy diagram for R-410A (heating mode). (Courtesy of RSES)

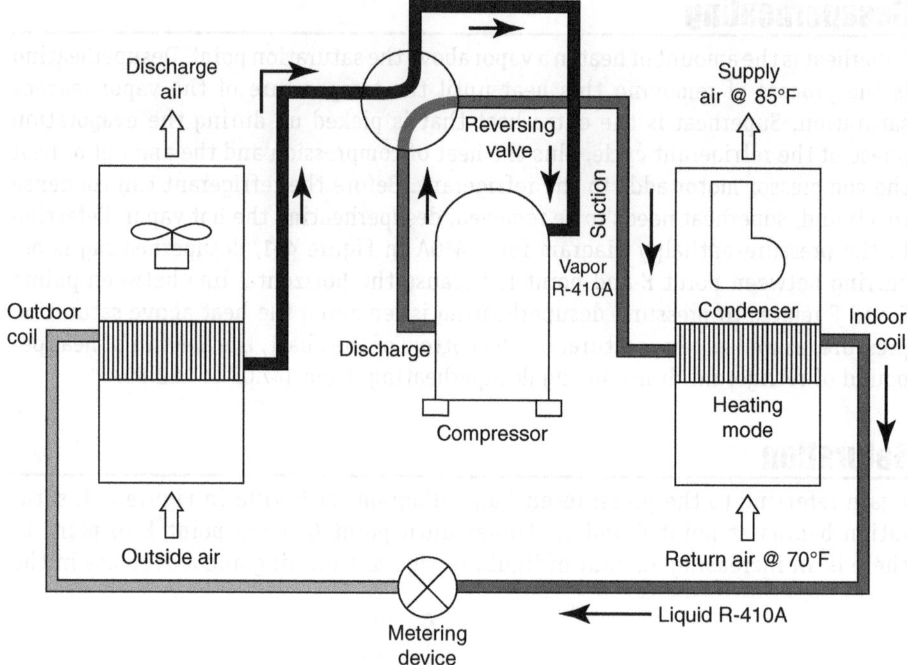

Figure 4–2
The indoor coil rejects heat to the indoor air as it is drawn through the coil. Indoor air is warmed as vaporous refrigerant condenses, changing state, while heat is removed from the refrigerant. Formed liquid refrigerant is sent to the metering device. (Courtesy of Delmar/Cengage Learning)

amount of vapor. At point F, there is 100% vapor. At point G, there is 100% liquid. The refrigerant changes state and the amount of heat per pound of refrigerant drops. This heat is being rejected from the refrigerant to the indoor air, warming the indoor air.

Subcooling

Subcooling is the process of cooling a liquid refrigerant below the saturation pressure/temperature point. At this point, there is 100% liquid and the entire amount of vaporous refrigerant has condensed. Referring to the pressure-enthalpy diagram for R-410A in Figure 4–1, subcooling is occurring from point G to point A. Subcooling occurs when the liquid refrigerant is able to lose additional heat to a point lower than saturation. If the indoor air temperature does not subcool the liquid refrigerant as it moves out of the indoor coil (condenser in the heating mode), some subcooling may occur as the liquid refrigerant moves from the indoor coil to the expansion device (Figure 4–2).

HEAT PUMP PERFORMANCE

An air-source heat pump is dependent on the air temperatures encountered at the functioning evaporator. In the heating mode, the evaporator is the outdoor coil. Outdoor air temperature will fluctuate, and so will the temperatures and pressures of the refrigerant. Checking a heat pump during the heating mode requires the use of an accurate electronic thermometer. The check requires the technician to measure the hot vapor discharge (superheat temperature) at the compressor and the outdoor air temperature. Table 4–1 shows the result of outdoor air temperature change and hot vapor temperature. The hot vapor temperature is based on the outdoor air temperature and should read between +/−4°F of that measurement.

Table 4–1 **Outside Air Temperature Plus 110°F = Hot Vapor Discharge Temperature (+/−4°F)**

Outdoor Air Temperature, °F	Hot Vapor Temperature	Plus 4°F	Minus 4°F
50	160	156	164
45	155	151	159
40	150	146	154
35	145	141	149
30	140	136	144
25	135	131	139
20	130	126	134
15	125	121	129
10	120	116	124

Tech Tip

Keep in mind that the hot superheated vapor temperature at the compressor discharge is not the saturation pressure of the refrigerant.

Tech Tip

For a general idea of the possible pressures that could be generated on the high side of an R-410A heat pump, see Table 4–2.

Table 4–2 **Possible Pressures That Could Be Generated on the High Side of an R-410A Heat Pump**

Discharge Vapor Temperature, °F	Estimated Discharge Saturation Temperature, °F	Estimated Discharge Saturation Pressure, psig
160	150	611.9
155	145	575.1
150	140	540.1
145	135	506.9
140	130	475.4
135	125	445.4
130	120	416.9
125	115	389.9
120	110	364.1
115	105	340.0
110	100	317.0
105	95	295.0
100	90	274.0
95	85	254.0

Safety Tip

R-410A is a high-pressure refrigerant and requires a different set of gauges rated for high pressure. Do not use standard gauge manifolds or other line sets that are not rated for R-410A. If a line should rupture, it could cause injury. Always use appropriate tools and personal protective equipment (PPE) for the task being performed.

Outdoor Coil as a Heat Rejector (Cooling Mode)

This chapter has been devoted to the discussion of how the indoor coil rejects heat during the heating mode. It should also be mentioned that the outdoor coil rejects heat in the same way during the cooling mode. Rejection of heat occurs in both heating and cooling modes, but the coil that is rejecting the heat is in a different location. During heating, the inside coil rejects heat to indoor air. During the cooling mode, the outdoor coil rejects heat to outdoor air. In both cases the coil's function will be as a condenser. The condenser function is where heat is rejected by the refrigerant under high-temperature and high-pressure conditions. The condition of the ambient air (indoor or outdoor air) surrounding the condenser in either heating or cooling mode will dictate the temperature/pressure condition of the refrigerant on the discharge side of the compressor.

SUMMARY

In this chapter you have reviewed the basic heat pump refrigeration cycle as it applies to basic split air conditioning systems. The heat pump uses a condenser (indoor coil, heat rejector) and a blower to heat the inside of the home or building. The outside unit that houses the compressor and evaporator (outdoor coil, heat absorber) absorbs heat from the outside air. The compressor pressurizes the vaporous refrigerant and sends it to the indoor coil (heat rejector). The indoor coil condenses vapor to liquid, sending it to the metering device. The metering device forms a pressure differentiation point and controls the flow of liquid refrigerant to the outdoor coil. The outdoor coil allows the refrigerant to expand, absorbing heat from the outside air. Vaporous refrigerant is pulled into the compressor, is compressed, and is sent to the indoor coil (heat rejector), and the entire refrigeration process begins again.

REVIEW QUESTIONS

1. Describe how the indoor coil acts as a heat rejector.
2. Describe the process of desuperheating.
3. Explain what happens to the liquid/vapor of a refrigerant at saturation.
4. Explain how a refrigerant can be subcooled.
5. Describe what happens to the quantity of heat in the indoor coil as the outside temperature changes.

CHAPTER 5

The Heat Pump Thermostat

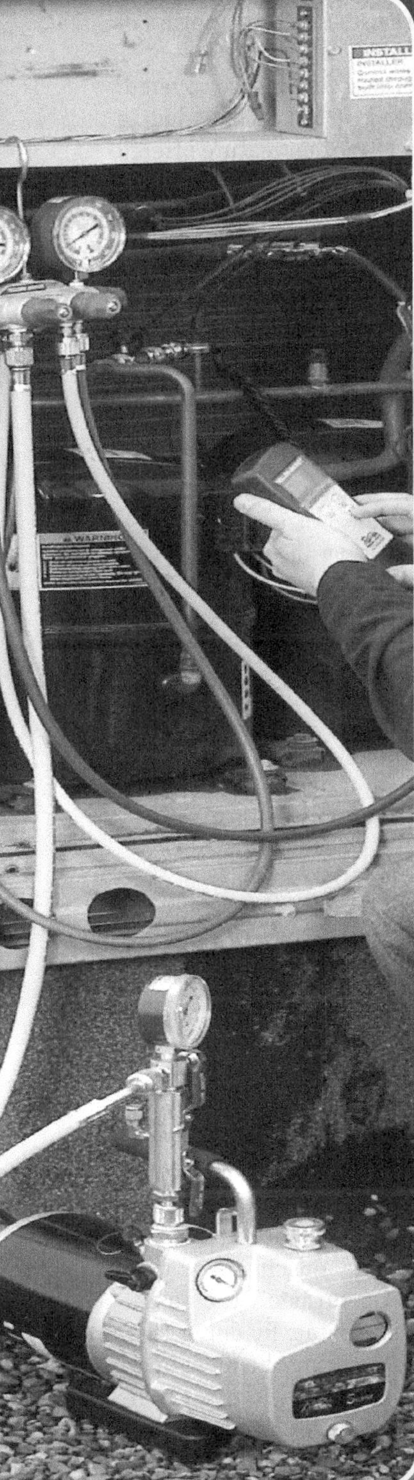

LEARNING OBJECTIVES

The student will:

- Describe what a single-stage thermostat can control
- Explain the difference between single-stage and multistage thermostats
- Describe what a multistage thermostat controls in a heat pump application
- Explain the difference between four-wire and seven-wire thermostats
- Describe what a programmable thermostat can do as compared to a nonprogrammable thermostat
- Explain what a touchscreen thermostat can provide for a user
- Describe night set-back and the optimum temperature and time settings for energy savings
- Describe the difficulty of dehumidification using a thermostat and a whole-structure cooling system

INTRODUCTION

The thermostat is the one control that the user interacts with on a daily basis. In many cases it is the only visible feature that the owner can point to as a symbol of system value. The more complex the system, the more complex the thermostat must be. The complexity of a thermostat should be in the control and not in the user interface. One mark of a good thermostat is that it is easy to use.

This chapter discusses thermostats, starting with a basic, commonly used thermostat, and then explores thermostats that can control more than one heating or cooling source. A heat pump two-stage heating, one-stage cooling thermostat is featured, along with discussion of generic multistage thermostats. The attributes of electronic programmable and nonprogrammable thermostats and the option of touchscreen models are also discussed.

Night set-back and dehumidification options for thermostats are also covered. Night set-back has been shunned in the past, but new information has led to reapplying this option for heat pumps. Humidity control is another feature offered by some thermostat manufacturers. The incorporation of humidity control would seem to be an added benefit, but with that benefit, there are some other design conditions that should be considered.

Field Problem

On a scheduled maintenance call, the customer asked the technician if additional energy savings could be obtained with his heat pump installation. He related to the technician that he had been warned away from set-back thermostats, so he was asking if there were any other options. The technician engaged the customer in a short conversation and asked him to list those things that he might like to have in an energy-saving system. The customer immediately said that he was very happy with his energy bills, but felt that there were times when he had little control over the operation of the system. In particular, he regularly went away from home and would manually lower the temperature, but would come home to a cold house that would stay cold for several hours as he slowly increased the temperature, fearing to operate supplemental heating.

The technician related that some studies have indicated that set-back thermostats are beneficial if the system is set back 8 hours or more, even if the supplemental heat was turned on to boost the temperature. Also, it is possible to have a programmable thermostat bring up the temperature incrementally over several hours to reduce the need to operate supplemental heat. Impressed, the customer asked if he could be shown more thermostat features and scheduled for a thermostat upgrade. As the technician finished the maintenance call, he did two things. First, he checked the thermostat cable for the number of conductors. Second, he called the company and scheduled a sales representative to call on the customer. He checked the wiring to determine if the cable would need to be upgraded to handle a new thermostat with multiple features and gave that information to the sales representative. In this case, there was a seven-wire cable at the thermostat, so there were plenty of wires to upgrade the thermostat.

SINGLE-STAGE THERMOSTATS

Thermostats are available in many different styles and with varying functions. Most will control temperature within +/−0.5°F of setpoint. If the setpoint is 70°F, then the thermostat would turn on the heating system when the temperature drops to 69.5°F and turn off the system when the temperature reaches 70.5°F. Room thermostats often have another built-in device called a "heat anticipator" that helps to prevent temperature swing (overheating). The anticipator is used to fine-tune the thermostat to the system. The anticipator turns off the burner slightly before the real temperature of the room reaches setpoint. By doing this, the amount of residual heat in the heat exchanger is used to bring the room up to setpoint instead of overshooting (going higher than) the setpoint. In older thermostats the anticipator was a heater that increased the temperature close to the thermostatic sensor (usually a bimetallic coil). In electronic thermostats, there may not be an anticipator option, or it may be a timer or a differential adjustment (depending on manufacturer) that is activated to turn off the burner a set amount of time before the thermostat would normally turn off. Always read the setup instructions for electronic thermostats.

Single-stage thermostats operate one heating and cooling source. If a single-stage thermostat were connected to an air-source heat pump (ASHP), it would operate the heating or cooling mode of a single-speed compressor and reversing valve. The thermostat serves to switch the system from heating to cooling by operating the reversing valve. While in heating or cooling, the thermostat turns the compressor on or off depending on the need for conditioning. However, all heat pump single-stage thermostats must have the ability to change the reversing valve in the cooling mode and turn on emergency heat in cases where the heat pump fails. For this reason, a standard, single-stage thermostat used for fossil-fuel heating would not be appropriate for heat pump applications. However, for review purposes, reviewing single-stage thermostat operation is a good place to start.

A simplistic, low-voltage, single-stage thermostat incorporates a thermostatically operated switch that closes to send power to either cooling or heating controls. Some single-stage thermostats can automatically switch from heating to cooling (Figure 5–1). Well-designed single-stage thermostats incorporate automatic switchover as well as system fan control, and more. Electronic thermostats can be programmed to operate at one temperature part of the day and another temperature at other times during the same day. This option is called set-back.

Tech Tip

There are many different types and styles of thermostats. When the HVAC system gets more complicated than just heating or just cooling, selecting the right thermostat becomes more complicated. Here are a few of the options available:

- Programmable versus nonprogrammable
- Digital versus mechanical
- Type of digital display (backlit, large, etc.)
- Single-stage versus multistage
- Automatic changeover from heating to cooling
- Touchscreen versus buttons
- Shape, style, color

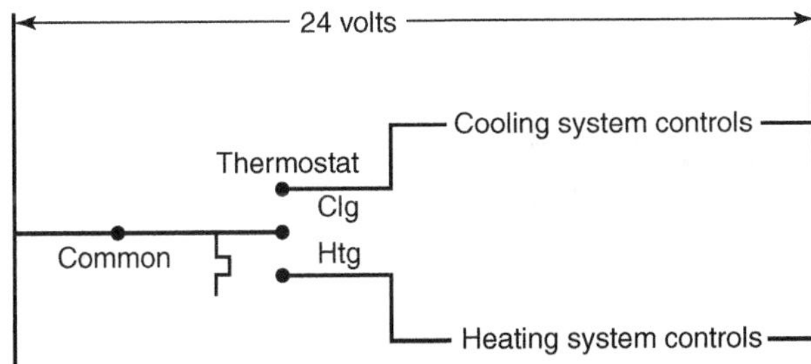

Figure 5–1
The simple single-stage thermostat closes a thermally operated switch to either heat or cool. As the temperature rises, it closes to cooling system controls. As the temperature falls, it closes to heating system controls. (Courtesy of Delmar/Cengage Learning)

MULTISTAGE THERMOSTATS

Multistage thermostats are controls that can operate more than one heating or cooling source. A simple way to think about the operation of a multistage thermostat (Figure 5–2) is a situation where two air conditioning systems and two furnaces are attached to one thermostat. The first air conditioning system is used most of the time, but the second supplements the first during periods of excessive heat load to maintain setpoint. The same concept applies to the furnaces. The second furnace only operates at times when the first cannot keep up with the heating load and the indoor temperature drops below setpoint. If the first furnace is only designed to meet the heating need to an outdoor temperature of 30°F, the first furnace would operate 100% of the time at 30°F outside. As the outside temperature drops, the second furnace is cycled to maintain the interior setpoint temperature.

An air-source heat pump (ASHP) requires a two-stage thermostat. Electric heat strips are used in many systems to augment the heating requirements when the heat pump operates below the balance point. As the outdoor temperature drops below the thermal balance point, the heat pump cannot supply enough heat to maintain the indoor setpoint temperature. The multistage thermostat will, at that point of need, energize the supplemental heat in addition to the compressor to maintain the indoor temperature setpoint. In installations where auxiliary heat is brought on in stages, the thermostat may be designed to control more than two stages or multiple stages of heat strips as the outdoor temperature drops. In this way, the amount of electric strip heat brought on is matched to the incremental needs of the building as the outside temperature drops. This design is more efficient and reduces the amount of electricity used.

Tech Tip

Reversing the functions of the indoor and outdoor coil can only happen if the flow of refrigerant is changed. Discharge vapor must be directed from the compressor to the coil functioning as the condenser (heat rejection). Suction vapor must be directed from the coil functioning as the evaporator (heat absorption) to the compressor. To accomplish the change of flow, a special valve is needed called a reversing valve (also called a changeover valve or four-way valve).

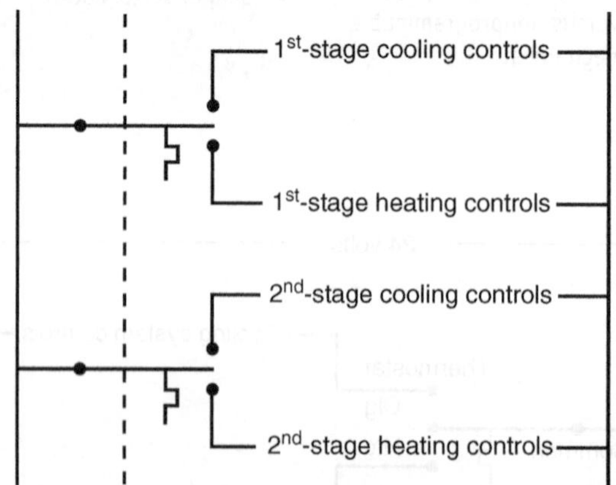

Figure 5–2
The simple multistage thermostat closes a thermally operated switch as the first stage to either heat or cool. As the temperature drops, the second stage closes to the heating system controls; or, as the temperature rises, the second stage closes to the cooling system controls. Each stage is 1.5°F apart from the other. Room temperature must drop 1.5°F below setpoint to activate the second stage of heat. (Courtesy of Delmar/Cengage Learning)

THERMOSTAT WIRING

Most installed thermostats are 24 volts. Our discussion of thermostat wiring will focus on 24-volt, low-voltage installations. Also, because most multistage thermostats are two-stage thermostats, multistage thermostat wiring will be confined to seven-wire installations (R, G, W, W_1, Y_1, Y_2, 0).

Single-Stage Wiring

The basic single-stage thermostat uses three wires: R, W, Y (red, white, and yellow). R is the 24-volt power wire from the transformer. W is the wire that connects to the heating controls. Y is connected to the cooling controls. If the indoor blower is controlled, there is a G (green) wire. Many times, G is assumed and a single-stage thermostat is referred to as a four-wire thermostat (R, G, W, Y; Figure 5-3).

Tech Tip

Some thermostats require a common terminal/wire to complete the power circuit for loads in the thermostat. These loads could be status lighting, backlighting and memory circuits. Care must be taken at installation to ensure there are enough conductors to the thermostat. Other options for thermostats may be those that are power-stealing or battery-powered.

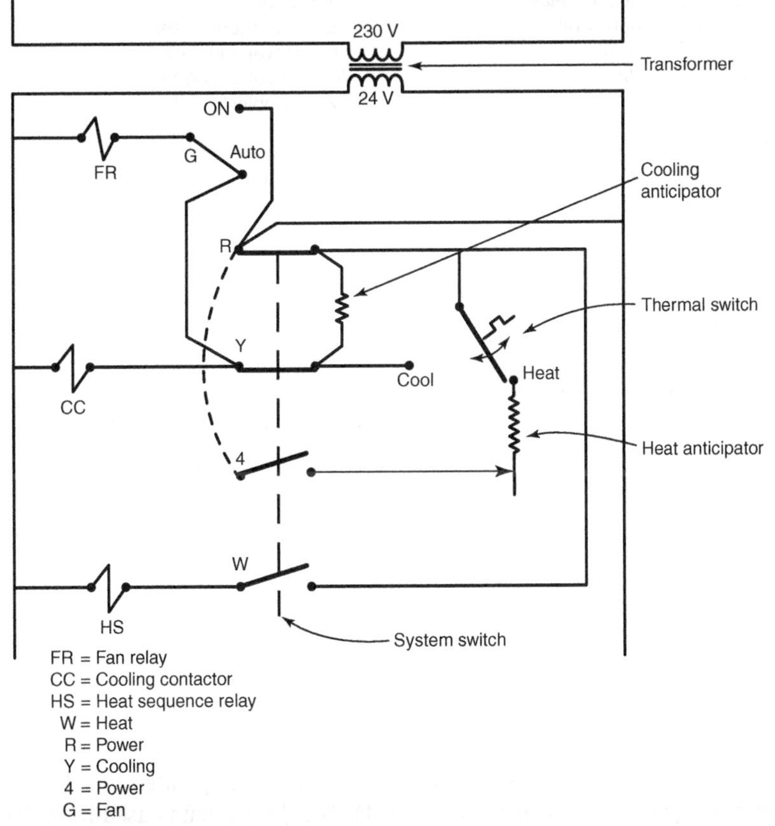

Figure 5-3
A single-stage heating and cooling thermostat. This is a typical four-wire thermostat for heating and cooling. (Courtesy of Delmar/Cengage Learning)

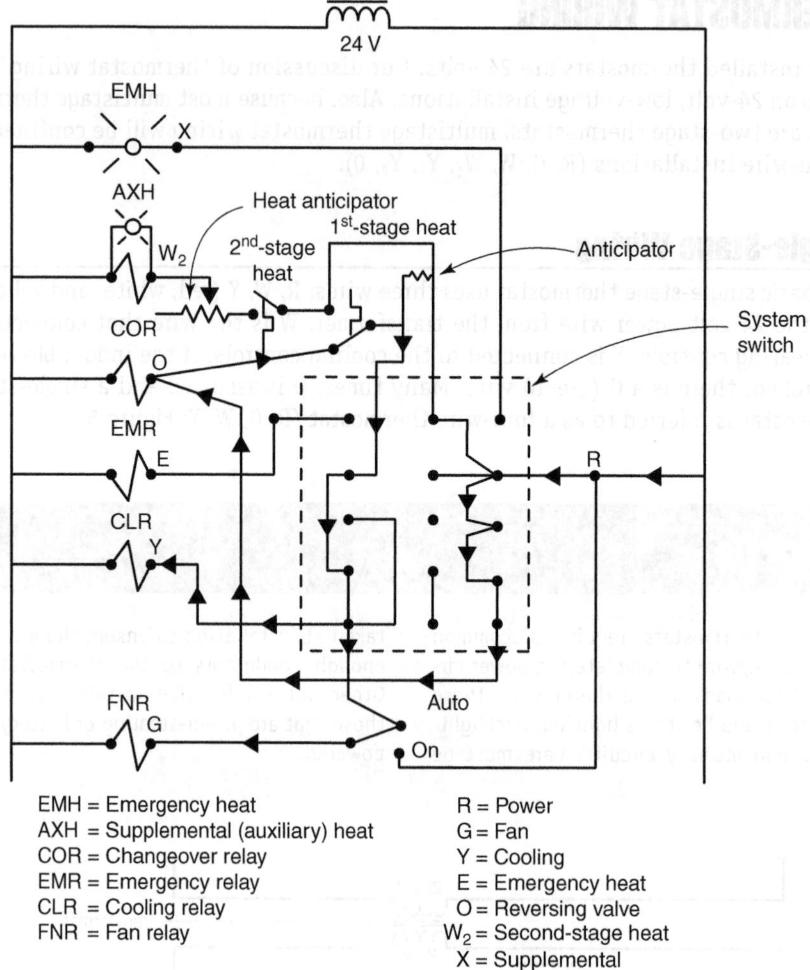

Figure 5–4
A generic heating/cooling thermostat for a heat pump is shown operating in the cooling mode. (Courtesy of Delmar/Cengage Learning)

Multistage Wiring

The multistage, or two-stage, thermostat can have many applications (Figures 5–4 through 5–7). If it is not for a heat pump, the first- and second-stage connections are labeled W_1 and W_2. The cooling connections are similarly marked, Y_1 and Y_2, for the two stages of cooling. This thermostat is also considered to be a seven-wire thermostat (R, G, W, W_1, Y_1, Y_2, O). When used for a heat pump, the thermostat is designed to operate a reversing valve, and the lettering for the valve connection is marked O (or B) for changeover.

A heat pump default mode is typically to heat (for some manufacturers this is different). In order for the heat pump to cool, the function of the heat exchangers needs to be changed. The changeover valve (reversing valve) is operated during the cooling mode to switch the function of each exchanger. We need to remember, too, that the compressor is operating in both the heating and cooling modes. When the thermostat operates the cooling system, the reversing valve and the compressor operate at the same time. In the heating mode, the compressor operates, but the reversing valve does not (with the exception of defrost). The Y terminal is used in cooling and heating modes. Refer to Figure 5–8, which shows the wiring of a two-stage heating, single-stage cooling system. When the cooling mode is selected at the system switch (and if cooling is called), the cooling relay and the reversing valve are operated at the

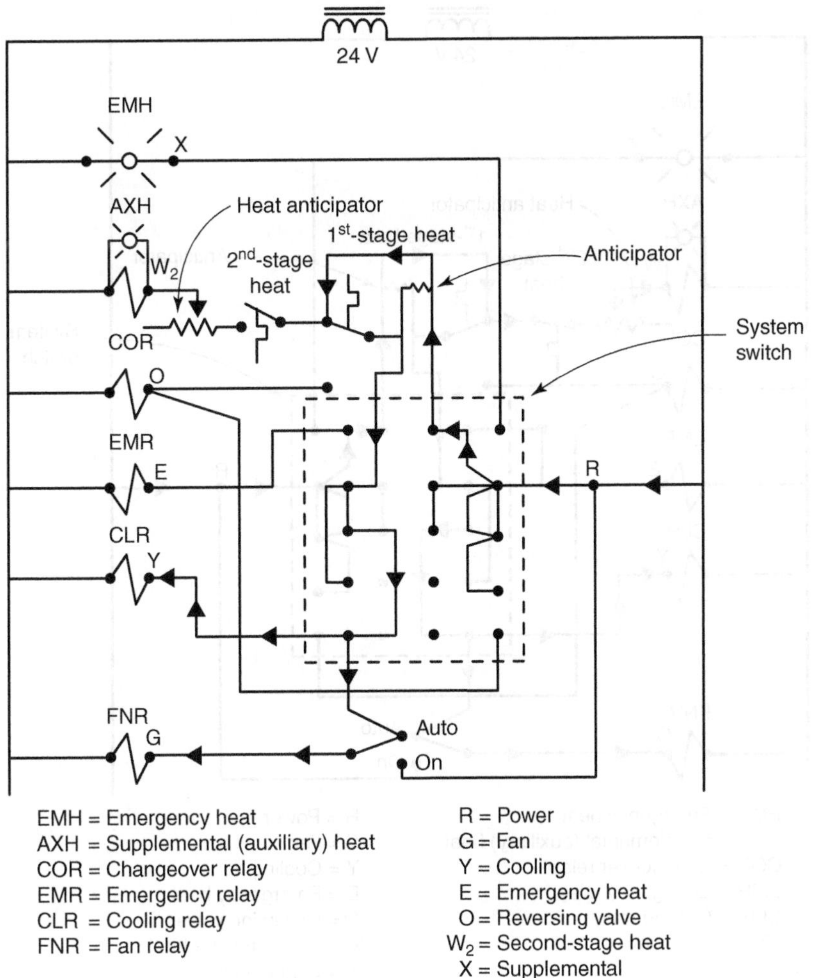

Figure 5–5
A generic heating/cooling thermostat for a heat pump is shown operating in the first stage of the heating mode. The heat pump is heating the structure without supplemental heat. (Courtesy of Delmar/Cengage Learning)

same time. The blower switch is selected for AUTO, which also brings on the blower. In the heating mode (with the system switch selected for HEAT), the first stage of heating is given to the heat pump. The cooling relay will operate the compressor, but the reversing valve is not operated. The blower will operate as in cooling if selected for AUTO. As the temperature drops below the first thermal balance point, the indoor temperature will drop when the heat pump capacity cannot supply all of the heating need. At that point, the second stage of heat will be operated by HEAT 2 (W_2) and will operate the supplemental (auxiliary) heat relay (AUX HT RELAY).

Tech Tip

The first thermal balance point occurs when the heat pump can no longer maintain the internal temperature setpoint for the structure. At this point the outdoor temperature has dropped to a point where an ASHP cannot pull enough heat from outdoor air to satisfy the heat loss of the structure. The second balance point is where the heat pump and the first stage of supplemental heat are not able to maintain indoor air temperatures.

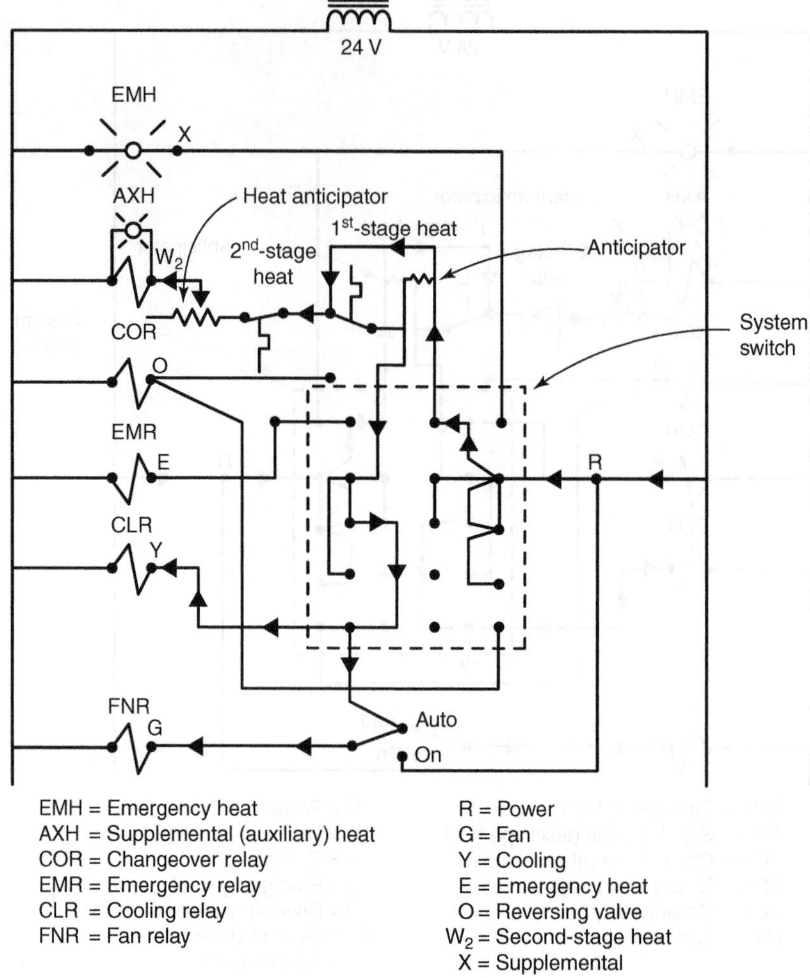

Figure 5–6
A generic heating/cooling thermostat for a heat pump is shown operating in the first and second stages of the heating mode. The heat pump alone is not able to maintain indoor setpoint conditions in the first stage. The second stage comes on as the indoor temperature drops and brings on supplemental heat. (Courtesy of Delmar/Cengage Learning)

Tech Tip

Some thermostats incorporate a B and/or O terminal that is used to control the reversing valve in a heat pump application. The typical heat pump energizes the reversing valve for the cooling mode and the typical thermostat energizes the O terminal for a call for cooling. Some heat pumps (manufacturer specific) energize the reversing valve in the heating mode, in which case the B terminal would be used for this type of application. Some digital programmable thermostats have only one terminal (B or O), and it is programmed in the setup menu to be energized by cooling or heating, depending on system requirements.

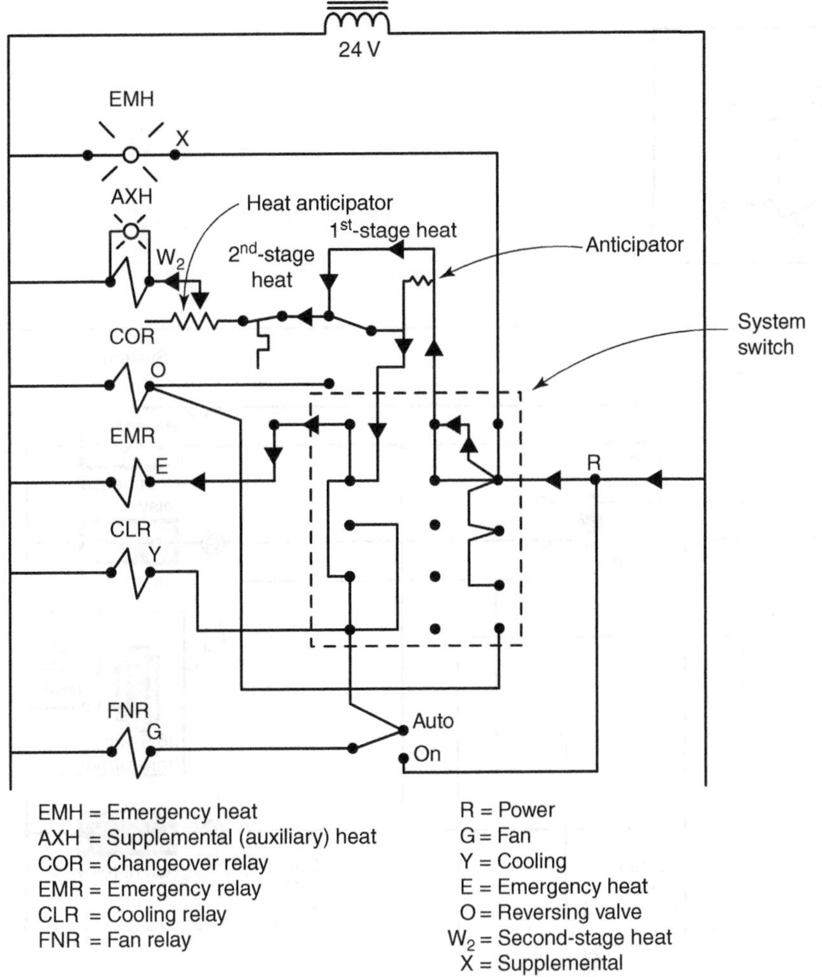

Figure 5–7
A generic heating/cooling thermostat for a heat pump is shown operating in the emergency heat mode. (Courtesy of Delmar/Cengage Learning)

Tech Tip

Thermostat cable is available in several configurations: two-wire (two wires surrounded by a plastic outer covering), five-wire, seven-wire, and so on. When selecting the wire option to use, it is advantageous to select more conductors than needed. A five-wire thermostat wire could be pulled for a simple three-wire thermostat to anticipate added system features and upgrades of the thermostat. This installation eliminates the need to pull extra wire at another time. If a heat pump is being installed, seven-wire should be considered as the minimum number of conductors installed, and eight or ten may be preferred. Additionally, if a wire is damaged and there are unused wires available, one other unused wire can be used.

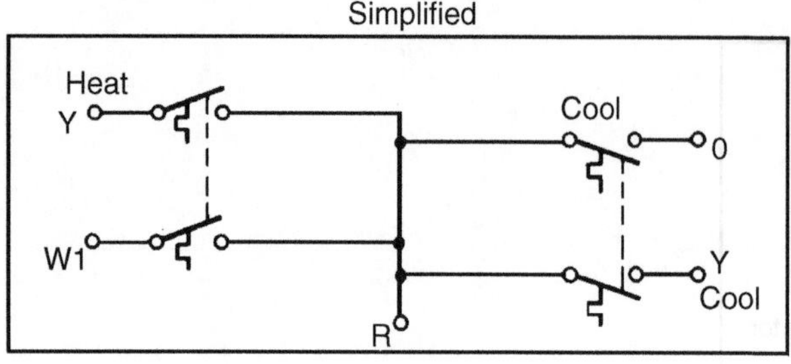

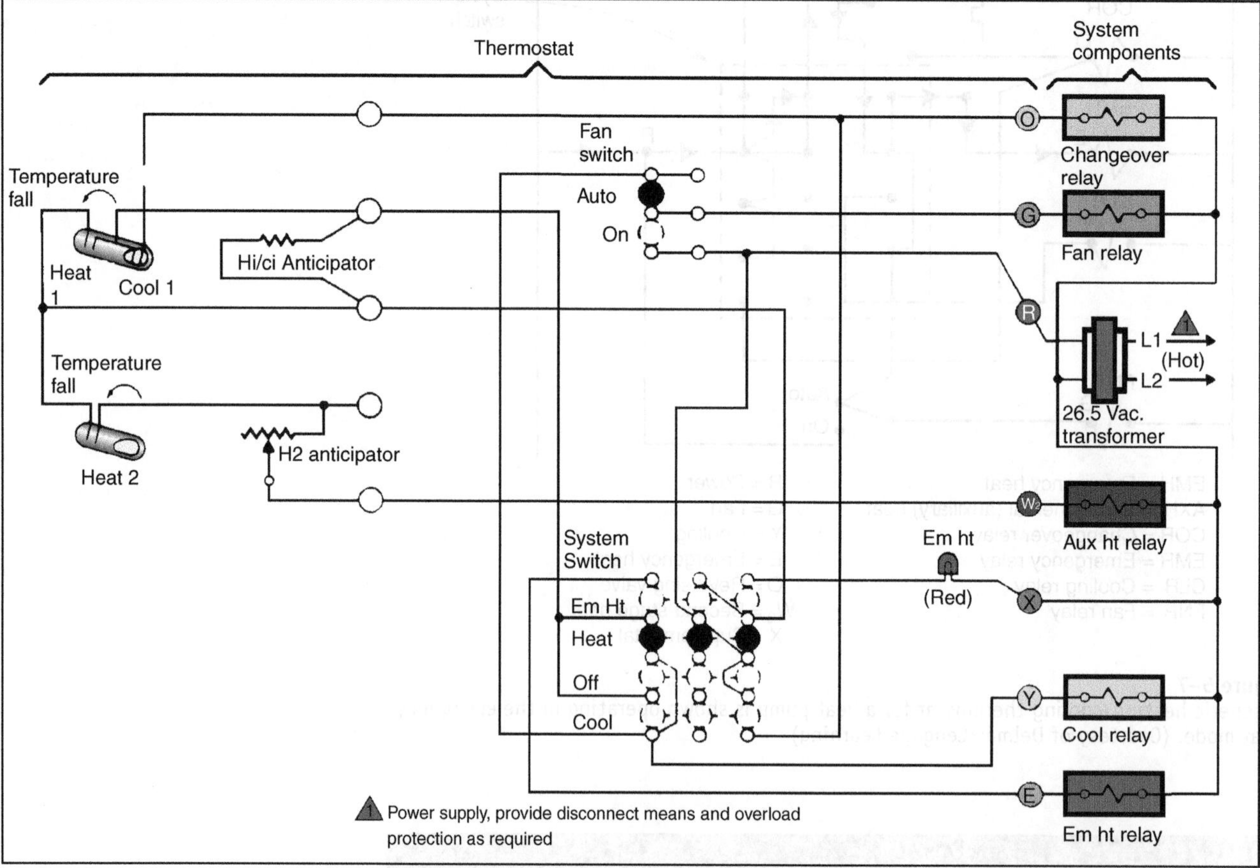

Figure 5–8
A two-stage, or multistage, thermostat for a heat pump with two stages of heating and one stage of cooling. Also note the emergency heat switch and LED. The customer can manually switch to emergency heat if necessary. (Courtesy of Honeywell, Inc.)

ELECTRONIC THERMOSTATS

Electronic thermostats (sometimes also referred to as digital thermostats) are available in programmable and nonprogrammable varieties. Set-back is the key option that moves a person to select a programmable thermostat.

Programmable Thermostats

Programmable thermostats allow the user to set a different indoor temperature for several different times in one day or consecutive days (Figure 5–9). For instance, the thermostat could be programmed to set back to 65°F during the night, bring the temperature up to 70°F at 6:30 AM, set back to 67°F degrees during the day, bring the temperature up to 72°F at 4:30 PM, and set back to

SIMPLICITY

1. **For quick reference,** easy-to-follow instructions are posted inside the door.

2. **Simple programming** is assured by duplicating the previous day's schedule with the Copy Previous Day button.

3. **To enjoy clean air,** the Reset Filter button lets you know when the air filter needs to be cleaned or changed. Timer length can be adjusted to your home's conditions.

4. **Make life easier** by setting the mode button to AUTO to automatically change the operating mode between heat and cool.

5. **Response** is quick and easy with large buttons.

6. **Override the system** when special occasions occur by utilizing the Hold button to temporarily adjust the temperature without reprogramming your entire schedule.

Largest Backlit LCD on the market displays large, easy-to-read numbers with a backlighting feature that is activated by the touch of a button.

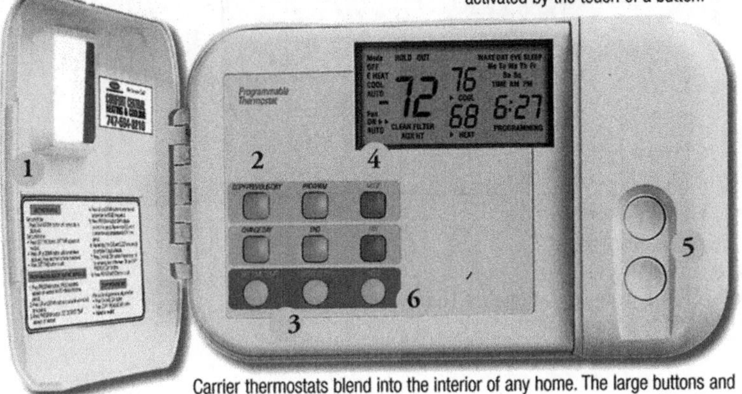

Carrier thermostats blend into the interior of any home. The large buttons and streamlined appearance make these thermostats very homeowner friendly.

Battery-free operation
Even in the event of a power outage, program settings are stored indefinitely and the clock will remain functional for 72 hours.

Figure 5–9
A typical electronic programmable thermostat and features. Note #2, the simple programming feature. (Courtesy of Carrier Corporation)

65°F at 11:30 PM. This program could be the same for 5 days during the week and have a different program for Saturday and Sunday. A thermostat chosen for this type of program would be referred to as a 5-&-2 programmable. If the desired program was different for each day of the week, a 7-day programmable would be chosen.

Some programmable thermostats will also control auxiliary heat during defrost for heat pumps. This model of thermostat will keep the electric resistance strip heaters off while defrosting the outdoor coil and only bring them on if the indoor temperature drops to a preprogrammed level.

1 **Easy operation** of the mode button selects between OFF, HEAT, COOL and AUTO operations. Heat pump thermostat models also include an EMERGENCY HEAT mode.

2 **Airflow is monitored** by the fan button. You can use this function to choose between ON or AUTO fan operations.

3 **Simple maintenance reminder** of the clean filter indicator ensures that you keep your system operating at peak performance and efficiency.

4 **Added convenience** is offered by the outdoor temperature sensor. This optional feature displays the outdoor temperature on the LCD readout.

Backlit LCD displays large, easy-to-read numbers with a backlighting feature that is activated by the touch of a button.

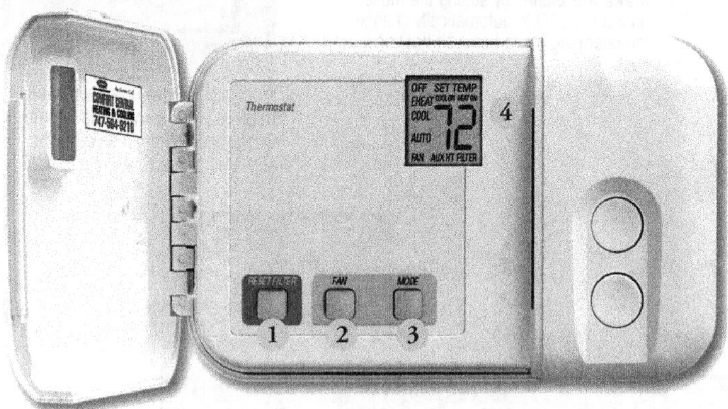

Figure 5–10
A typical electronic nonprogrammable thermostat and features. (Courtesy of Carrier Corporation)

Nonprogrammable Thermostats

A nonprogrammable electronic thermostat is designed to allow changes in setpoint, but not automatic set-back. The thermostat has all of the other features of a programmable thermostat, but not the option to change the setpoint several times during one day. It may, however, have a lower temperature setting that can be selected automatically. A simple nonprogrammable electronic thermostat should be easy to read and operate (Figure 5–10).

Touchscreen Thermostats

Touchscreen thermostats use a flat liquid crystal display (LCD) panel for user input (Figure 5–11). The user is presented with relevant information and new sets of buttons for each feature of the thermostat. The ability of the touchscreen to show the user specific information and the changing user input buttons allow the thermostat to be more user-friendly and reduce the number of manual buttons that need labeling.

These thermostats also have these features:

- adjustable backlighting
- large temperature displays for the customer to read the setpoint and temperature
- system diagnostics
- capability to widen the differential to reduce short cycling
- temperature range limit controls
- setback for multiple programmable occupied/unoccupied fan programming
- filter, UV lamp, and humidifier pad reminders
- ability to accommodate single- or multiple-stage systems

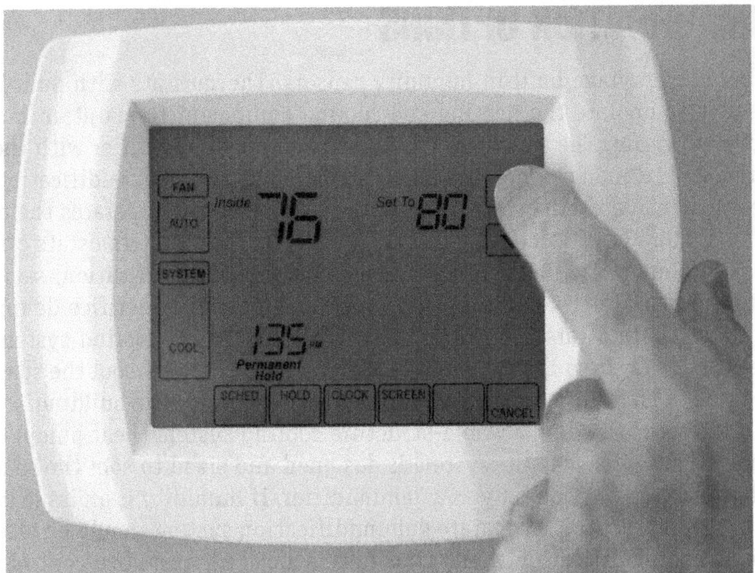

Figure 5–11
Touchscreen thermostat. (Courtesy of iStockphoto.com)

- multiple power options
- dehumidification mode
- 5-minute lockouts for compressor protection
- ability to monitor outdoor temperature
- ability to manage remote and/or averaging sensors
- adjustable fan off delay
- heat pump outdoor lockout temperature control
- dual-fuel capabilities
- auto-changeover
- keypad lockout

NIGHT SET-BACK CONSIDERATIONS

In the past, heat pumps were considered as systems that should not have set-back thermostats. It was felt that the heat pumps could not recover from a set-back condition easily and would require electric heat strips to be used, defeating the savings that could be obtained during set-back. In recent years, computer simulation and actual studies have indicated that savings through set-back can be achieved. A night set-back of 7°F below the normal daytime setpoint for time periods of 8 hours or more in length can result in energy savings.

Tech Tip

With all of the different models and features, such as compressor time delays, the technician should have a discussion with the customer and review system requirements to ensure that the selected thermostat is correct for the application and before programming the thermostat. The discussion should include things like: display size, backlighting, power source, backup/battery source, precision, ease of customer use (instructions), programmability needs, electronic reminders (filter, etc.), and more as additional features are offered by thermostat manufacturers.

DEHUMIDIFICATION OPTIONS

Some thermostats have built-in humidity sensors. Thermostats with built-in sensors are able to operate the heating and cooling equipment to maintain humidity levels in a structure. The thermostat is able to operate a humidifier with the heating system and turn on cooling when the structure requires dehumidification. With some less expensive thermostats, the thermostat specifically operates the cooling system to dehumidify, which may lead to overcooling. Some thermostats offer two settings to dehumidify. One setting is to maximize dehumidification, sacrificing comfort. The other setting is to optimize comfort level and sacrifice dehumidification. Other thermostats that are coupled with multistage cooling systems may operate a smaller system that will reduce moisture, but not overcool the space. The second stage of the thermostat handles the cooling load for the building.

The problem with using a whole-structure cooling system (heat pump) to control humidification is that the system is designed and sized to cool the space and is not designed to function only as a dehumidifier. If humidity is an issue in addition to cooling the space, a separate dehumidification system should be offered as a solution. Dehumidification would thus have a separate humidity control sensor and control circuit.

Tech Tip

System design and equipment selection must be carefully considered when utilizing dehumidification options, especially in northern climates.

SUMMARY

In this chapter we have discussed thermostats, both three-wire and four-wire. Basic thermostats are also known as single-stage controls that operate a single source of heating or cooling. Multistage thermostats can control more than one source of heating or cooling. Multistage thermostats are needed for heat pump applications, because a heat pump has supplemental heat that is used to augment the amount of heat needed when the heat pump experiences conditions below the thermal balance point.

Thermostat wiring can be more complicated as the number of stages and system features increase. Thermostats may require seven or more wires when they are used with heat pumps to handle all of the stages and features. Thermostats that are multistage and applied to heat pumps need to be matched to the manufacturer. Some manufacturers design the reversing valve energized for heating. Other manufactures have the default as cooling. The thermostat must be selected to operate the specific heat pump and the default mode.

Electronic thermostats are offered as programmable and nonprogrammable. Programmable thermostats allow for several different temperature setpoints per day. Both programmable and nonprogrammable thermostats are offered as touchscreen models. The touchscreen provides a variety of information and a changing user interface, while eliminating external buttons. The touchscreen tends to reduce user confusion and increase confidence.

Night set-back is no longer considered taboo for heat pump systems. It has been shown that a modest set-back for an extended period of time will result in energy savings. This means that programmable thermostats can be specified and installed for heat pump applications.

Humidity control is a feature that some thermostats can provide. A humidifier can be operated when the heating system is operating. During the dehumidification process, the cooling system is operated. Operating the cooling system with a priority on dehumidification may lead to overcooling. If dehumidification is an issue as well as cooling, a separate dehumidification system and sensing control should be considered.

REVIEW QUESTIONS

1. Describe what a single-stage room thermostat can control.
2. Explain the difference between single-stage and multistage thermostats.
3. Describe what a multistage thermostat controls in an air source heat pump application.
4. Explain the application of single-stage and multistage thermostats.
5. Describe what a programmable thermostat can do as compared to a nonprogrammable thermostat.
6. Explain what a touchscreen thermostat can provide for a user.
7. Describe night set-back and the optimum temperature and time settings for energy savings.
8. Describe the difficulty of dehumidification using a thermostat and a whole-structure cooling system.

CHAPTER 6

Heat Pump Components

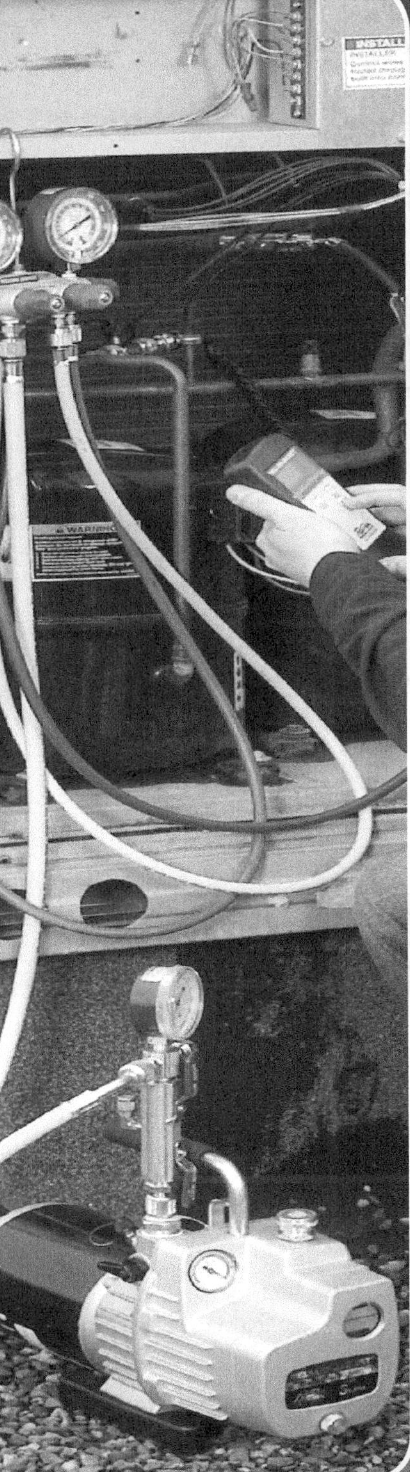

LEARNING OBJECTIVES

The student will:

- Describe the operation of the reversing valve
- Describe the function of the pilot operator of a reversing valve
- Explain how a "default" mode or position for a reversing valve relates to the mode (heating or cooling) of the heat pump
- Describe the use of two metering devices in one heat pump
- Explain how single-direction filter driers can be used in a reverse-flow heat pump system
- Describe the construction of a biflow filter drier
- Describe the purpose and operation of an accumulator

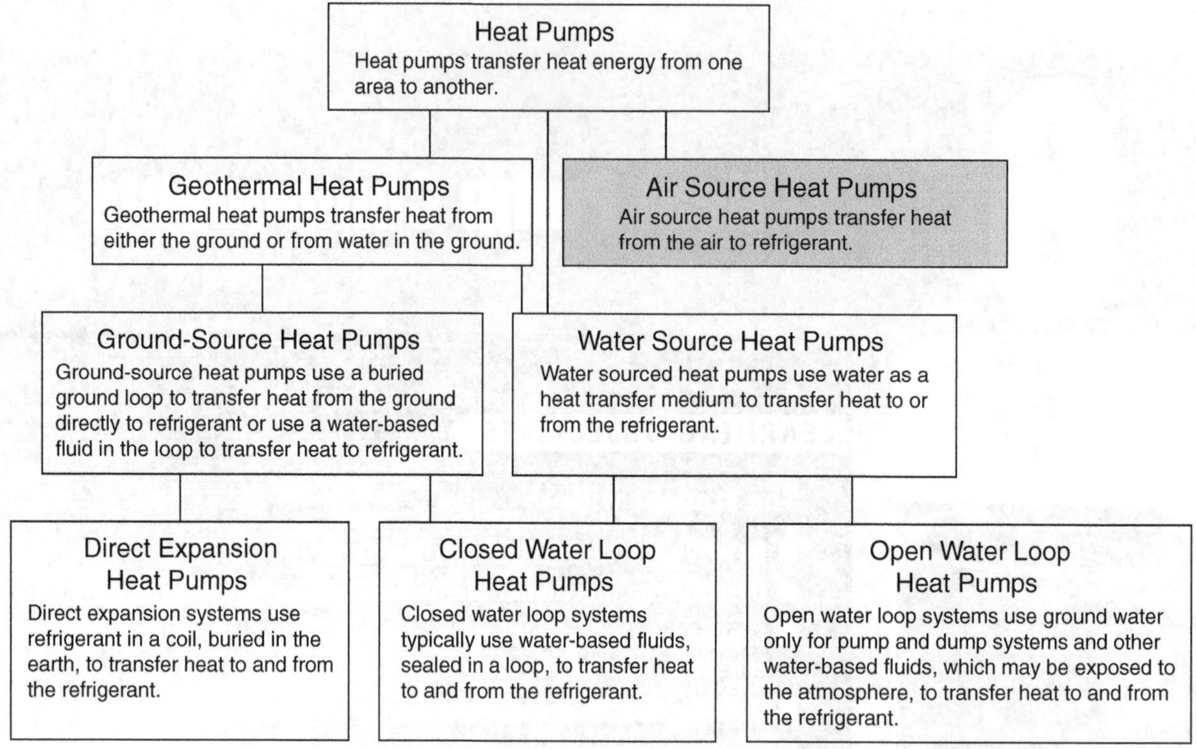
Heat Pumps
Heat pumps transfer heat energy from one area to another.
Geothermal Heat Pumps
Geothermal heat pumps transfer heat from either the ground or from water in the ground.
Air Source Heat Pumps
Air source heat pumps transfer heat from the air to refrigerant.
Ground-Source Heat Pumps
Ground-source heat pumps use a buried ground loop to transfer heat from the ground directly to refrigerant or use a water-based fluid in the loop to transfer heat to refrigerant.
Water Source Heat Pumps
Water sourced heat pumps use water as a heat transfer medium to transfer heat to or from the refrigerant.
Direct Expansion Heat Pumps
Direct expansion systems use refrigerant in a coil, buried in the earth, to transfer heat to and from the refrigerant.
Closed Water Loop Heat Pumps
Closed water loop systems typically use water-based fluids, sealed in a loop, to transfer heat to and from the refrigerant.
Open Water Loop Heat Pumps
Open water loop systems use ground water only for pump and dump systems and other water-based fluids, which may be exposed to the atmosphere, to transfer heat to and from the refrigerant.

INTRODUCTION

All heat pumps have common components, and every compression refrigeration system requires four components: compressor, condenser, metering device, and evaporator (in that order). Heat pumps have an indoor and outdoor coil (evaporator or condenser), compressor, and at least one metering device. In most cases, the heat pump has two metering devices. In addition, sometimes two filter driers are required along with two check valves. Another device may be necessary to protect the compressor from liquid refrigerant. This device is called an accumulator and acts as a refrigerant-flow safety device to the compressor.

Field Problem

A business owner requested service of his heat pump. The heat pump was not working in the cooling mode. This was the first time that the system was operated in the cooling season this year. The day was warm and the owner reported that the unit had worked fine during the heating season. The technician assigned suspected the reversing valve, because this heat pump did not operate the reversing valve during heating (the default mode). During cooling, the valve needs to move to the new position to reverse the refrigerant flow.

Symptoms: System blower operated; outside unit was operating; indoor temperatures were high.

Possible Causes: Reversing valve stuck; thermostat not set correctly for cooling mode; control wiring problems; reversing valve solenoid is bad.

Placing the system in cooling mode, the technician made a simple test of the solenoid coil that operated this piloted-operated reversing valve. Placing a small screwdriver on the coil, he determined that there was no magnetism felt (no buzzing or vibrating felt in the screwdriver). Conducting a continuity check, he confirmed that the coil was open. After getting approval from the owner, he performed the work. Replacing the coil and placing the system into cooling, he reported to the customer that the problem had been eliminated.

Tech Tip

Always get the owner's approval before doing any repair.

REVERSING VALVE

Central to a heat pump is the reversing valve (also known as a changeover valve), shown in Figure 6–1. This valve is responsible for changing the flow (Figure 6–2) to the indoor and outdoor coils. Changing the direction of refrigerant flow means that the indoor and outdoor coils change function (Figures 6–3 through 6–5). When the flow changes, it also changes the direction of flow through the other refrigerant components: drier, metering device, check valves, and so forth.

Tech Tip

A leak test is done on a reversing valve to determine if the valve is leaking from discharge to suction. Temperature measurement is used to determine leakage (Figure 6–4). The valve may also stick, not closing completely, causing the leakage. Other problems with a valve are: sticking in one position (not moving) and a defective solenoid coil.

Figure 6–1
A four-way reversing valve (also known as a changeover valve). (Courtesy of Delmar/Cengage Learning)

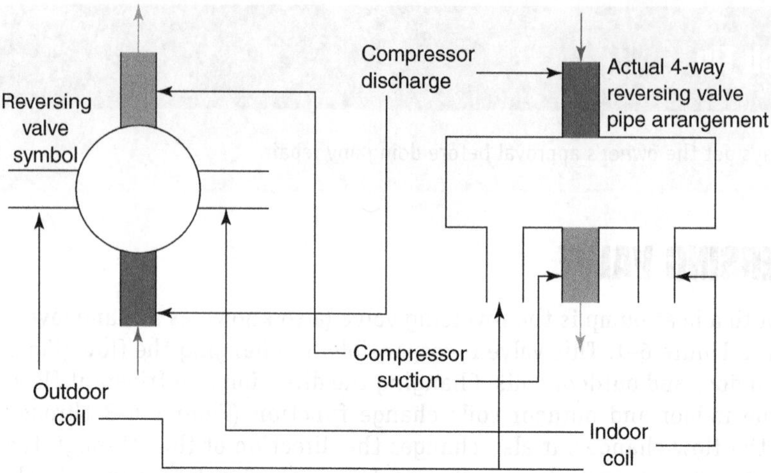

Figure 6–2
The symbol we are using for the reversing valve is labeled to show the relationship to the actual piping arrangement of the four-way reversing valve. We are using the symbol to simplify the understanding of fluid flow through the reversing valve. (Courtesy of Delmar/Cengage Learning)

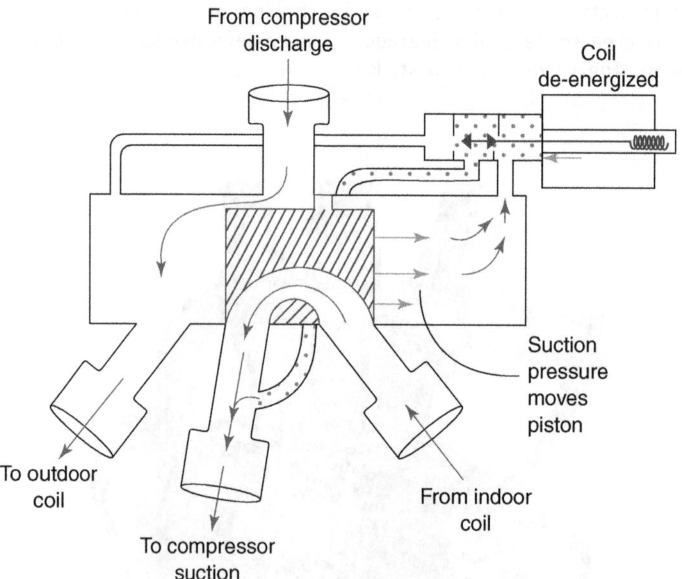

Figure 6–3
This reversing valve is operating in the heating mode. In most cases, the solenoid is de-energized as shown, which is the "default mode" for most ASHPs. (Courtesy of Delmar/Cengage Learning)

Tech Tip

A system operating with a leaky reversing valve will have the same symptoms as a system with an inefficient compressor.

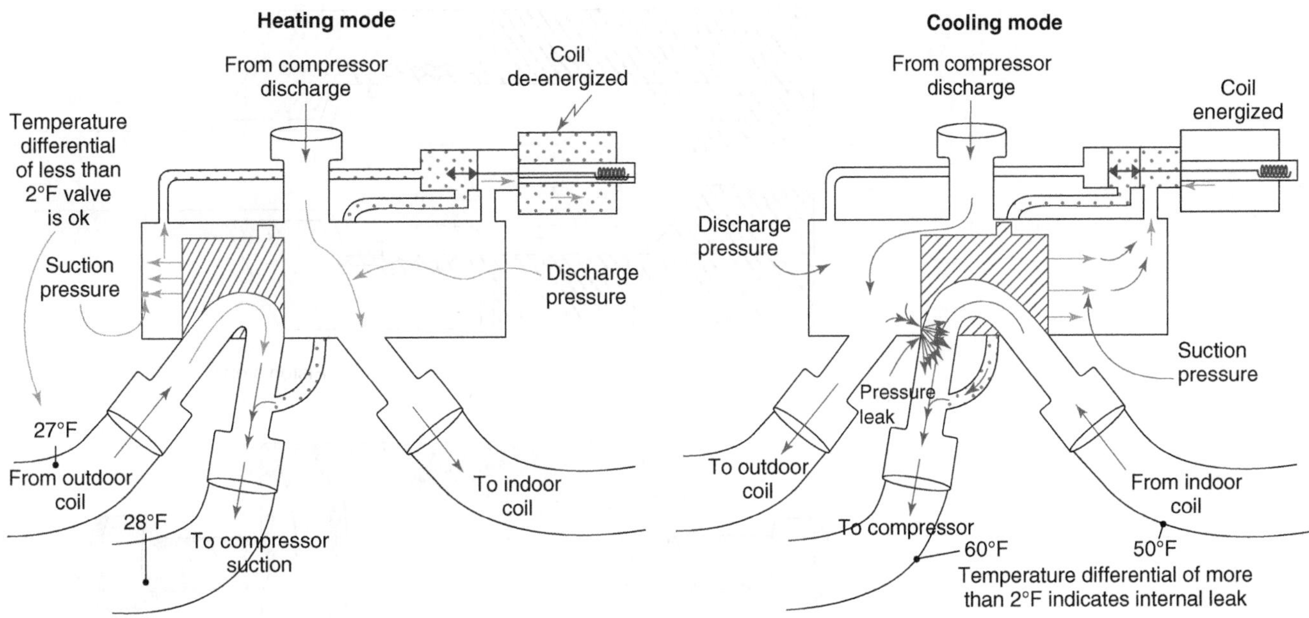

Figure 6–4
Checking reversing valve for leakage from high side to low side. Valve may leak in one mode and not the other. (Courtesy of Delmar/Cengage Learning)

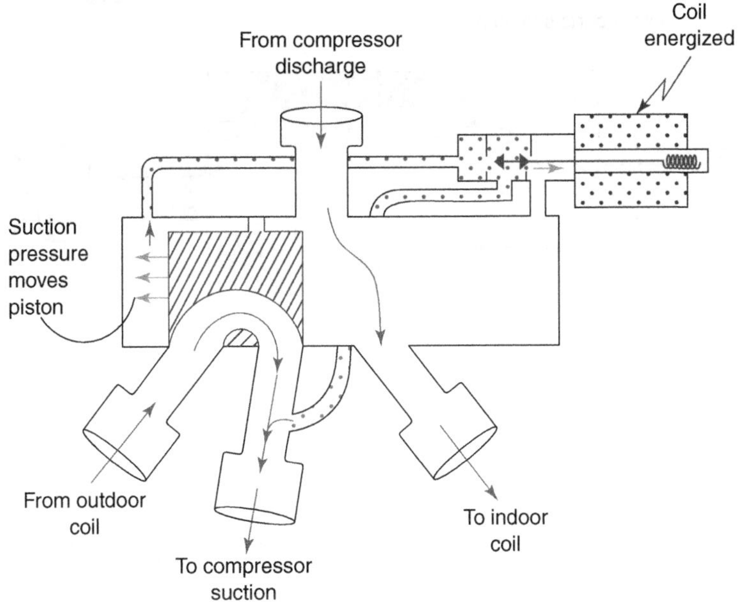

Figure 6–5
This reversing valve is operating in the cooling mode. The solenoid is energized as shown, which moves the valve from the heating default mode to the cooling mode. (Courtesy of Delmar/Cengage Learning)

METERING DEVICES

Heat pumps use several types of metering device configurations. Some use a metering orifice, which is basically a plate with a small, precisely drilled hole or a piston with an orifice. See Figures 6-6 through 6-8 for an example of how a metering piston operates.

Some systems use two capillary tubes and check valves to provide metering for both cooling and heating modes. See Figures 6-9 and 6-10 for examples of this in both heating and cooling modes.

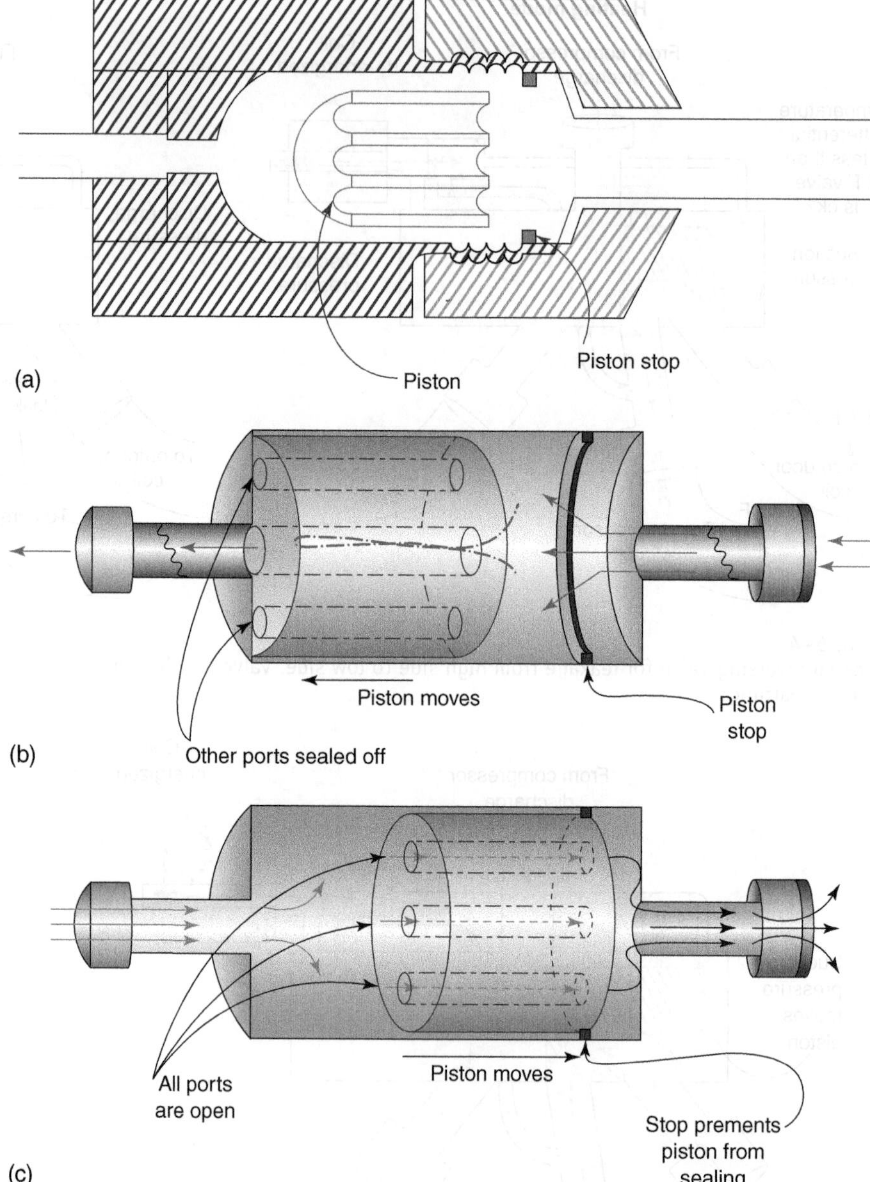

Figure 6–6
(a) This is the construction of a metering orifice in a piston. The piston is behind a retainer assembly that can be removed for service. (b) The illustration shows the metering piston in the cooling position and metering liquid refrigerant to the indoor coil. In the heating position. (c) The piston allows unrestricted flow of refrigerant, essentially moving the metering orifice out of operation for the heating mode. (Courtesy of Delmar/Cengage Learning)

Thermostatic expansion valves (TXVs) are also used and could replace one or both of the capillary tubes shown in the figures. A popular combination is the use of a TXV for the heating mode (see Figure 6–11a) metering device and a capillary tube for the cooling mode metering device.

Single electronic expansion valves are commonly used in high efficiency heat pumps because the valve can be modulated in either direction (see Figure 6–11b). The sensing element can be installed on the suction of the compressor, after the reversing valve. In this way the expansion valve can sense the suction vapor in either heating or cooling mode.

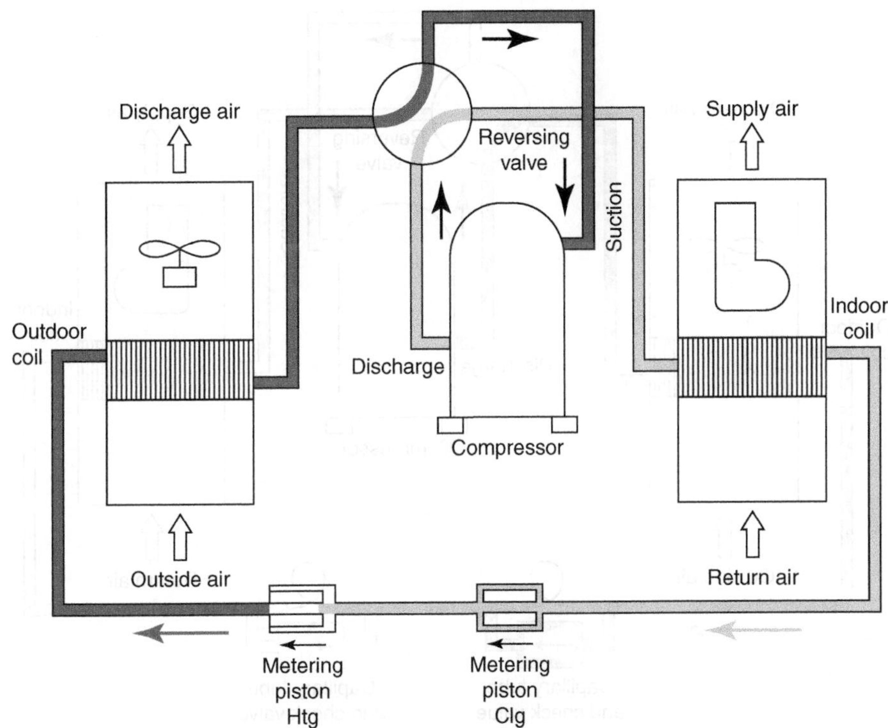

Figure 6–7
The heating (htg) metering piston is moved out of the way for cooling, and the cooling (clg) piston is moved into the metering position. The cooling metering piston is metering the flow of liquid refrigerant to the indoor coil. (Courtesy of Delmar/Cengage Learning)

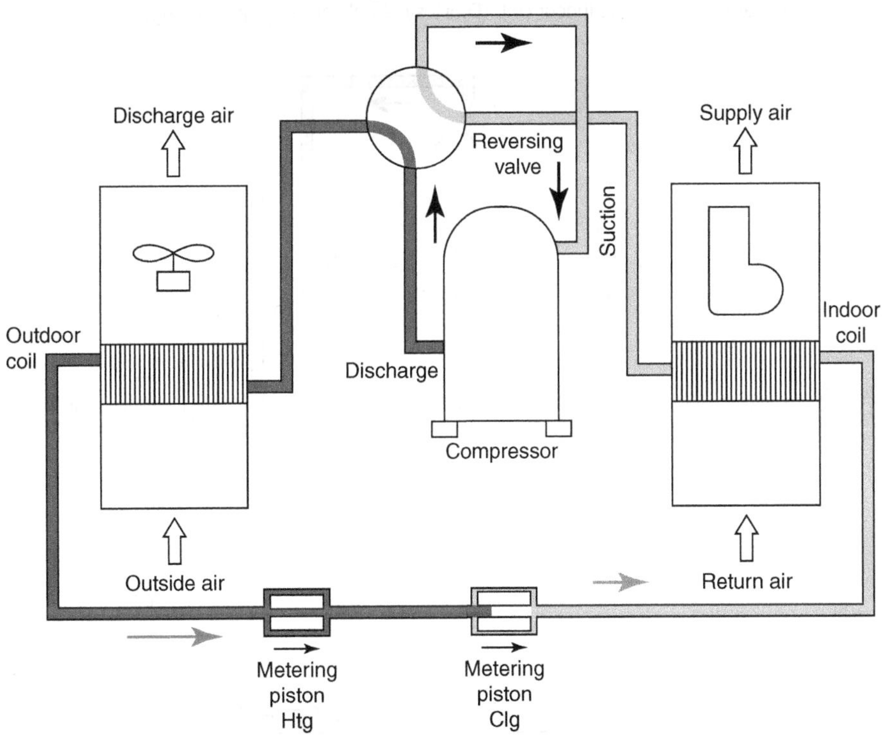

Figure 6–8
The cooling (clg) metering piston is moved out of the way for cooling, and the heating (htg) piston is moved into the metering position. The heating metering piston is metering the flow of liquid refrigerant to the outdoor coil. (Courtesy of Delmar/Cengage Learning)

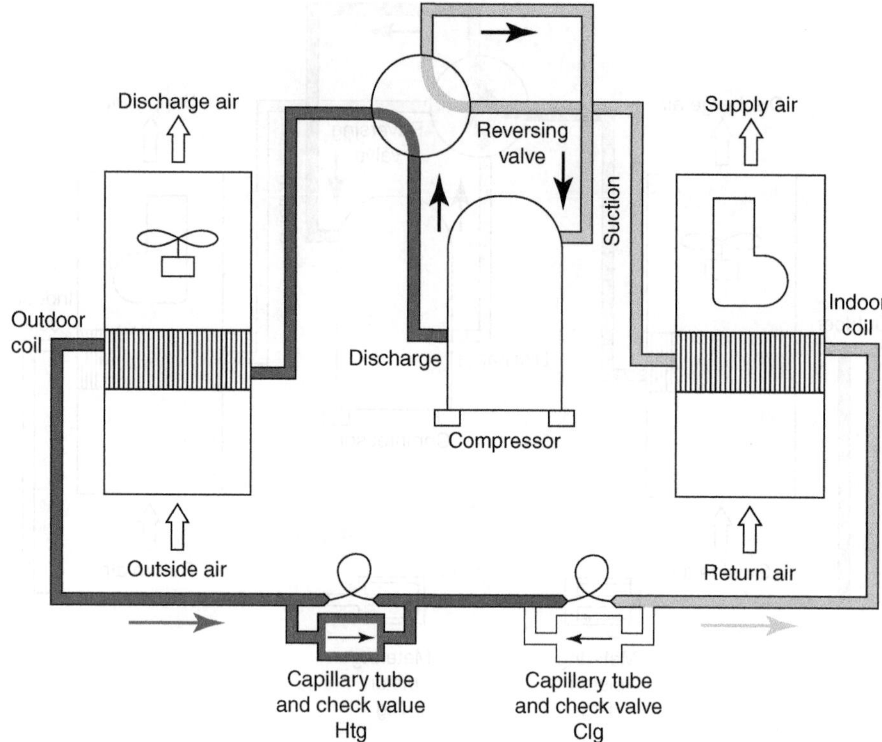

Figure 6–9
In this diagram, two capillary tubes and check valves are used for each mode—heating and cooling. Notice that the check-valve arrow points in the direction of "full flow" or unrestricted flow. If the heat pump is in the cooling mode, the cooling check valve is closed to fluid flow. Liquid will only flow from the outdoor coil, through the heating check valve (open to flow; in the direction of the arrow), and then through the cooling capillary tube (cooling check valve is closed to flow; against the direction of the arrow) and to the indoor coil. (Courtesy of Delmar/Cengage Learning)

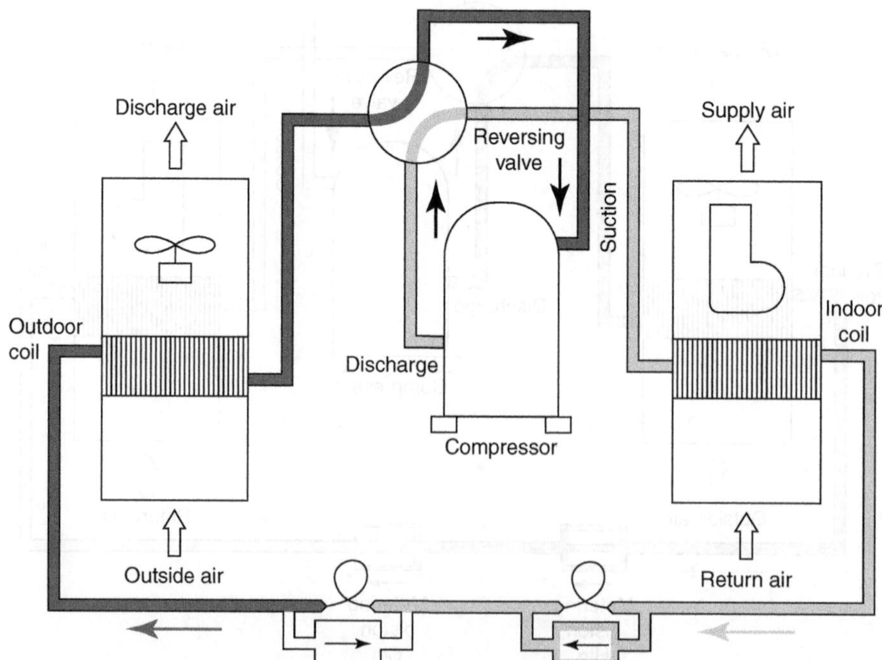

Figure 6–10
The reverse would be true for the heating mode. The liquid would flow from the indoor coil through the cooling check valve (open) and then through the heating capillary tube (heating check valve is closed). (Courtesy of Delmar/Cengage Learning)

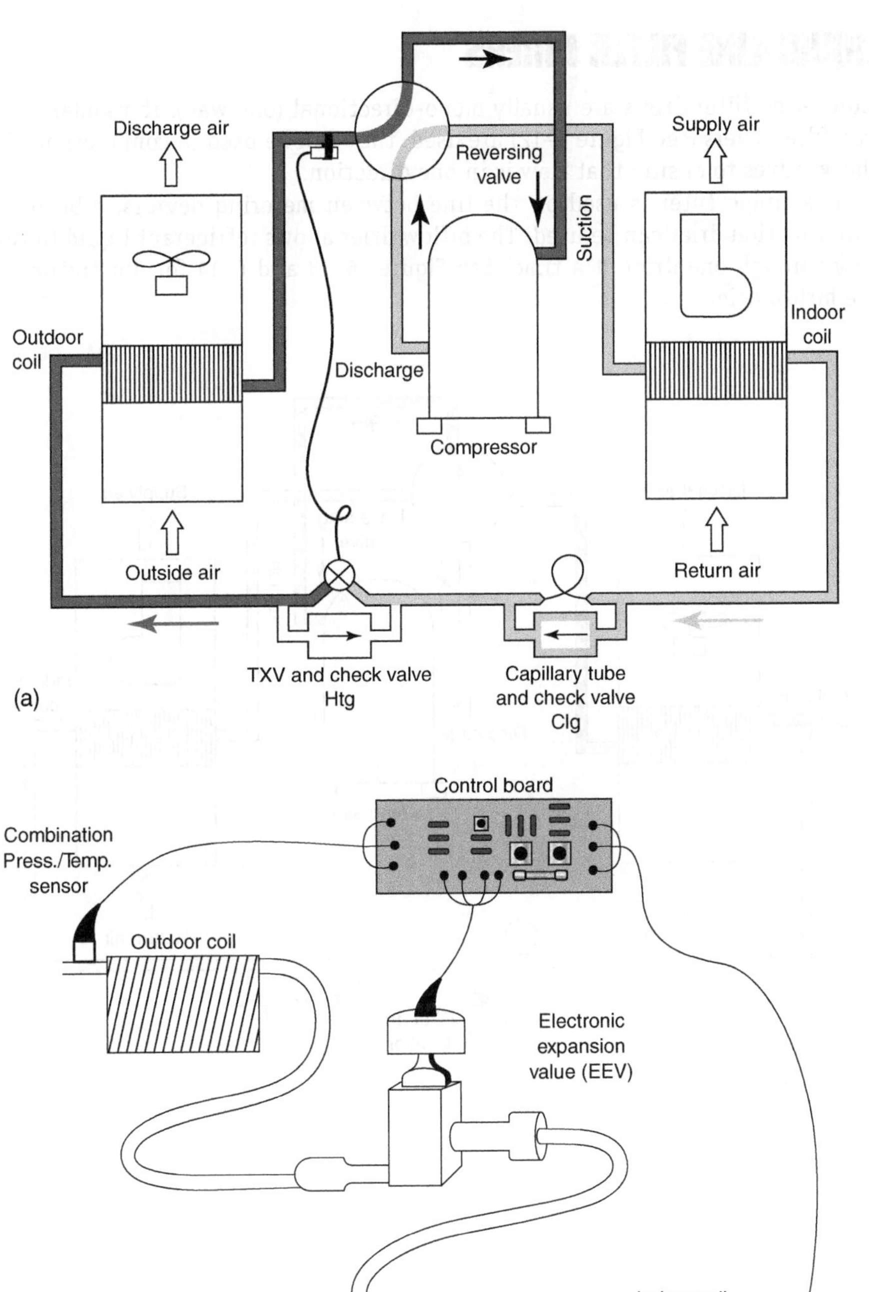

Figure 6–11
(a) Some systems use a combination of metering devices. This diagram shows the use of a thermostatic expansion valve (TXV) for heating mode and a capillary tube for cooling mode. Check valves are used to ensure flow to the correct metering device. A single electronic expansion valve with a sensing element connected to the suction of the compressor can control the flow of liquid refrigerant in either heating or cooling mode. The diagram shows the heating mode. (b) This diagram shows an electronic expansion valve (EEV). Electronic temperature sensors are wired to the control board, monitoring both temperature and pressure. The EEV is placed between the indoor and outdoor coil and can control the flow of refrigerant in either direction, depending on whether the system is in cooling or heating mode. Both electronic sensors and the EEV work to maintain the right amount of refrigerant flow to either the indoor or outdoor coil. (Courtesy of Delmar/Cengage Learning)

LIQUID-LINE FILTER DRIERS

Liquid-line filter driers are usually mono-directional (one way). If standard one-way filter driers (see Figure 6–12) are used, they can be used in conjunction with check valves to ensure that flow is in one direction.

If a single filter is used on the line between metering devices, a biflow or two-direction drier can be used. The biflow drier allows refrigerant liquid to flow only through one drier at a time. See Figures 6–13 and 6–14 for illustrations of the biflow drier.

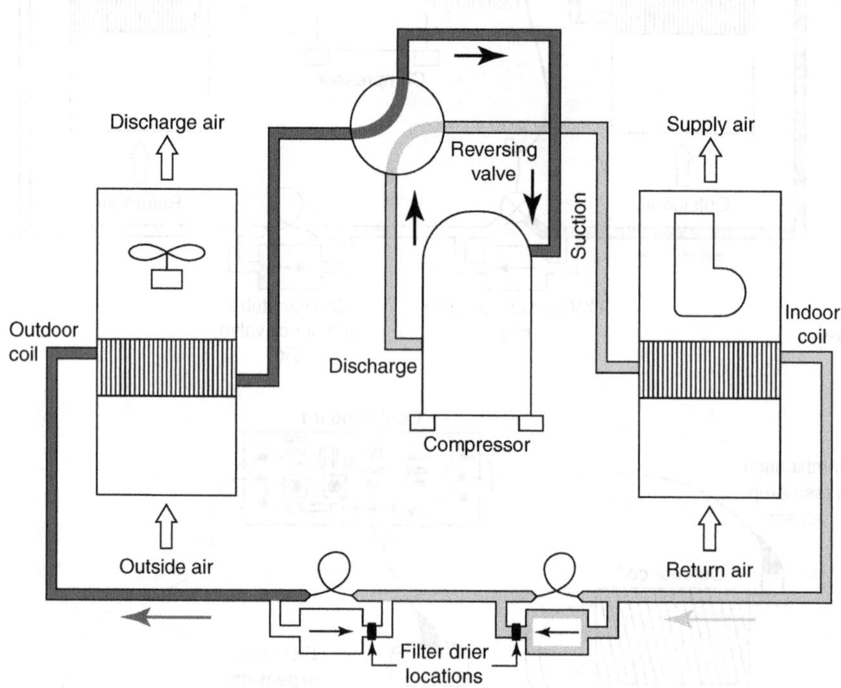

Figure 6–12
Filter driers are placed in the check valve circuit. Check valves direct flow for each liquid-line drier. (Courtesy of Delmar/Cengage Learning)

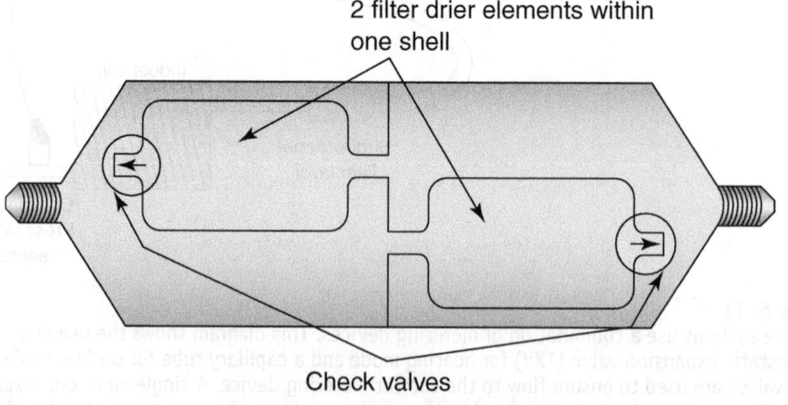

Figure 6–13
Refrigerant flow can be in either direction. A biflow drier allows liquid to flow through only one drier at a time, because of built-in check valves. Essentially, two individual driers with check valves are housed within a single shell with refrigerant connectors at either end. The drier appears to be a single filter drier from the outside. (Courtesy of Delmar/Cengage Learning)

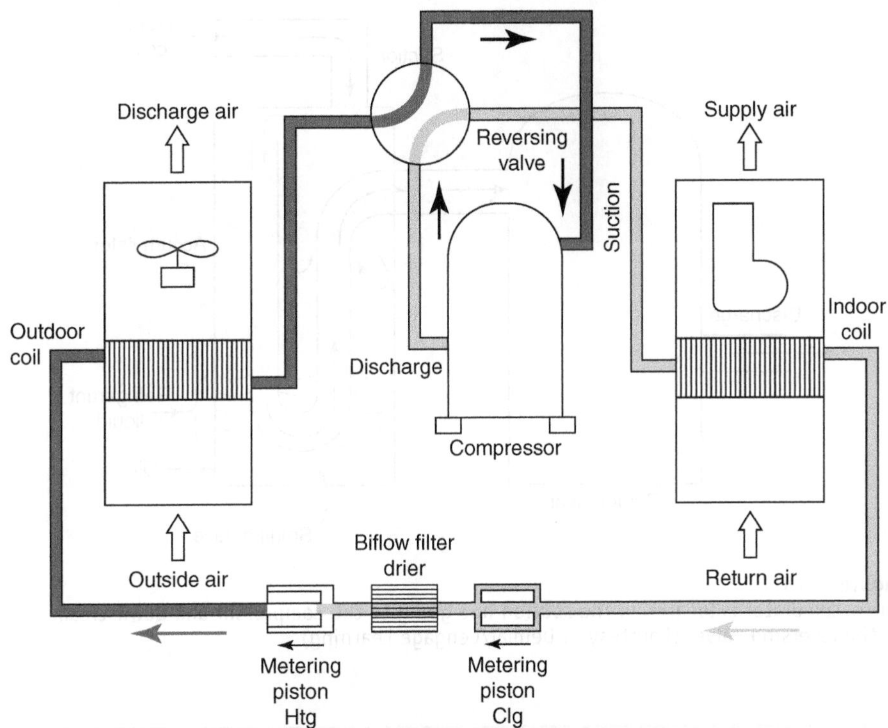

Figure 6–14
The biflow filter drier is positioned between both metering devices. Shown is a metering piston system with a biflow drier between both metering devices. (Courtesy of Delmar/Cengage Learning)

> **Tech Tip**
>
> When a filter drier becomes plugged, the pressure will drop from inlet to outlet. When this happens, a small portion of liquid refrigerant may vaporize. The filter drier will become cooler than the liquid line entering the drier. This temperature difference can be felt and measured. If any temperature difference is measured from the inlet to the outlet of the drier, the drier should be replaced.

Accumulator

An accumulator ensures that all of the refrigerant returning to the compressor is vapor. Any drops of liquid refrigerant are given a chance to evaporate within the accumulator. The piping configuration of the accumulator only allows vaporous refrigerant to be pulled from the top of the tank. The velocity of the refrigerant in the lines is very high, but in the tank the velocity or speed of the refrigerant slows. Liquid droplets lose their speed and slow down, falling out of the vapor. Liquid droplets, because they are heavy, will drop to the bottom of the tank.

The accumulator provides a means of protection for the compressor. Liquid refrigerant entering the compressor could damage the compressor, because liquid cannot be compressed. As it protects the compressor from receiving liquid refrigerant, the accumulator could also cause oil entrapment. To prevent this from happening, a small orifice is placed at the bottom of the "U" to cause any trapped oil at the bottom of the accumulator to move up the suction line and back to the compressor. Figure 6–15 showing the accumulator does not show the orifice at the bottom of the "U," for simplicity.

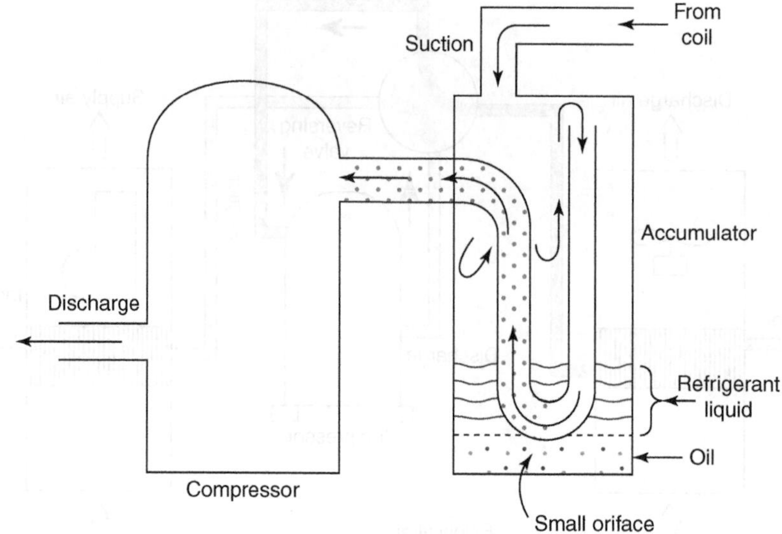

Figure 6–15
The accumulator is located in the suction line going to the compressor and downstream of the reversing valve. (Courtesy of Delmar/Cengage Learning)

Tech Tip

During compressor burnout, the orifice in the accumulator may become plugged. There is no good way of checking for blockage or to remove a blockage. Replacement of the accumulator when a burned-out compressor is replaced is suggested.

Tech Tip

As liquid accumulates within an accumulator device, a line of condensation or ice indicates the level of liquid. When removing refrigerant from the system, the accumulator may need to be heated to evaporate liquid refrigerant.

SUMMARY

In this chapter we have discussed the common refrigerant components found in an ASHP. The reversing valve is the functional "heart" of the system, determining the mode of the system—heating or cooling. The reversing valve is also called a changeover valve. The operation of the valve is to switch the connection of the compressor to the indoor and outdoor coils. When the reversing valve operates, the function of each coil switches, changing the mode from heating (default) to cooling in most ASHPs.

Several different styles of metering devices can be used in ASHPs. The simplest are capillary tubes or metering orifices because they do not have moving parts. A hybrid of this is the metering piston that moves to a new position, depending on the mode—heating or cooling. Thermostatic expansion valves provide higher efficiency and can be used in combination with capillary tubes. The electronic expansion valve is unique, because it has the ability to control liquid flow in either direction. This means that a single valve may be used instead of two metering devices.

Filter driers are usually single-direction devices (mono-directional) and a heat pump system will typically require two filter driers. There is also a biflow drier that looks like a single device, but has within it two driers with check valves to divert the flow as the direction of flow changes.

An accumulator is used in most heat pump designs to contain liquid refrigerant that may flow back to the compressor. This device allows liquid refrigerant to drop to the bottom of a tank so that only vaporous refrigerant is returned to the compressor. The accumulator acts as a refrigerant-flow safety device for the compressor.

REVIEW QUESTIONS

1. Describe the operation of the reversing valve.
2. Describe the function of the pilot operator of a reversing valve.
3. Explain how a "default" mode or position for a reversing valve relates to the mode (heating or cooling) of the heat pump.
4. Describe the use of two metering devices in one heat pump.
5. Explain how single-direction filter driers can be used in a reverse-flow heat pump system.
6. Describe the construction of a biflow filter drier.
7. Describe the purpose and operation of an accumulator.

CHAPTER 7

Motors

LEARNING OBJECTIVES

The student will:

- Describe one type of high-start-torque motor
- Describe one type of low-start-torque motor
- Explain why a high-start-torque, single-phase motor would require a start capacitor
- Describe two types of motor-starting devices
- Explain why blowers and fans use a certain type of motor, and name one motor type
- Describe how to identify common, start, and run motor terminals
- Describe how to electrically check motor windings
- Explain what "ECM" stands for and how these motors work

INTRODUCTION

Heat pumps use electric motors to move heat. These motors are used in a variety of ways. Some motors are used to drive compressors and will be studied in more detail in another chapter. Other motors move both high- and low-temperature fluids, such as air and water, through heat exchangers. Motors account for a significant amount of energy usage. Some new motors, such as the electronically commutated motor (ECM), reduce the amount of energy needed to move fluids. This chapter describes types of motors and their uses in heat pump systems.

Field Problem

An air-source heat pump (ASHP) was not cooling as required. The outside fan of the ASHP was not running, the system was locking out, and the customer was continually resetting the system at the thermostat. The customer reported this to the repair company, which dispatched a repair technician to remedy the problem. When the technician arrived, he met with the owner and used good customer relations to determine the symptoms. The technician went to the outside unit and observed that the fan was stalled.

Symptoms: Customer resetting; no cooling; outdoor fan not running; system locking out on a safety device.

Possible Causes:

1. Bearings
2. Broken wires to the fan
3. Motor terminals at the contactor
4. Motor run capacitor
5. Burned-out motor windings

Using this list, the technician eliminated the most obvious possible causes. Turning off the power to the outside unit, the technician tried moving the blades and found that the motor turned freely. Visually checking the motor terminals helped to establish that the terminals were securely connected and there was no visual discoloring or oxidation. The motor and wiring ohmed-out correctly. Checking the capacitor, the technician found that the capacitor was open (see Figure 7–1). After replacing the capacitor, the motor started and the rest of the system checked out for correct operation.

Figure 7–1
This PSC fan motor shows three leads (lower right), with two leads to the capacitor on the lower left and a green ground wire from the motor to the blower housing. The capacitor leads were removed to check the capacitor, while the capacitor remained mounted to the blower housing. The technician would have also checked between motor leads and ground to be sure the motor was not grounded. (A.O. Smith Electrical Products Company)

MOTOR REVIEW

Electric motors are essential components of all heat pump systems. This section is a review of common electric motors and their purposes.

Types of Motors

Both single-phase and three-phase motors are used in HVAC systems. Within the single-phase category, there are electric motors used for heat pumps called split-phase motors. Split-phase motors have a run and start winding that is physically out of phase with the other. This means the motor stator (stationary windings) will have run and start windings that have electromagnetic poles offset from each other. There are two types of split-phase motors:

1. High-start-torque, single-phase motors, used for compressors
   a. Capacitor start, induction run (CSIR)
   b. Capacitor start, capacitor run (CSCR)
2. Low-start-torque, single-phase motors, used for fans, pumps and blowers, and compressors
   a. Permanent split capacitor (PSC)

High-start-torque motors are required to start under mechanical, or fluid (gas or liquid), pressure. These motors will use starting devices to get the motor rotating from a stalled position. The starting devices include:

1. Start relays
   a. Current magnetic (current relays)
   b. Voltage magnetic (potential relays)
   c. Solid state (noncontact/electronic relays/power transistor)
2. Capacitors
   a. Start capacitors
   b. Run capacitors

Contactors are used to connect power to high-draw electric motors that are operated by a control circuit. The control circuit could be line voltage (the voltage being applied to the unit), but in heat pump applications, it is usually low voltage (voltage that has been stepped down using a transformer). Control voltage that has been stepped down is typically 24 volts.

Tech Tip

Because contactors are used to start and stop compressors and fan motors, they are susceptible to wear. Contacts of the contactor should be checked for wear during scheduled maintenance; if worn, replace the contactor.

Tech Tip

Contacts should never be sanded or filed. Sanding or filing may cause the contacts to fail sooner. Contacts should always be replaced (if they are replaceable), but more commonly the entire contactor or relay needs to be replaced. The technician should never spend service time trying to repair a contactor or relay by sanding or filing the contacts (see Figures 7-2 and 7-3).

Figure 7–2
Two-pole contactor. (Courtesy of Delmar/Cengage Learning)

Figure 7–3
(a) Three-phase motor starter. (b) Overload element. (Courtesy of Delmar/Cengage Learning)

Motor nameplates will state the motor phase type, voltage(s), rotational direction, starting amperage (locked-rotor amperage, or LRA), and maximum running amperage (full-load amperage, or FLA). They may also list the frame size, RPM, whether or not the motor has internal thermal protection, safety factor (SF), and maximum temperature rise. Motor safety controls monitor the amperage of the motor and disconnect motors having high-amperage-draw problems. Safety controls (overloads) take the form of heat-sensing devices that can be externally mounted on the motor or internally imbedded in the windings. In both types of overloads, the device is sensing the amperage draw and the generated heat. If either high amperage or high heat is sensed, the overload opens and disrupts electrical flow to the motor.

Tech Tip

Service factor (SF) is a measure of the amount of electrical overloading that an electrical motor can be subject to without damaging the motor. Service factor is a multiplier which, when applied to the rated FLA, indicates a permissible short-term loading which may be carried under the conditions specified for that service factor.

Multiplying the full load amps on the data plate by the SF shows how much overcurrent can be done without overheating and damaging the motor. Example: A 1-HP motor with an SF of 1.15 can operate to the level of 16 amps × 1.15, or 18.4 amps (115% load) momentarily. A motor operating at full load will have a given temperature rise. As the SF rises, so will the rating for temperature rise. Both of these are indicators of how the motor will survive when overloaded or overheated.

Motor-Starting Devices

Current relays are commonly used for fractional motors of 1/8 to 3/4 horsepower and applied to systems that start in an equalized state (Figure 7–4). Current relays sense the total amperage (amp) draw of the run windings. When high-amp (locked-rotor) conditions are sensed, the relay closes the switch to the start windings. As the amperage drops and the motor reaches 85% of its rated running speed, the relay drops open the switch and de-energizes the start windings. This type of relay is normally open (NO) and position sensitive (see Figure 7–5)—it must be mounted in the upright position (as identified on the relay). Test the contacts in the de-energized condition using an ohmmeter.

Tech Tip

Current relay coils seldom burn out. There is usually no need to test the operator. Contacts usually burn up or get stuck. The best way to check contacts is to remove the relay and turn it over (upside down). Contacts will fall together and can be checked with an ohmmeter.

Potential relays are used for 3/4 horsepower to 3 horsepower single-phase motors to energize the start winding. As the motor starts, the start winding acts as an alternator, creating an opposite voltage to the voltage being used. This is referred to as "counter emf," or reverse voltage. For example, a 220-volt motor starts and produces a counter emf in the start winding that creates a magnetic field in the potential relay. The magnetic field is greater than the line voltage and is strong enough to open the start switch when the voltage reaches a counter emf of about 340 volts and closes the switch when the counter emf drops to 290 volts. Both the operator (coil) and the switch are susceptible to damage. The operator should be checked for continuity and the contacts, which are normally closed (NC), can be checked using an ohmmeter in any position.

Figure 7–4
A current relay showing the operator coil made of heavy wire. (Courtesy of Delmar/Cengage Learning)

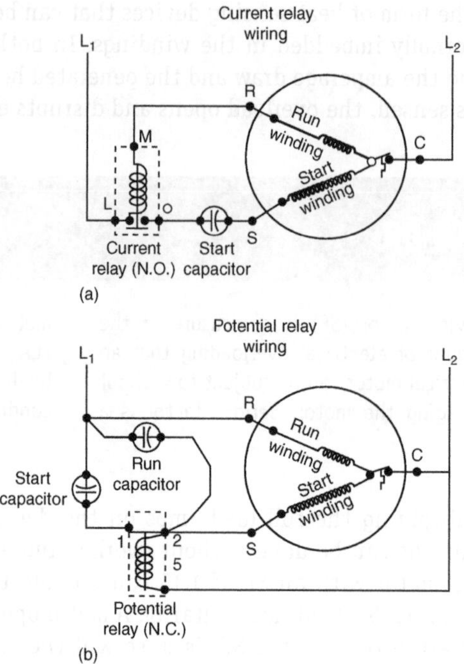

Figure 7–5
Current and potential start relays shown wired without start or run capacitors. Notice that the current relay is a NO switch and the potential relay is NC. The potential relay operator (coil) is wired in parallel with the start windings. (a) Motor shown wired with current relay and start capacitor. (b) Motor shown wired with potential relay and start and run capacitor. (Courtesy of Delmar/Cengage Learning)

Solid-state motor starters—more often referred to as positive temperature coefficients, or PTCs—have no moving parts and one relay that may fit a number of compressors over a horsepower range (Figures 7-6 and 7-7). When the solid-state material is cold, there is very little resistance through it. As amperage flows through the material to the start winding, the solid-state relay begins to heat and the resistance increases; amperage (current) flow is cut off to the start winding and the motor continues to operate on the run windings. A small amount of electrical leakage (through the material) is required to maintain high temperature.

Fan and Blower Motors

Fan and blower motors are low-start-torque motors. This means that they are not designed to start against much force or pressure. All permanent split capacitor (PSC) motors (Figure 7-8) are wired with run capacitors. The capacitor may be mounted to the motor, adjacent to the motor, or remotely in the control cabinet.

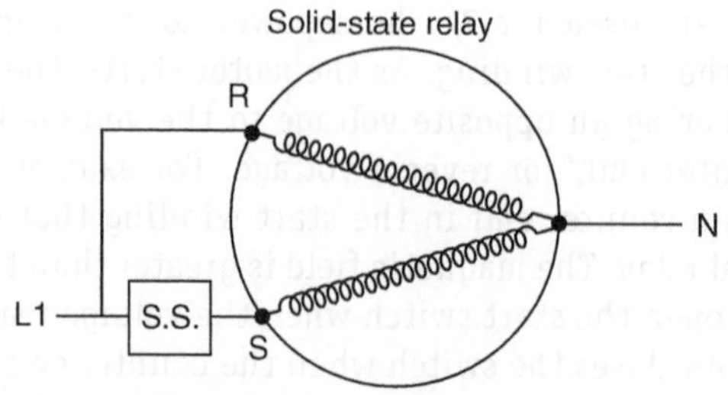

Figure 7–6
The solid-state device is wired in series with the start winding. (Courtesy of Delmar/Cengage Learning)

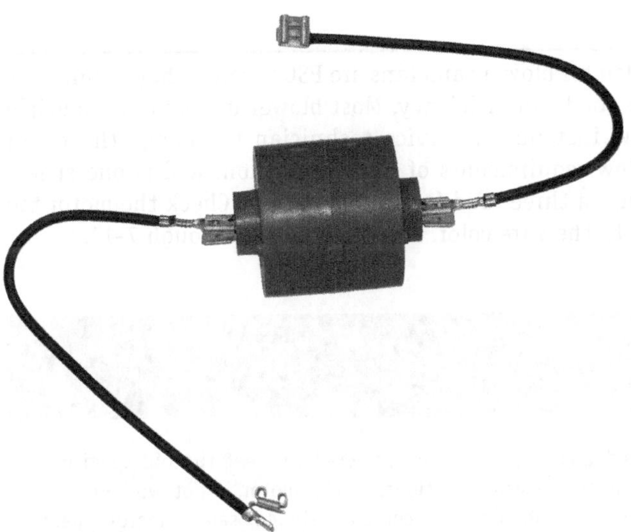

Figure 7–7
One type of solid-state motor starter, PTC (positive temperature coefficient) device, or thermistor. (Courtesy of Delmar/Cengage Learning)

Figure 7–8
A typical permanent split capacitor (PSC) motor used to turn a blower wheel or propeller fan. The run capacitor is external from the motor. (A.O. Smith Electrical Products Company)

Figure 7–9
When checking a bearing seal, look for the bearings. If you can see the bearings behind the seal, as in this picture, the seal is broken, worn, or missing. Sealed-for-life bearings have a basic weakness. Where the bearing seal rides against the moving bearing races, the seal wears. When the seal wears enough to allow dirt into the bearing, it fails at an increasing rate. Always check to see if the seal is intact. (Courtesy of Delmar/Cengage Learning)

Tech Tip

Most motors have sealed motor bearings and require no lubrication. During scheduled maintenance, motor bearings should be checked for broken seals and wear (Figure 7–9). Broken bearing seals should be treated as worn bearings and replaced before they fail. Motors with oil ports should be lubricated at the manufacturer's recommended periodicity with the proper oil.

Multispeed Motors

Most of the motors installed for blowers and fans are PSC because they require low starting torque and have good run efficiency. Most blower motors have multiple speed taps that allow the installer or service technician to change the motor speed to match the airflow requirements of the installation. Red is one speed, blue is another, and so on, on three- and four-speed motors. Check the motor tag for speeds corresponding to the wire color. See Figures 7-10 through 7-13.

Tech Tip

Before a motor is replaced and if it is still running, the system static pressure and cubic feet per minute (CFM) should be checked and recorded. The new motor should match the operating characteristics of the old motor. Motor speed taps are selected to meet the old specifications. If the motor is not working, the technician should select motor speed taps to meet manufacturer specifications for static pressure and CFM.

RUN CAPACITORS

Capacitors used with motors are classified as either start or run capacitors (Figure 7-14). Start capacitors are short-duration capacitors and have no means of dissipating heat. They are packaged in plastic and are typically round in shape. Run capacitors are designed to stay in the circuit as long as the motor runs. These capacitors are packaged in metal or plastic containers that are oval

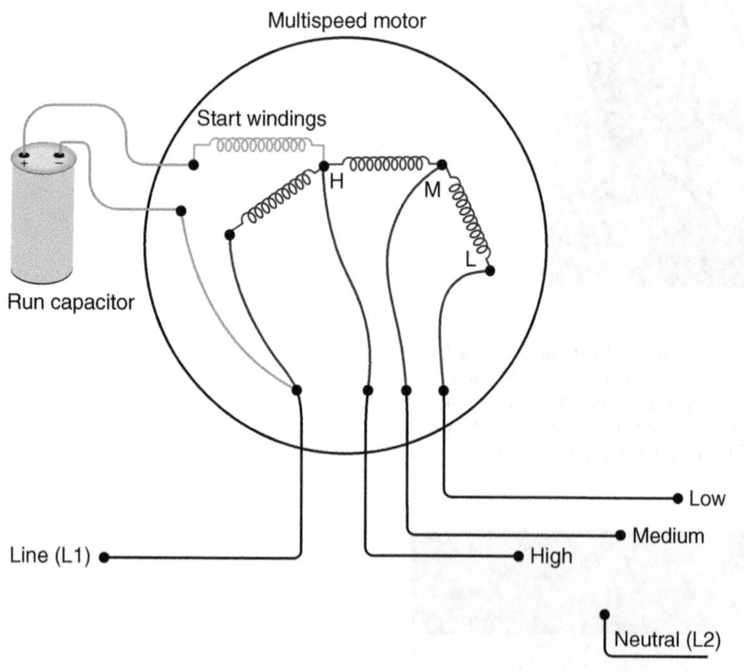

Figure 7-10
Diagram of a PSC multiple-speed motor. The electrical diagram shows how the windings are arranged. The tested electrical resistance increases as the motor winding taps decrease the speed of the motor. (Courtesy of Delmar/Cengage Learning)

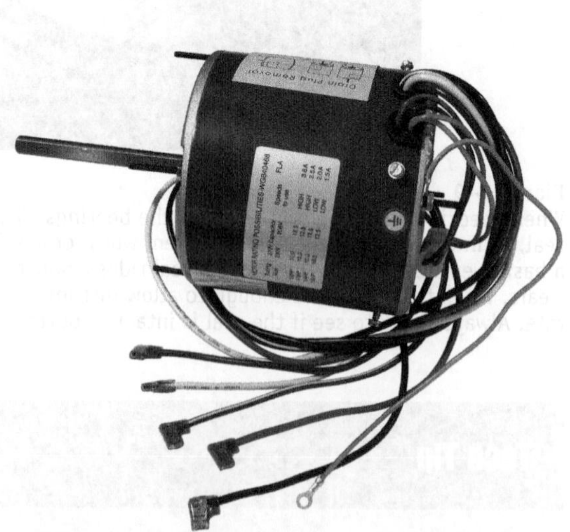

Figure 7-11
Multispeed motor showing the wiring connections used to change speed. The nameplate relates the colors of the wire to the motor speed. (Courtesy of Delmar/Cengage Learning)

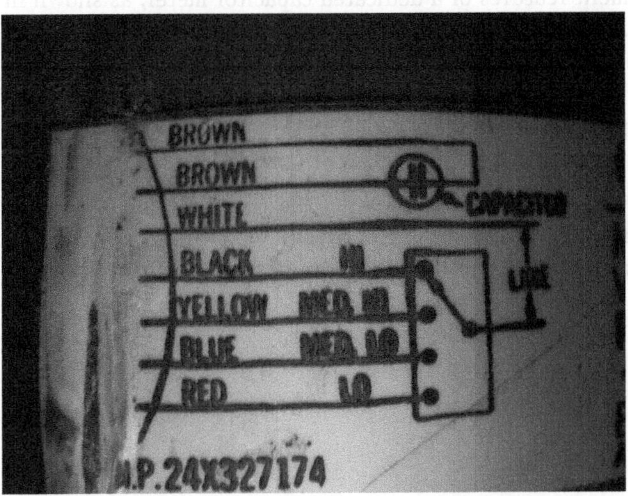

Figure 7–12
This portion of the PSC multispeed motor shows wire color coding for the run capacitor and motor-speed selection. This motor tag shows taps for four speeds: HI, MED HI, MED LO, and LO. (Courtesy of Delmar/Cengage Learning)

Figure 7–13
The capacitor shown is a three-tap, multi-tap capacitor. The center connections are common to each of the three capacitors. This example is designed for small PSC fan motors and has a 2.5-, 4-, and 5-mfd capacitor in one package. (Courtesy of Delmar/Cengage Learning)

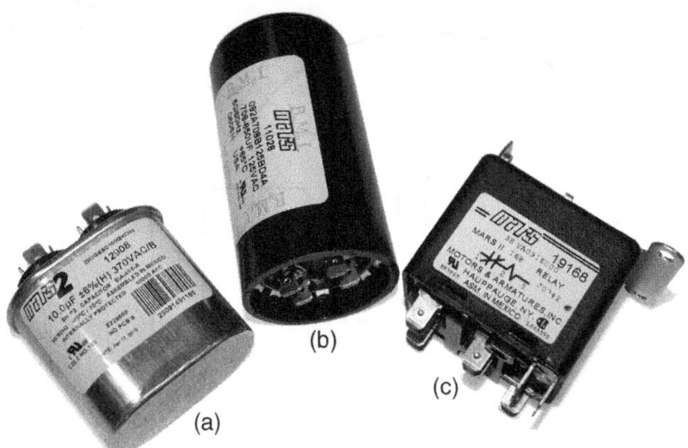

Figure 7–14
Three motor-starting components. (a) run capacitor; (b) start capacitor; (c) potential relay. (Courtesy of Delmar/Cengage Learning)

or round cylinders, with an internal heat transfer liquid that allows the capacitor to give up heat to the metal casing package. These capacitors are most often silver or gray in color. Many manufacturers use dual capacitors with two capacitors in one assembly with three terminals—one common and one for each of the two motors. Frequently, the terminal for the compressor is labeled "HERM," which is an abbreviation for *hermetic*, and the other terminal is labeled "FAN" for the outdoor fan.

Safety Tip

Oil-filled run capacitors made prior to 1978 could contain PCBs (polychlorinatedbif-enals). These capacitors must be treated as hazardous waste. Fluid leaking from these capacitors is hazardous to your health. If the capacitor has no date stamp, looks old, or is leaking, treat the capacitor as if it contains PCBs.

Multi-Tap Capacitors

Multi-tap capacitors are very valuable to service technicians. This type of capacitor can be "tapped," or wired, to match the capacitance requirements of nearly any piece of equipment. Several sizes of capacitor and microfarad ratings are available. The lowest and the combined total capacitance are marked on this type of capacitor. The service technician wires as many taps as necessary to achieve the necessary amount of capacitance. Table 7–1 provides a multi-tap capacitor chart.

Capacitors can be checked for microfarad (mfd) rating by using a digital multimeter with capacitance measurement features or a dedicated capacitor meter, as shown in Figure 7–15 (how to check for a good or bad capacitor). Some meters are auto-ranging. Auto-ranging meters have a selection switch to measure a certain capacitor range.

Table 7–1 Multi-Tap Capacitor Chart

To get the following microfarads (mfds), connect the leads in series to the Xs.	2.5-mfd connection	4-mfd connection	5-mfd connection	Common connection (connect one motor lead here)
2.5 mfd	×			×
4 mfd		×		×
5 mfd			×	×
6.5 mfd	×	×		×
7.5 mfd	×		×	×
9 mfd		×	×	×
11.5 mfd	×	×	×	×

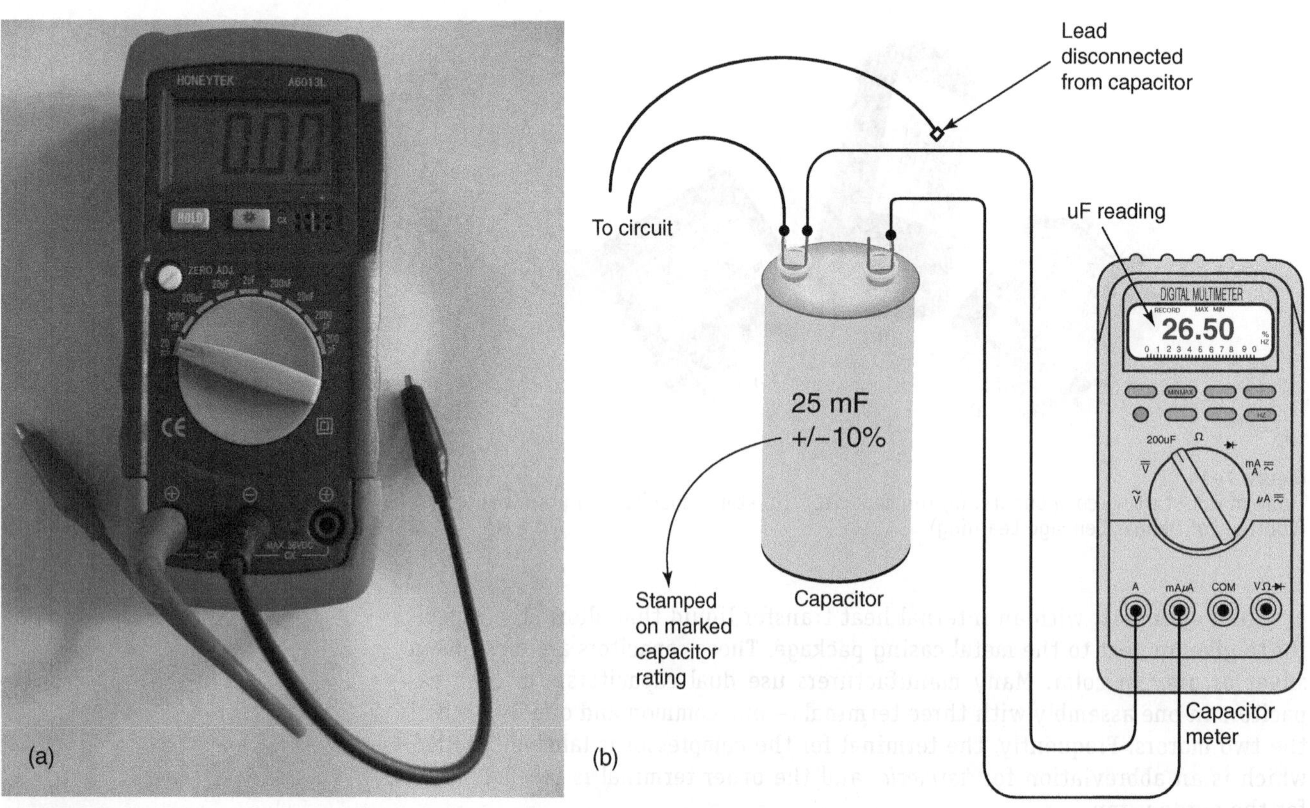

Figure 7–15
(a), (b) How to check for a good or bad capacitor. (Courtesy of Delmar/Cengage Learning)

To measure a capacitor, one side of the capacitor needs to be removed from the circuit so that there is no feedback through the circuit. The meter selector is set for the capacitor within the maximum capacitance range on the face of the meter. Fully discharge the capacitor; typically this is done with a 20,000-ohm resistor. If the meter reads "1," the capacitor is out of range of the meter setting. If this occurs, turn the meter selector to the next larger setting. If the display shows zeros, move the selector to the next smaller setting to improve the meter reading. If the capacitor is unmarked, testing should start at the highest meter selection (200 pF for this meter); change the selector to the next lower selection until a reading can be made. Shorted capacitors will show "1" or out of range for all meter selections. Figure 7–16 shows a permanent split capacitor (PSC) motor.

When reading capacitance, be sure to factor in the capacitor tolerance. A capacitor with a rating of 25 uF with a +/−10% (or +/−2.5 uF) tolerance may read from 22.5 to 27.5 uF. Figure 7–17 shows a run capacitor's label and specifications.

Safety Tip

Always handle capacitors as if they were fully charged. Capacitors can store voltage and may carry a charge. Always bleed off energy using a 20,000-ohm resistor. Never use a screwdriver to discharge a capacitor. Shorting may damage a capacitor.

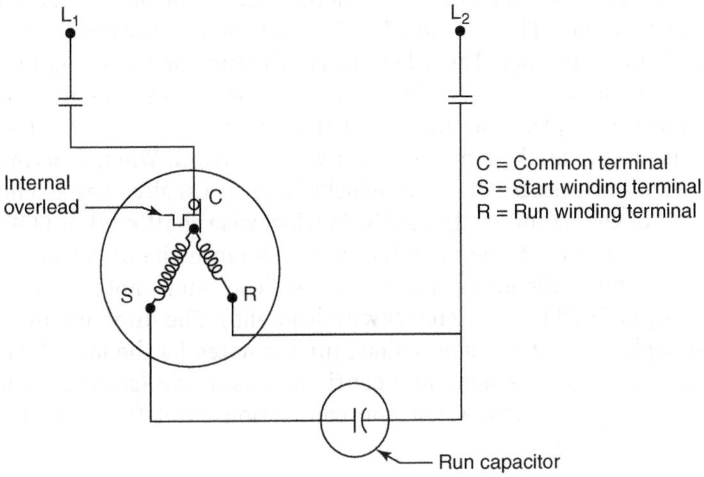

L_1 L_2

Internal overlead

C

S

R

C = Common terminal
S = Start winding terminal
R = Run winding terminal

Run capacitor

Figure 7–16
A schematic showing a PSC motor. (Courtesy of Delmar/Cengage Learning)

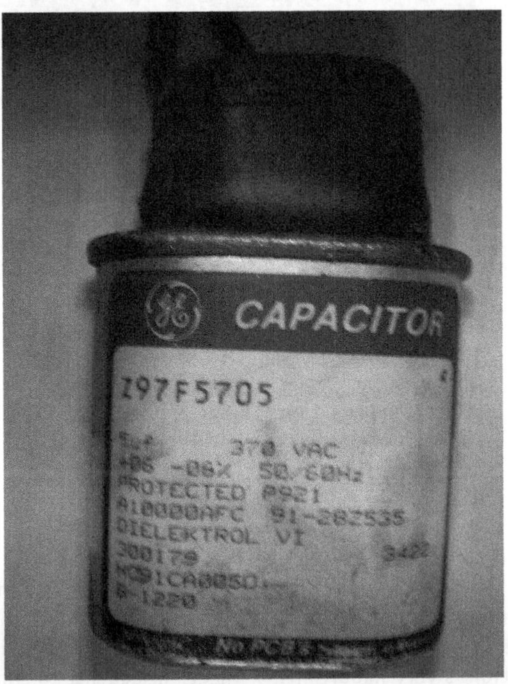

Figure 7–17
Small-run capacitor used with a PSC blower motor, showing label and specifications. (Courtesy of Delmar/Cengage Learning)

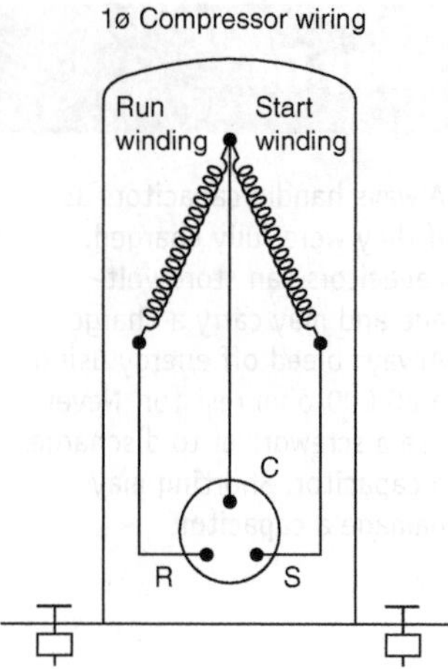

Figure 7–18
The split-phase wiring inside a compressor is the same for all split-phase motors. (Courtesy of Delmar/Cengage Learning)

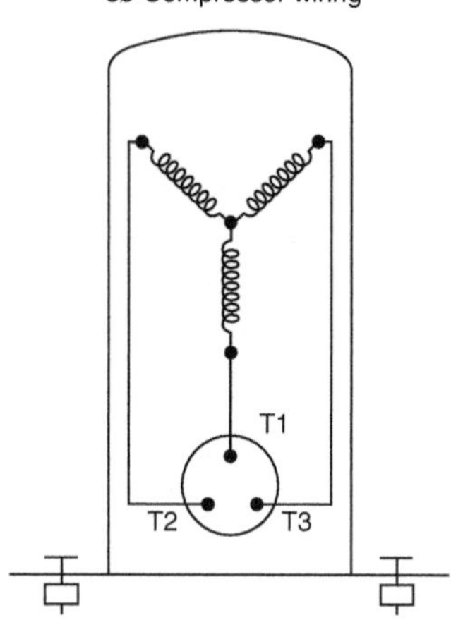

Figure 7–19
Three-phase wiring configuration. This configuration is the same for compressors and blower motors. (Courtesy of Delmar/Cengage Learning)

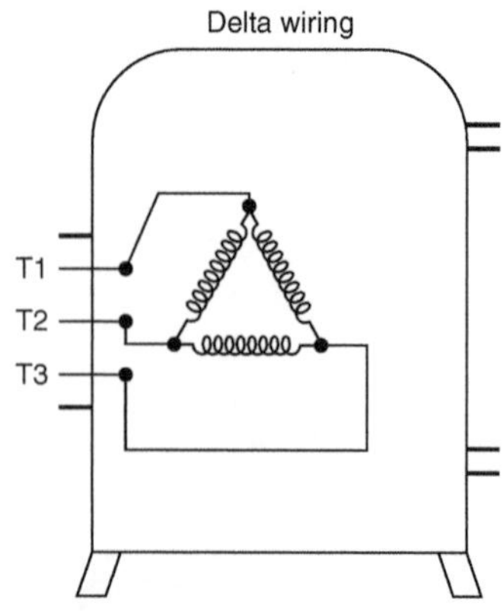

Figure 7–20
Some three-phase motors are wired in a delta pattern (triangle) where each winding is connected to the other two windings. (Courtesy of Delmar/Cengage Learning)

Tech Tip

Capacitance testers are readily available and inexpensive. Many times they can be one of the features of a multimeter. Look for test instruments that meet your needs on the job.

MOTOR WINDINGS

All single-phase motors fall into the category of split-phase motors. This means that the motor is designed with a set of start windings and another set of run windings. In the case of multiple-speed motors, there are speed taps off the run winding. This means that all split-phase motors have a common connection to the start and run windings (Figure 7–18). The C or "common" is the connection point of both run and start windings. The other ends of the run and start windings are designated R for run and S for start. Most motors operate in a single direction that is determined by the way the manufacturer wired the motor internally. The direction of rotation is usually marked on the motor with an arrow. When describing direction of rotation as clockwise or counterclockwise, it is usually done as viewed from the shaft end of the motor—noted as CWSE (clockwise shaft end) or CCWSE (counterclockwise shaft end). Some manufacturers reference the direction of rotation from the other end of the motor, known as the "lead" end—noted as CWLE (clockwise lead end) or CCWLE (counterclockwise lead end). The direction of rotation is determined by looking at the motor shaft (unless noted by the manufacturer).

Three-phase motors are generally configured as a wye (sometimes identified as a "star" or "Y") and have a common connection point for each of the three windings (Figure 7–19).

Tech Tip

In addition to a wye three-phase motor, there is also a delta. The delta is not as common as the wye motor (Figure 7–20).

Tech Tip

Phase sequence is related to three-phase power supplies and motors that are connected. Because the motor is designed to start and run using the out-of-phase delivery of this type of power, how the motor is connected will determine the direction of rotation (clockwise [CW] or counterclockwise [CCW]). When three-phase motors are connected to certain equipment, the direction of rotation is very important. Blowers and fans are affected when run in reverse of their intended rotation. See Figure 7–21.

Three-phase motor rotation can be easily changed by reversing the connection of any two of the three wires connected to the power source. When the direction of rotation needs to be right the first time it is connected, a phase sequence or phase rotation meter is needed. These meters identify the connection sequence to eliminate guesswork and ensure proper motor rotation.

Winding Identification

When determining the common, start, and run terminals on single-phase motors, use the procedure shown in Figure 7–22.

Electrically Checking Windings

There are a number of checks to be conducted on motor windings after the voltage has been verified:

1. Check start and run amperage (Figure 7–23).
2. Check for grounding or a short to ground (Figure 7–24).
3. Check for continuity or open circuit (Figure 7–25).
4. Check for internal protection (if the motor is hot and there is no continuity from common).
5. Perform routine check for winding insulation degradation with megohmmeter.

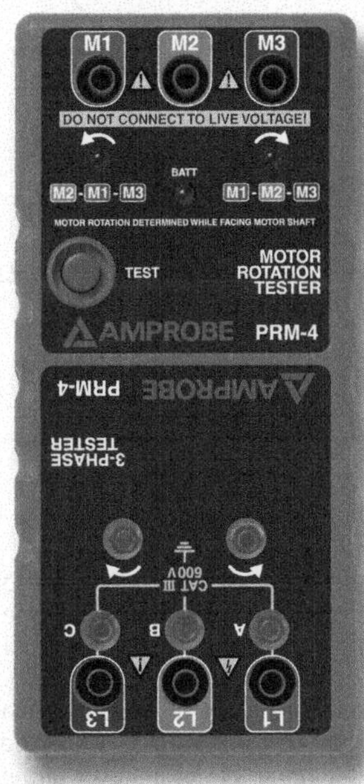

Figure 7–21
A phase sequence or phase rotation meter is connected to the three-phase power supply to determine which wire should be connected to which connectors to obtain the correct rotation for three-phase motors. (Courtesy of Amprobe)

Tech Tip

Always check motor volts and amps after replacing any motor! These checks should also be performed after adjusting or replacing pulleys or sheaves on belt-drive systems.

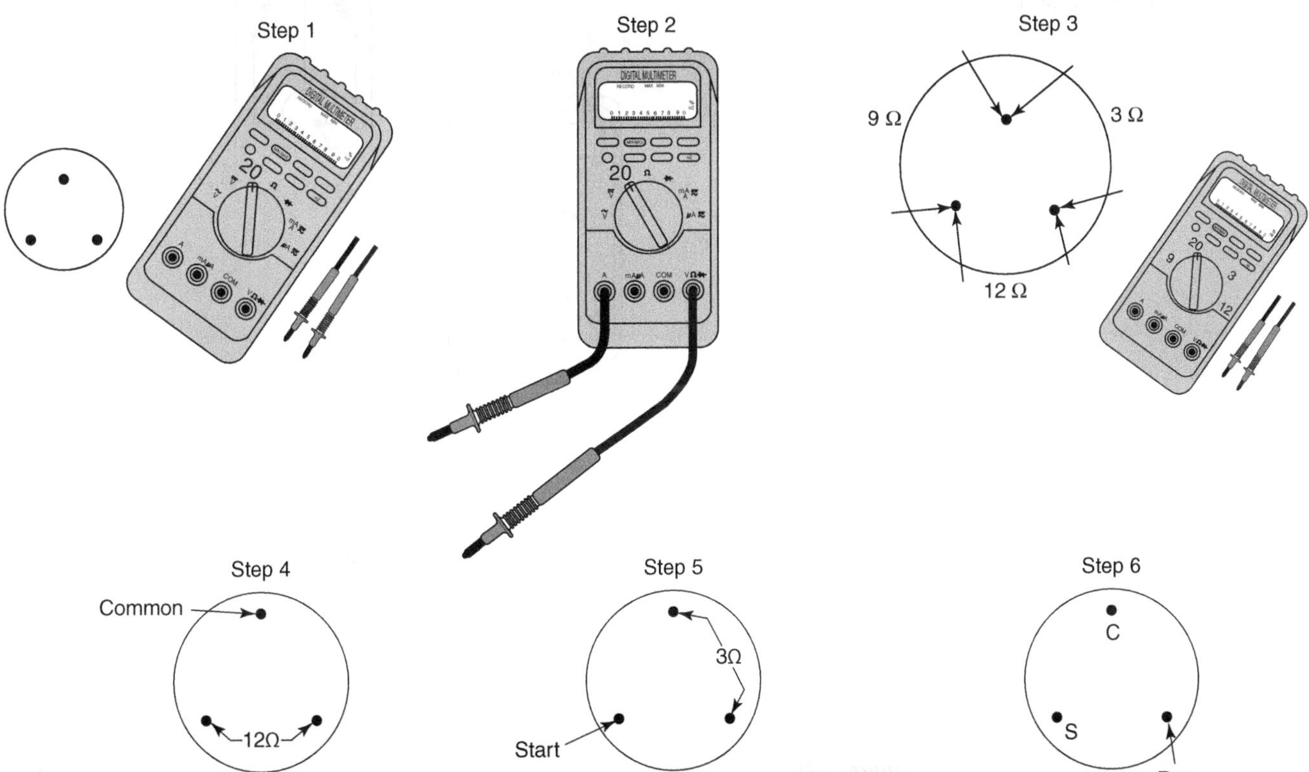

Figure 7–22
Winding identification procedure. (Courtesy of Delmar/Cengage Learning)

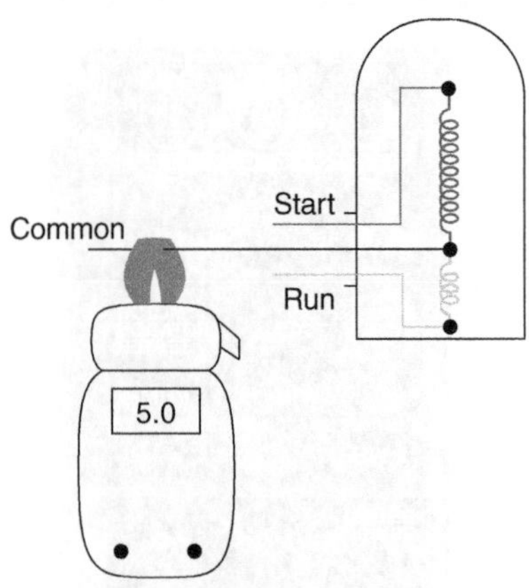

Figure 7–23
To check the motor for internal shorts, it is difficult to obtain a resistance specification of the motor. Instead, use the amperage that is marked on the motor as FLA (full-load amps) and LRA (locked-rotor amps). FLA is the running amperage of a motor under load. If the motor is not under load, the amperage draw should be lower while running. LRA is the starting amperage. Most ammeters have a maximum setting that will record the maximum amperage on start. If either of these readings is higher than marked on the motor and the motor is not under higher-than-normal load, the internal windings may be suspect. Both FLA and LRA should be taken on the common winding of the motor. (Courtesy of Delmar/Cengage Learning)

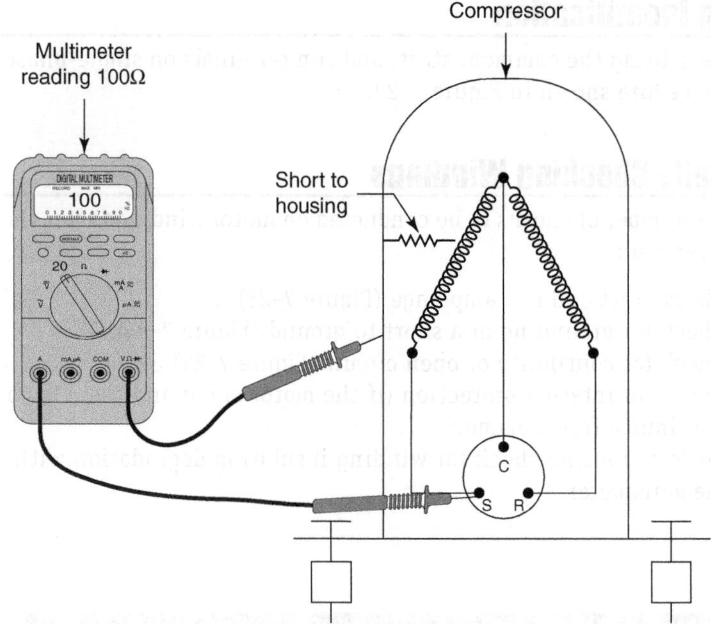

Figure 7–24
Checking for grounding or short to ground requires that the meter be set for the highest ohms scale available, usually a R × 10,000 scale, and that a test be done from any motor terminal to the motor case. If there is any indication of ohms on the meter, the motor should be suspected of having grounding problems. (Courtesy of Delmar/Cengage Learning)

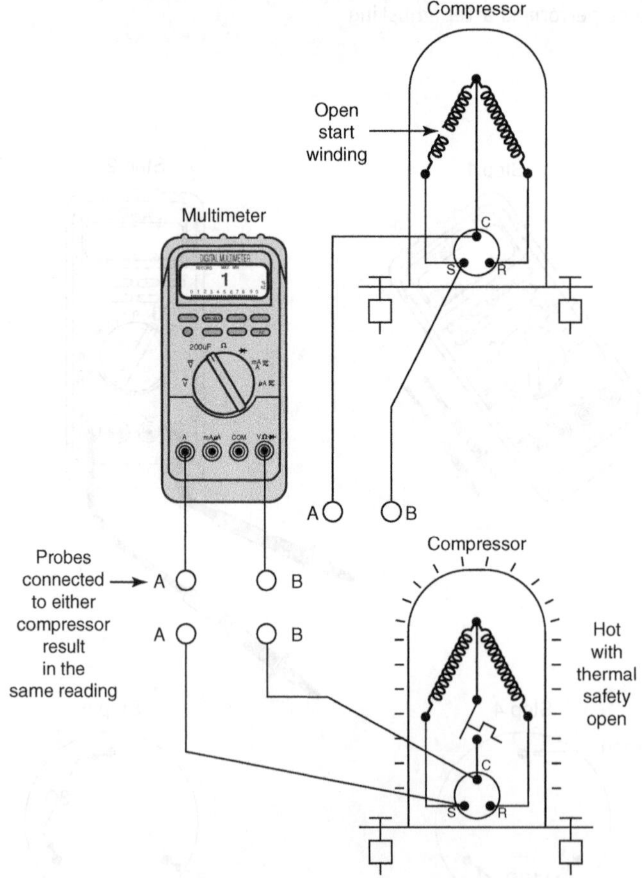

Figure 7–25
Motors that have open circuits will not show continuity (meter set for ohms) on all windings. If a motor is hot and has no continuity from the common to R or S, the cause may be an open internal motor protector. (Courtesy of Delmar/Cengage Learning)

All resistance (ohm) motor tests should be conducted with the power off and the motor terminals disconnected from the rest of the unit. Amperage checks must be done while the motor is connected and powered. It is suggested that the amperage check be done first and that the resistance checks be done following the amperage checks. As a final check to ensure that it was rewired into the circuit correctly, a voltage and amperage check should be done to complete the work.

ELECTRONICALLY COMMUTATED MOTORS

The electronically commutated motor (ECM) can be an induction motor or a permanent magnetic DC motor that uses an integrated circuit (micro computer chip) to measure and determine the position of the rotor. There is a broad range of application for ECM motors. These motors are being used in place of PSC motors for fans, blowers, pumps, and some compressor applications. Replacement of PSC motors ranging from 1/5- to 3/4-HP motors for both 115 volts and 230 volts can be done with two sizes of ECM motors (Figure 7–26). An ECM is capable of reducing electrical consumption and helping to increase system ratings for seasonal energy efficiency ratio (SEER) and heating seasonal performance factor (HSPF).

There are two advantages of ECM motors. The first is that these motors are more efficient than the traditional PSC motors by as much as 20%. The second is the microprocessor's ability to measure motor torque during operation, which directly corresponds to airflow. A quick look at the manufacturer's specifications for an indoor blower unit equipped with a PSC motor will show a given amount of CFM delivered for a specific total external static pressure for each of the given speed taps. As the total external static pressure becomes higher, the total CFM delivered decreases. If we now look at the specifications for a unit equipped with an ECM motor, the second big advantage becomes obvious. These specifications show that for a given ECM speed selected, there is a specific CFM delivered over a wider range of total external static pressures. This is due to the ability of the microprocessor to speed up the motor to compensate (within limits) for higher external static pressures.

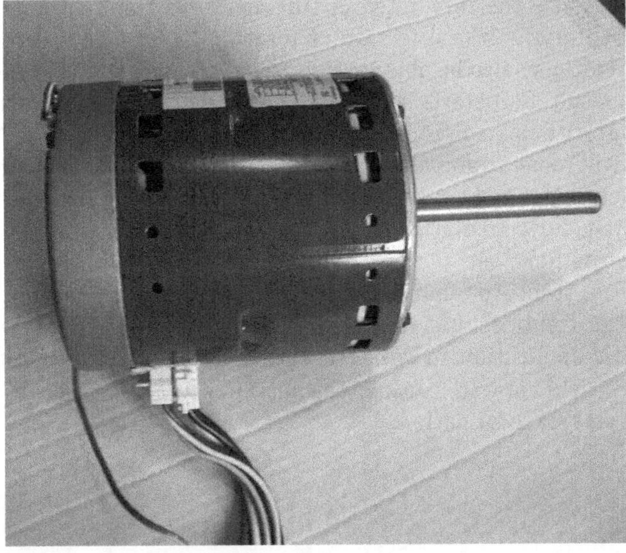

Figure 7–26
ECM replacement motor for a PSC. (Courtesy of Delmar/Cengage Learning)

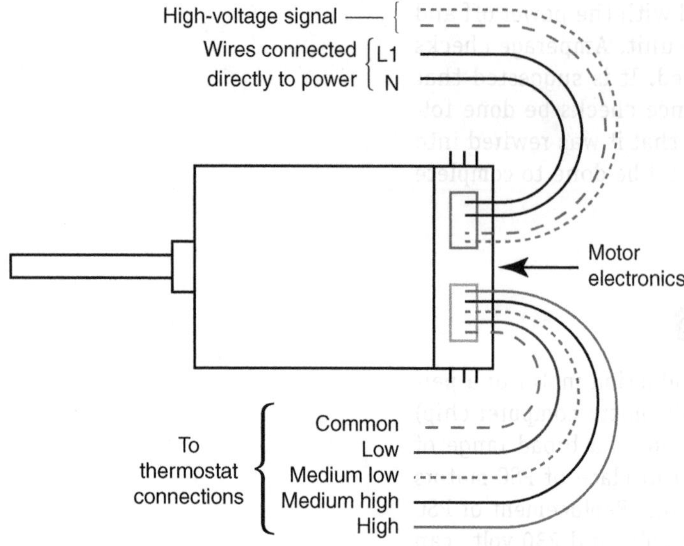

Figure 7–27
Motor connections differ from manufacturer to manufacturer. Power is connected directly to the motor. Signal wiring is for both high voltage and low voltage and controls the speed of the motor. (Courtesy of Delmar/Cengage Learning)

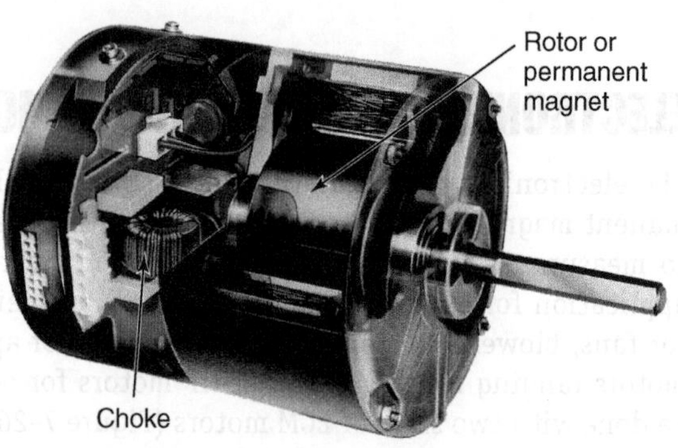

Figure 7–28
An ECM cutaway showing the electronic control area and the rotor. (Courtesy of General Electric)

ECM motor speed selection is often done through the thermostat instead of through a fan control. The ECM has low-voltage signal wires that accept signals from the thermostat for heating and cooling motor-speed switching. Line-voltage signal wires are also present to receive signals from the fan control center. The motor is connected directly to power through cabinet door switches to maintain electrical power to the integrated circuit (Figures 7-27 and 7-28). The original equipment manufacturer (OEM) literature should always be consulted because many of these motors are programmed by OEMs to match the equipment.

Some ECMs are able to determine the rotation electronically and fix the rotation in memory so that the motor starts in the same direction from the time of installation. Because the integrated circuit measures the amperage flow to the motor, it can determine the direction of rotation based on the motor load. The integrated circuit may also be able to "soft start" the motor and reduce noise and vibration as a result. Soft starts also reduce the amount of amperage used at start. Reduction of start amperage increases the efficiency of the motor. Additionally, these motors will ramp up and down at the beginning and end of their cycles. This has advantages for both the mechanical wear on the motor and the efficiency of the HVAC system. Low circulation of air during cooling mode, for instance, allows for more moisture removal.

There is no standard configuration for ECMs. They are programmed by the OEM or the field technician (in the case of universal replacement ECMs) for the specific application in which they are used. When encountering these motors in the field, always consult the manufacturer's literature for troubleshooting and service information.

SUMMARY

In this chapter we have discussed motors and starting devices for split-phase motors. We have also briefly named other motors, such as three-phase motors, that may be used in commercial heat pump systems that run on three-phase power. Both high- and low-start-torque motors are used. Many of the motors used are PSC (permanent split capacitor), but many manufacturers are using ECM motors in new heat pump systems in order to reduce energy consumption and increase the energy ratings of their systems.

Capacitors are an integral part of PSC motor circuits. Many PSC motors have multiple-speed capability. The speeds of these motors are chosen by selecting the color-coded wire identified with a particular motor speed. Motor speed is based on the need to supply air volume and the static pressure needed in a system installation.

The newest motor being used on high-efficiency heat pumps is the ECM (electronically commuted motor). This motor has higher operating efficiencies and a host of other functions that make it desirable to use in heat pump and other HVAC applications.

REVIEW QUESTIONS

1. Describe one type of high-start-torque motor.
2. Describe one type of low-start-torque motor.
3. List 3 types of motor-starting devices.
4. Explain why blowers and fans use a certain type of motor, and name one motor type.
5. Describe how to identify common, start, and run motor terminals.
6. Describe how to electrically check motor windings.
7. Explain what "ECM" stands for and how these motors work.

CHAPTER 8

Compressors

LEARNING OBJECTIVES

The student will:

- Explain the differences between reciprocating, rotary, and scroll compressors
- Describe how compressors can be staged
- Explain how compressors are "unloaded"
- Describe how a two-speed compressor operates electrically
- Describe how a variable-speed compressor operates
- Describe how to conduct a compressor efficiency check

INTRODUCTION

The compressor is considered by most to be the heart of the heat pump system. If the heart is not functioning well, the entire system suffers. If the compressor is inoperable, the system is dead. Reciprocating, rotary, and scroll compressors make up the bulk of compressors used in heat pumps. The classification of compressors and their relationship is shown in Figure 8–1. This chapter discusses the

Field Problem

The compressor was running, but the system was not performing in the same fashion that the customer had experienced in previous seasons. The technician was recalling the customer's complaints as he began his checks of the system. The customer had said:

1. "This thing has only run right once, and only after a technician made it work."
2. "The heat pump is only 3 years old."
3. "During this past cooling season it worked okay, but it doesn't put out the heat."

Symptoms: The system was operating; operated during the cooling season; is not putting out enough heat in the heating mode.

Possible Causes: Refrigerant charge; inefficient compressor (compressor mechanical problems, compressor electrical problems); leaking four-way valve.

Starting with the outdoor unit, the technician did the sensory checks (visual, sound, smell, and touch) and observed that:

1. There were no visual signs of oil (indicating a leak) in the outdoor unit.
2. The filter drier had been changed (signs of brazing).
3. There was a hot smell to the compressor.
4. The top (discharge) of the compressor was very hot.
5. The compressor had an unusual "whining" sound that was not characteristic of a scroll.

The technician made a mental list of possible causes:

1. Refrigerant charge
2. Inefficient compressor
   a. Compressor mechanical problems
   b. Compressor electrical problems
3. Leaking four-way valve

System pressures needed to be checked, so the technician decided to conduct a compressor efficiency check at the same time, which includes the measurement of pressure. This check also requires the comparison of compressor electrical data with measured operating data using the manufacturer's charts. For reference, the technician recorded the compressor nameplate information:

- Locked-rotor amps (LRA)—300 amps
- Rated-load amps (RLA) (or full-load amps [FLA])—90 amps
- Voltage—230 volts

The technician's readings, using accurate voltage and amperage test instruments, were:

- Running amps 40 amps
- Voltage—233 volts
- Saturated discharge temperature—110°F
- Saturated suction temperature—50°F

The compressor was operating at lower-than-rated amperage (this test is explained later in this chapter) and system pressures were unusual given the operating conditions. Based on the compressor noise, the high temperature, the amp test, a temperature test of the four-way valve, and a check for correct charge, the technician concluded that the compressor was inefficient (had internal mechanical problems). With this diagnosis, the technician discussed the options with the customer.

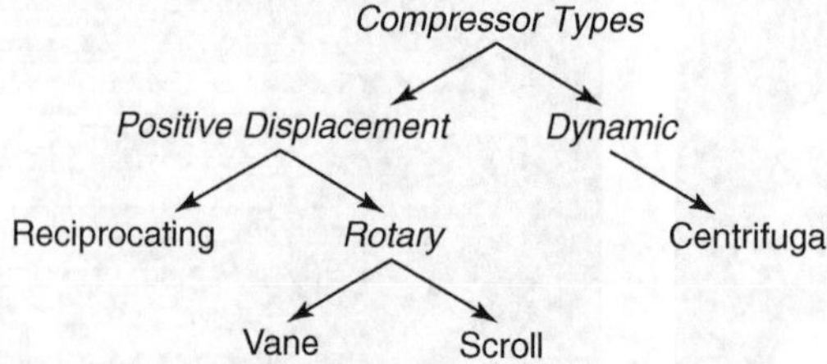

Figure 8–1
Compressors are first classified as either positive displacement or dynamic. "Positive displacement" means that if the discharge of the compressor was completely blocked, the compressor would eventually stop. "Dynamic" means that if the discharge was completely blocked, the compressor motor would still turn. Most compressors used for heat pumps are positive-displacement compressors. (Courtesy of Delmar/Cengage Learning)

design and characteristics of compressors. Additionally, the concepts of staging, unloading, and compressor speed are discussed as methods of compressor capacity control to save energy.

RECIPROCATING COMPRESSORS

Reciprocating compressors (see Figure 8–2) have been consistently used in the HVAC industry and will continue to be used in many applications. The purpose of all compressors is to increase pressure of vapor refrigerant.

Most compressors used for heat pumps are welded hermetic compressors. The compressor and motor are sealed inside a shell. This means that the internal parts of compressors are not serviceable. The technician must determine if the compressor is functioning mechanically and electrically by taking readings from the outside (Figure 8–3).

Refrigerant is pulled into the cylinder on the intake stroke of the piston, pulling suction vapor through the suction valve. As the piston compresses on the upstroke (compression), it forces refrigerant vapor through the discharge valve. Reciprocating compressors leave a small amount of vapor in the cylinder after each compression stroke. This is necessary to have enough clearance for the piston to come to the top and not hit the cylinder head. This small amount of vapor re-expands and accounts for additional superheat.

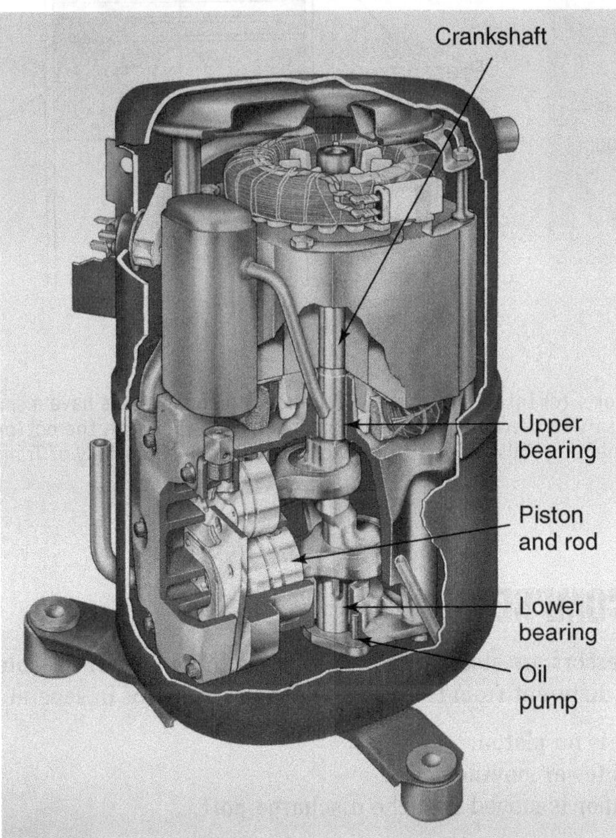

Figure 8–2
The internal workings of a welded hermetic, reciprocating compressor.
(Courtesy of Tecumseh Products Company)

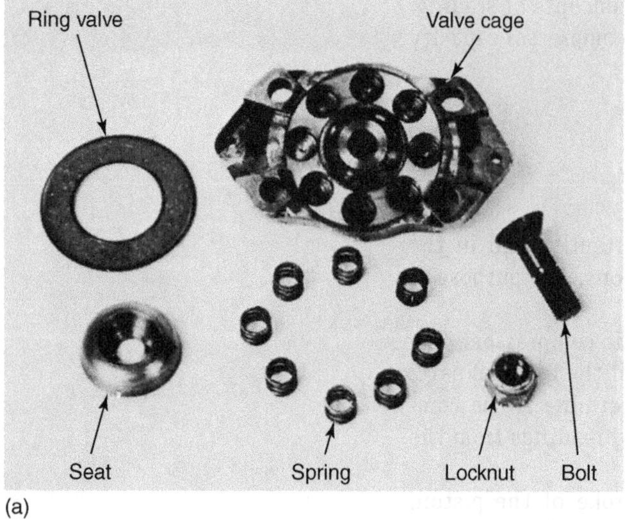

(a)

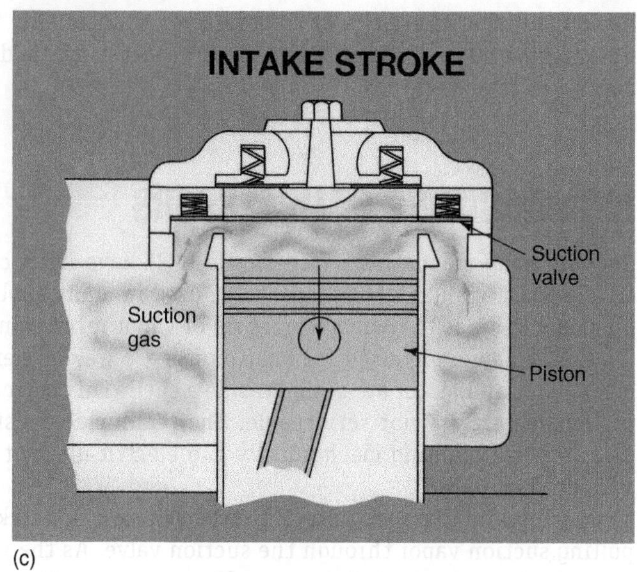

(c)

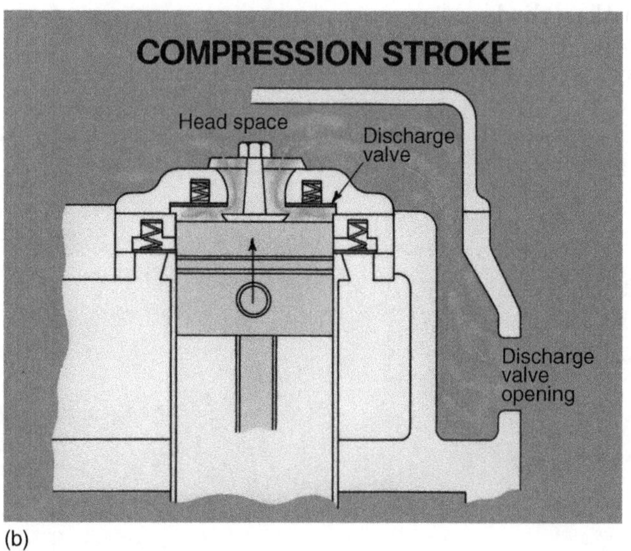

(b)

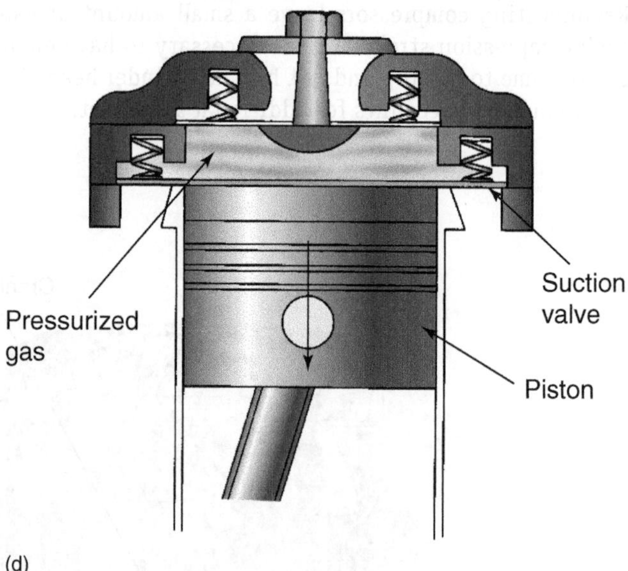

(d)

Figure 8–3
(a) Compressor valve plate and parts. (b) Compression stroke. (c) Intake stroke. (d) Reciprocating compressors have a small portion of vapor left in the top of the upstroke, and this is necessary to prevent the piston from making contact with the bottom of the valve plate. Without this clearance volume, damage to the connecting rods and wrist pins would also result. (Courtesy of Trane)

ROTARY COMPRESSORS

Rotary compressors are also housed in a shell and are not serviceable. The rotary compressor is different from the reciprocating compressor in several ways:

1. There is no piston.
2. It has fewer moving parts.
3. All vapor is moved into the discharge port.

The rotary compressor sweeps vaporous refrigerant around the cylinder, pulling new vapor through the suction intake, compressing the vapor, and sending it through the discharge valve. Because this is rotary motion instead

Figure 8–4
A hermetically sealed rotary compressor with a suction accumulator. The rotary compressor has an eccentric rotor that wipes the cylinder wall as it rotates. The sliding vane traps suction vapor and moves it around the cylinder, compressing it as it leaves through the discharge valve. (Photo by Susan Brubaker)

of reciprocating, the compressor is known as a rotary compressor. Some rotary compressors have stationary vanes and others have vanes that are mounted in the rotor (see Figure 8-4).

SCROLL COMPRESSORS

The scroll is the most recent design in compressors and has been widely accepted in the industry. It is technically considered to be a type of rotary compressor, but in the industry is not referred to as a rotary compressor. It is always referred to as a scroll compressor.

Like the rotary compressor, the scroll moves in a circular motion, pulling in vaporous refrigerant and compressing it with one quick motion. The orbital action of the movable or driven scroll against the stationary scroll creates the suction and compression actions of the compressor. There are no other moving parts. Figures 8–5 through 8–7 show scroll compressors.

Scroll compressors have several advantages. When oil or particles move through the compressor, the scroll can move, without binding, to allow the oil or particles through the compressor. This means that some liquid slugging can be tolerated by the scroll compressor. This compressor does not need valves. By eliminating the suction (and discharge valves in some compressors), the compressor further reduces the number of working parts. There are no volume losses caused by residual refrigerant vapor; all of the vapor is moved through the compressor, like in a rotary compressor. The design of the scroll separates the suction and discharge vapors by a greater distance than either the reciprocating or rotary compressor. By separating the suction and discharge vapors, the scroll reduces the amount of heat transfer that occurs in the compressor. The scroll compressor creates less noise, resulting in a quieter compressor operation, and operates with increased efficiency.

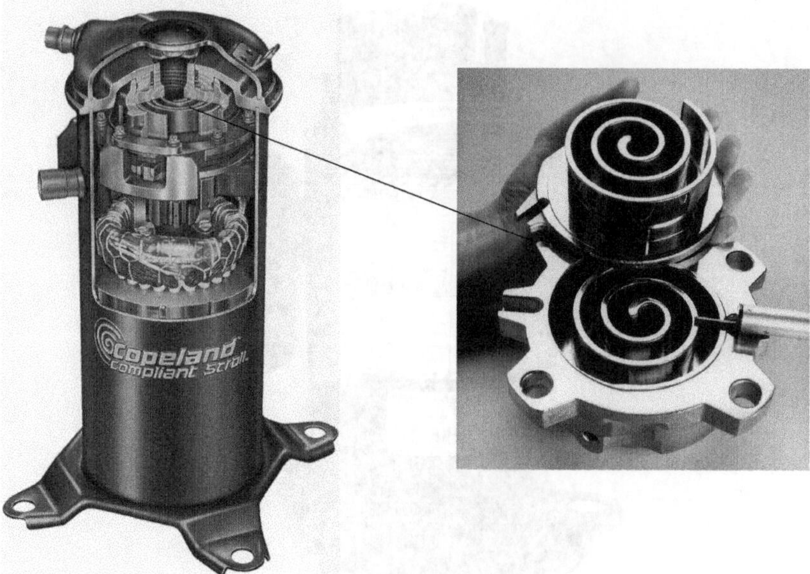

Figure 8–5
A hermetically sealed scroll compressor showing the stationary and movable scroll. (Courtesy of Emerson Climate Technologies)

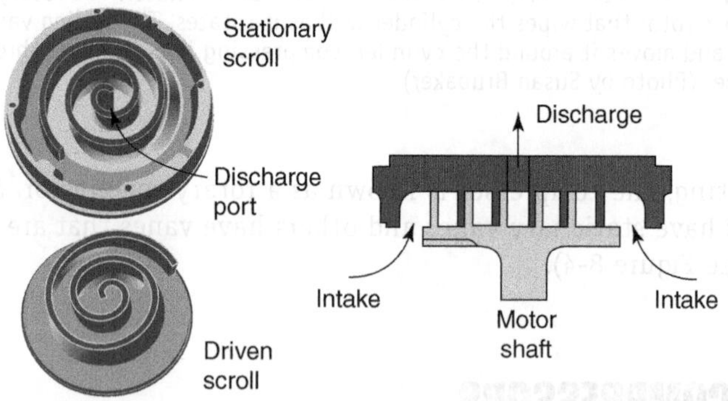

Figure 8–6
Stationary and movable (driven) scrolls. The driven scroll is connected to the motor shaft. The stationary scroll sits above the driven scroll and drive motor. Vapor discharges through the center of the stationary scroll. (Courtesy of Trane)

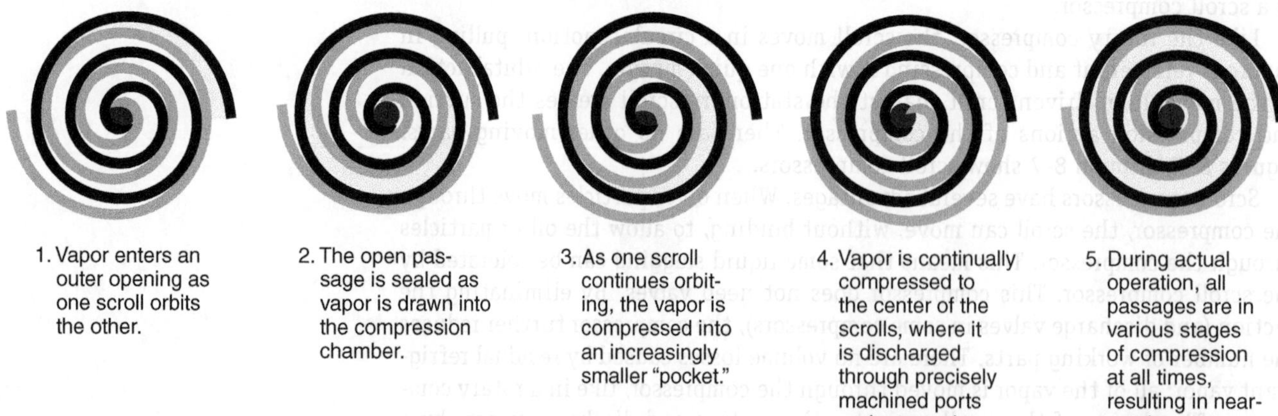

1. Vapor enters an outer opening as one scroll orbits the other.
2. The open passage is sealed as vapor is drawn into the compression chamber.
3. As one scroll continues orbiting, the vapor is compressed into an increasingly smaller "pocket."
4. Vapor is continually compressed to the center of the scrolls, where it is discharged through precisely machined ports and returned to the system.
5. During actual operation, all passages are in various stages of compression at all times, resulting in near-continuous intake and discharge.

Figure 8–7
This series of images shows the movement of the scroll and how the refrigerant vapor enters, compresses, and discharges through the scroll compression process. (Courtesy of Copeland Corporation)

STAGING

Compressor staging for some heat pump systems consists of two compressors in parallel that are operated at various conditions to increase the heating or cooling capacity of the heat pump and to minimize cycling. Generally, most staged compressor systems will operate one compressor for the first period of time (first stage). During this period of time, the control system monitors to determine if the desired conditions (set conditions) are met. If the system cannot meet the desired conditions, the second stage will operate in conjunction with the first stage to meet the condition. After the desired condition has been met, the second stage is powered off and the first stage continues to operate as needed or necessary to maintain the desired or set condition (Figure 8–8).

By using two compressors and staging their operation, less power can be used during those times when there is less demand. This is in contrast to using one large compressor and cycling it more often. Usually, the cycling and the required starting amperage will cause the single compressor system to use more energy. By using controls to match the heating or cooling needs of the building, the control system can operate a smaller single compressor to meet the need and only bring on a second compressor when required.

In some systems, the two compressors used for staging are different sizes. The smaller compressor is operated most of the time and the larger compressor is only operated during times of peak demand. Peak demand is generally periods of low or high ambient temperature. Because peak temperature occurs less frequently, most of the heating or cooling season can be handled by the smaller compressor. When only the smaller compressor is used, energy can be saved, compressor cycling is reduced, and more moisture is removed in the cooling mode.

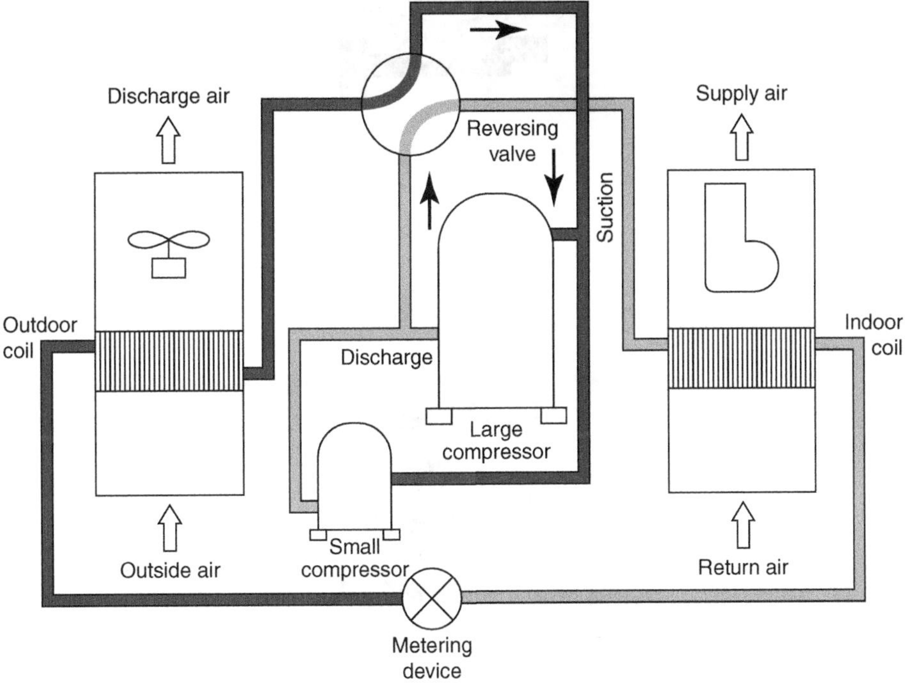

Figure 8–8
This diagram shows the smaller compressor operating during periods of lesser demand. The larger compressor would be staged to operate during times when the smaller compressor needs assistance. Note: This type of staging is represented by a larger compressor that typically handles 60% of the load. The smaller compressor typically represents 40% of the load. Even when it is a single compressor that has internal staging, the ratio of 60% and 40% still generally applies. (Courtesy of Delmar/Cengage Learning)

UNLOADING

Another way to control system capacity is through a mechanical process that reduces the pumping capacity of the compressor. This process is called "unloading" and involves holding a valve open or closed and allowing the compressor to pump without compressing vaporous refrigerant for a portion of the compression cycle (Figure 8–9). In a single-scroll compressor, bypass ports are held open, thus allowing only part of the compression path to function.

There are two different ways to stop the cylinder from compressing: (1) hold the suction valve closed (or close the suction line) or (2) hold the suction valve open (on a reciprocating compressor). See Figure 8–10. Either configuration will effectively eliminate one cylinder from pumping. If the compressor has four cylinders, each of the other three cylinders could have independent unloading capabilities.

Figure 8–9
Some scroll compressors have unloading or capacity control ports. When these ports are open, the scroll works at reduced capacity. When the ports are closed, the scroll works at 100% capacity. (Courtesy of Delmar/Cengage Learning)

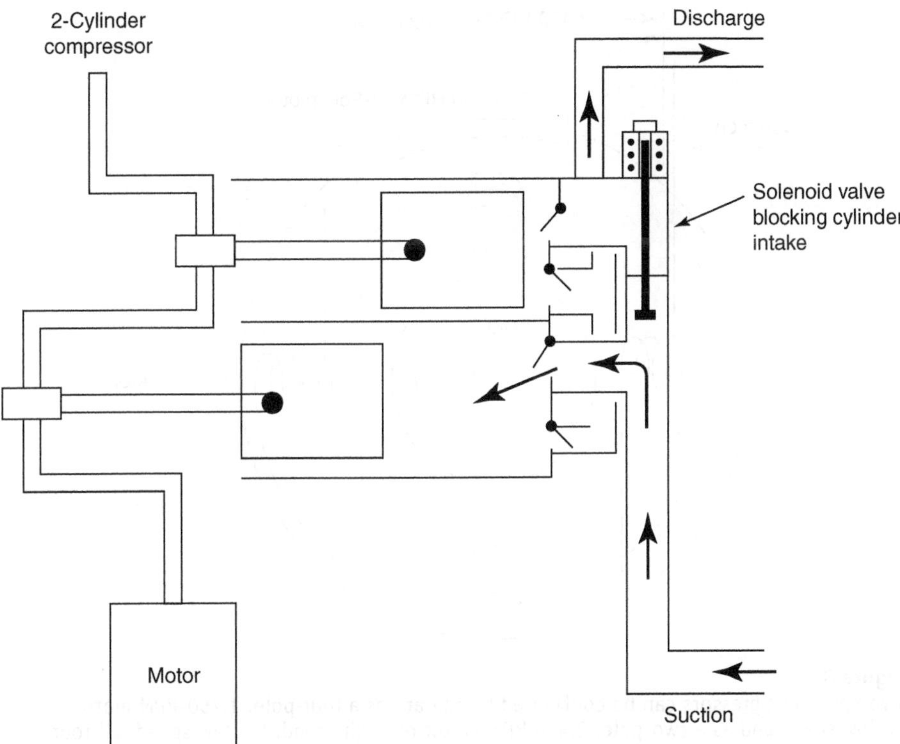

Figure 8–10
Blocking the flow into one cylinder of a two-cylinder compressor is one way of unloading the compressor or reducing the compressor capacity. Another method is to hold the suction valve open on a reciprocating compressor. (Courtesy of Delmar/Cengage Learning)

VARIABLE-SPEED/MULTISPEED COMPRESSORS

Variable-speed compressors are in the same classification as frequency drive and electronically commuted motors. Multispeed compressors are wired to run at two or more known speeds. Both the variable-speed and multispeed configurations operate the compressor at reduced capacity during low demand. During high demand, the entire capacity or highest speed is engaged to meet peak demand.

Two-Speed Compressors

Most two-speed compressors are configured to run at two different RPMs. Most applications involve operating as a four-pole motor on low speed and a two-pole motor on high speed. Low-speed operation would be used during low demand. High speed would be used only when low-speed capacity could not keep up with demand. Remember that motor speed is the frequency times 120, divided by the number of poles.

Two-speed compressors can also be operated as a two-stage compressor (Figure 8-11). The first stage of the thermostat would operate the low-speed winding configuration of the compressor. The second stage would operate the high-speed configuration.

Variable-Speed Compressors

When comparing standard, single-speed compressors to variable-speed compressors, significant energy savings can be obtained. Variable-speed compressors operate at lower speeds on start-up and during times of low demand. By continually

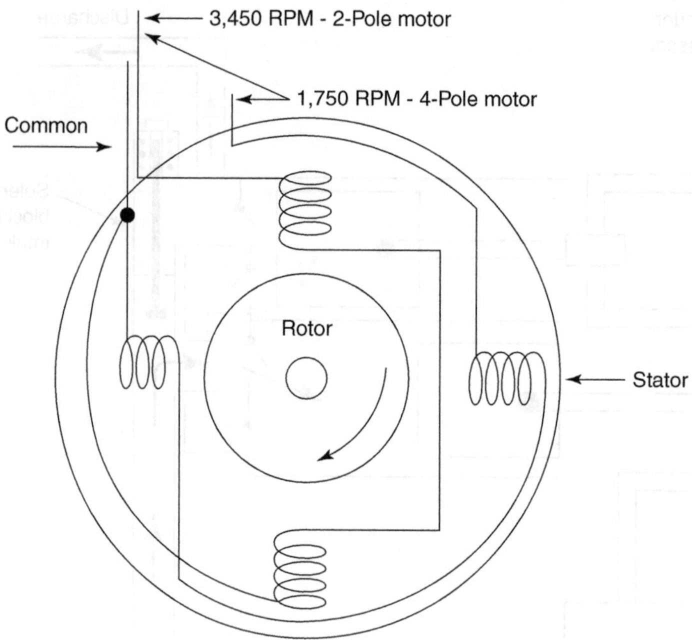

Figure 8–11
Two-speed compressors can be configured to operate as a four-pole, 1,750-RPM motor on low speed and as a two-pole, 3,450 RPM motor on high speed. For low speed, all four poles of the windings would be connected. High speed would use only use two poles. (Courtesy of Delmar/Cengage Learning)

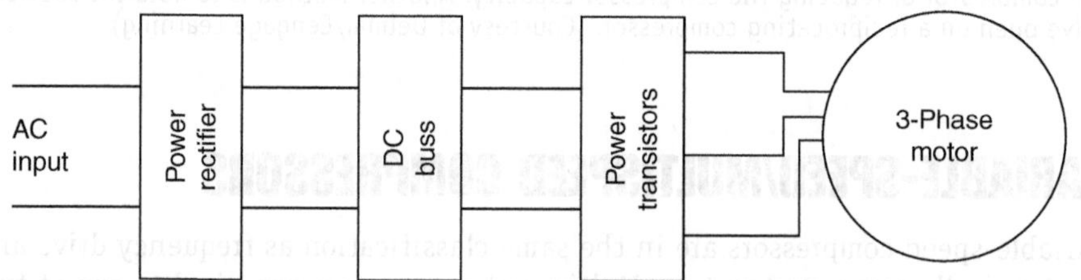

Figure 8–12
Variable-frequency drive motors have complicated control systems attached to single- or multiple-speed three-phase motors. As the frequency changes, the speed of the motor changes. (Courtesy of Delmar/Cengage Learning)

matching the load to the speed of the compressor, the compressor will operate almost constantly. By operating more of the time, cycling is reduced, increasing the life of the compressor and improving operating efficiency. Varying the speed also means that the compressor can operate at lower noise levels, making the variable-speed compressor quieter than a single-speed compressor. These compressors can also be more compact in size and weight. Using a variable-speed compressor also has the advantage of reducing other types of controls, manifolds, and extra circuits associated with other forms of capacity control.

Variable-speed compressors can have either commutated DC or frequency drive motors. Both technologies have been employed. Commutated DC motors are also referred to as electronically commutated motors (ECMs). Frequency drive motors use electronics and changing frequency to vary the speed above and below standard 60-cycle power. Figure 8–12 shows a variable-frequency drive motor.

Control systems that monitor and adjust variable-speed compressors (Figure 8–13) can also be connected to variable-speed indoor blowers and outdoor fans. In this way, the capacity of the compressor is being matched to the capacity of the indoor and outdoor coils.

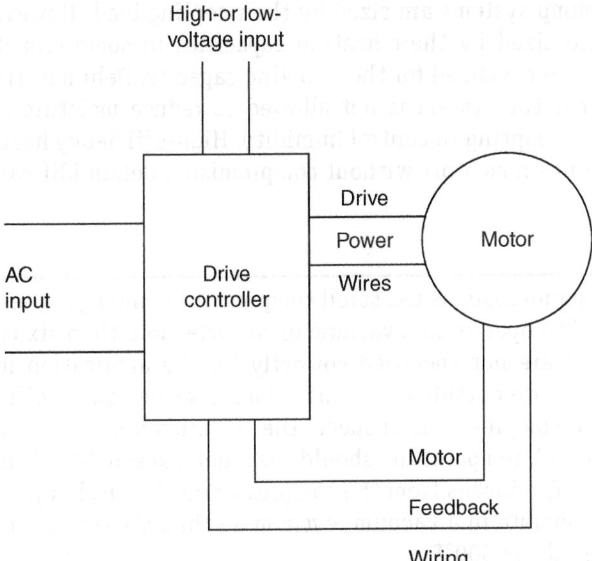

Figure 8–13
Variable-speed motors employ a motor drive controller that provides power to the motor and receives feedback to monitor motor amperage and speed. There are many variations of this general configuration. ECM motors are among the motor and control types that fall within this configuration. (Courtesy of Delmar/Cengage Learning)

EFFICIENCY

The efficiency of heat pumps has nearly doubled since their early days. Improvements in efficiency are a result of improved controls and advanced compressor technology. Unloading, staging, variable-speed motors, and multispeed motors, electronic expansion valves (EEVs), as well as more efficient heat transfer coils, continue to increase efficiency.

Heat pump efficiency can be judged using the seasonal energy efficiency ratio (SEER) rating for air conditioning mode. Higher SEER ratings can be related to heating mode and tend to mean that heating seasonal performance factor (HSPF) ratings are also higher. Heat pump SEER ratings increase as more sophisticated technology is used. The basic trend for greater efficiency includes such components as: electronic expansion valves, variable-speed motors (blower and compressor), capacity control compressors (staging, variable speed, and unloading), and capacity monitoring controls and sensors. The more sophisticated the technology, the higher the cost of the heat pump, but the greater the efficiency. Greater efficiency means that there is a lower operating cost. All of these factors need to be weighed with the customer when selecting a heat pump.

COMPRESSOR DOs AND DON'Ts

Heat pump compressors are continuing to change. This is due to the fact that higher efficiencies are being strived for by all manufacturers. This means that there will be continual innovation and change in the industry for years to come. The following are some general guidelines and reminders about heat pump compressors.

Dos

Ultra-efficient heat pumps may have unloading capabilities, variable speed, or staging. Whatever the system, all of these must begin a cycle unloaded, at low speed, or at low capacity. None of these systems should be operated directly from no power to full power, speed, or capacity.

Most heat pump systems are sized by their cooling load. However, heat pump systems are also sized by their heating capacities in some climates. This may mean that they are oversized for their cooling capacity. Dehumidification or comfort may suffer if the system is not allowed to reduce moisture sufficiently or overcool while attempting to control humidity. High-efficiency heat pump system designs can maintain comfort without compromising dehumidification.

Don'ts

Numerous heat pump systems use scroll compressor technology. These compressors are not designed to operate in a vacuum or to cycle more than six times per hour. Heat pumps that are not specified correctly for the application may cause the compressor to operate outside of the manufacturer's design specifications. If the operation of the compressor is suspect, the technician should monitor and test the compressor. Oil temperature should also not exceed 300°F, measured as a maximum of 225°F, 6 inches from the compressor on the discharge line. Do not let the compressor operate in a vacuum, cycle more than six times per hour, or have oil temperatures above 300°F.

Tech Tip

Discharge temperature, as measured 6 inches from the discharge of the compressor, should never exceed 225°F. The temperature of 225°F indicates a maximum temperature internal to the compressor of 300°F. Above this temperature, oil in contact with these parts begins to break down, causing acid and sludge, which then go on to cause breakdowns in the chemical composition of the refrigerant. These changes in chemical composition of the refrigerant form acids which will, over time, break down copper ions in the system causing copper plating. Acids also will attack motor insulation winding, resulting in failed motors.

EFFICIENCY CHECKS

An accurate set of gauges, clamp-around ammeter, and manufacturer specifications are needed to conduct efficiency checks of any compressor. Manufacturer data sheets and specifications are available from most manufacturers and are often available on their websites.

The first step is to measure and record the pressure at the suction and discharge service connections. Convert the suction and discharge pressures to their related saturation temperatures. Plot these temperatures on the manufacturer's specification graph within the data sheet (see Figure 8-14) and determine if the temperatures are within the operating range specified.

Next, measure the operating amperage and voltage of the compressor. Both of these readings should be taken at the load side of the contactor. Amperage will be related to the suction and discharge pressure produced by the compressor. The amperage should match or fall within the operating specifications for the compressor.

Amperage higher than the manufacturer's performance data could mean:

1. Low voltage
2. Weak or malfunctioning run capacitor (single-phase compressors)
3. Too much oil
4. Internal mechanical problems (replace the compressor)
5. Internal winding-to-winding shorts (replace the compressor)

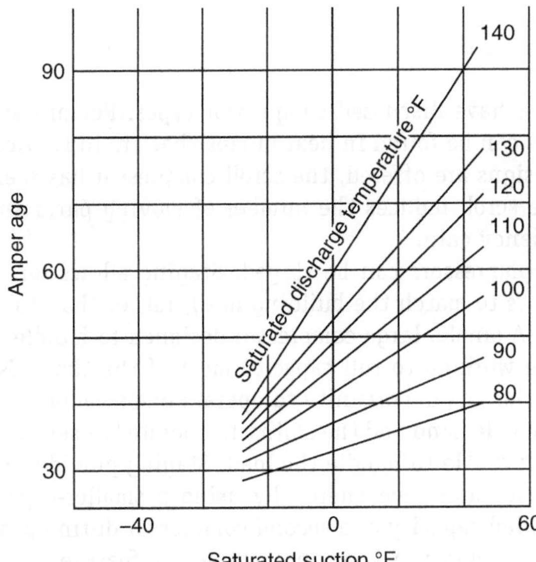

Figure 8–14
This is a basic example of the specifications that can be found in a manufacturer's information about the compressor. This type of chart is used to determine if the compressor is working within tolerances specified by the manufacturer. Saturated suction temperature, saturated discharge temperature, and amperage should all coincide (cross) or be very close if the compressor is operating according to specifications. (Courtesy of Delmar/Cengage Learning)

Or, it could be other mechanical or system issues that have nothing to do with the compressor. It is imperative that a full set of readings be taken when evaluating the performance of the compressor.

Amperage lower than the manufacturer's performance data could mean:

1. Rods are broken (replace the compressor)
2. Valves are faulty (worn, bent, or broken; replace the compressor)
3. Piston rings or wear surfaces of rotaries or scrolls are worn (replace the compressor)
4. Internal leakage in the head or mating surfaces (replace the compressor)

Or, again, it could be other mechanical or system issues that have nothing to do with the compressor. It is imperative that a full set of readings be taken when evaluating the performance of the compressor.

Tech Tip

It is strongly recommended that the technician obtain the manufacturer's compressor performance data. This information is available online and in pamphlet form. There are a lot of "condemned" compressors that are still good. Care needs to be taken to be sure that the compressor is faulty before replacing the compressor. It is highly recommended that technicians attend the manufacturers' schools.

SUMMARY

In this chapter we have discussed compressor types. Reciprocating, rotary, and scroll compressors can be found in heat pumps that are in service today. As new, more efficient designs are offered, the scroll compressor has seen increased use. The design of the scroll reduces the number of moving parts and the operation provides an efficiency gain.

All types of compressors can be staged. Staging allows for the operation of smaller compressors to match the building need, rather than to operate a single, large compressor. A single, large compressor designed to handle all of the design load would not be working to full capacity most of the time. Using two smaller compressors provides the opportunity to operate one smaller compressor to meet most of the seasonal demand and the ability to operate the second compressor only when the first is not able to handle the load. Staging provides a way to increase equipment longevity and save energy by using a smaller-capacity compressor, but having the added capacity of a second compressor during peak demand.

Unloading of a compressor is an internal design feature that reduces the capacity of the compressor in order to reduce amperage and save on power consumption. "Unloading" means that the lbs/min flow is reduced. Such designs effectively cut off the amount of compressed vapor produced and reduce the compressor capacity. Compressor capacity is reduced during times of low demand to save energy, increase efficiency, and minimize cycling.

Variable-speed and two-speed compressors are other ways of matching compressor capacity during times of low demand. Both of these methods require compressor motors to be designed in particular ways. Two-speed compressors are designed to run as two- or four-pole motors. Variable-speed compressors use electronics to increase or decrease the speed of specially designed motors.

Whatever the compressor technology, the technician is concerned with the operation of the compressor within a system. Compressor efficiency checks using manufacturer specifications are the only way to determine if the compressor is working properly. In order to accomplish these checks, the technician must have a set of high-quality diagnostic tools and use them to make accurate measurements of pressure, amperage, and voltage. The technician needs to compare these measurements and the compressor specifications to the manufacturer's operating specifications.

REVIEW QUESTIONS

1. Explain the differences between reciprocating, rotary, and scroll compressors.
2. Describe how compressors can be staged.
3. Explain how compressors are "unloaded."
4. Describe how a two-speed compressor operates electrically.
5. Describe how a variable-speed compressor operates.
6. Describe how to conduct a compressor efficiency check.

CHAPTER 9

Specific Defrost

LEARNING OBJECTIVES

The student will:

- Describe the reason why air-source heat humps (ASHPs) require defrost systems
- Describe how the outdoor coil is warmed to melt frost
- Describe the reason why certain operations need to occur to have a good defrost cycle
- Describe the operation of a pressure-initiated/temperature-terminated defrost system
- Explain how electronic controls have helped improve heat pump defrost
- Describe the benefits of demand defrost controls

INTRODUCTION

Air-source heat pumps (ASHPs) require that the outdoor coil be defrosted because the saturation temperature is commonly below freezing. When defrost should occur and how much defrosting will be necessary depends on the outdoor ambient condition. Since outdoor ambient conditions change and because outdoor ambient is different (meaning having more or less moisture) in certain parts of the country, defrosting the outdoor coil becomes a challenge (Figure 9-1).

If the frost accumulation is not removed from the outdoor coil, the efficiency of the unit goes down. Frost acts as an insulator, inhibiting the movement of heat from the outdoor air to the colder refrigerant. The effect is cumulative. If defrost cycles are not initiated correctly or the unit stays in defrost mode longer than necessary, the heating seasonal performance factor (HSPF) will be affected. Manufacturers want the HSPF to be high to attract new buyers. To keep the HSPF high, manufacturers try to match the defrosting needs with the installation by providing different ways to defrost and adjustable controls to meet defrosting needs for all climate conditions.

All manufacturers have adopted the "hot gas defrost" method. This means that the reversing valve is used to place the heat pump in the cooling mode during the defrost cycle. Hot vapor is diverted from the indoor coil to the outdoor coil to melt frost buildup. When the defrost cycle ends, the reversing valve is allowed to move to the heating mode position. This chapter provides information on the various defrost initiation (start) and termination (stop) methods. Many methods have been used in previous years, but the methods presented here are those currently used for heat pumps in the field.

Field Problem

The business owner reported that his heat pump was not heating adequately and seemed to be running all of the time. The technician noted that the day was not below 40°F and the heat pump was only 5 years old. Arriving at the scene, the technician noted that the outdoor fan was not running and that the outdoor coil was covered with ice (Figure 9-2).

Shutting down the outdoor unit, the technician removed the electrical panel cover. Reading the schematic, he noted that the defrost control directly operated both the outdoor fan and defrost relay. Possible causes were a defective defrost control or a defective outdoor fan. He pulled the fan wire and defrost relay wire off the defrost control. Using his ohmmeter, he checked for continuity from the power connector to the fan connector—there was no continuity. He checked for continuity from the power connector to the defrost relay connection—there was no continuity. He disconnected the power connector and taped all leads to protect against arcing. Placing the heat pump manually into the cooling mode, he defrosted the outdoor coil while maintaining an ohmmeter connection between the power and fan connections on the defrost control. He also monitored the temperature sensor on the defrost control with an electronic thermometer. When the electronic thermometer read 40°F and the coil was clear, he looked for continuity at the defrost control and found none. Shutting down the outside unit a second time, the technician verified his finding by connecting the outdoor fan directly to the power wire to test the fan—the fan operated.

The technician replaced the defrost control and conducted a defrost test to determine that the heat pump was functional before reporting his finding and repairs to the business owner.

DEFROST SEQUENCE

Several requirements and conditions need to occur for good defrost (Table 9-1). Each of these must happen or defrost problems will occur. Some of the operations occur at nearly the same time, but each is dependent on the other.

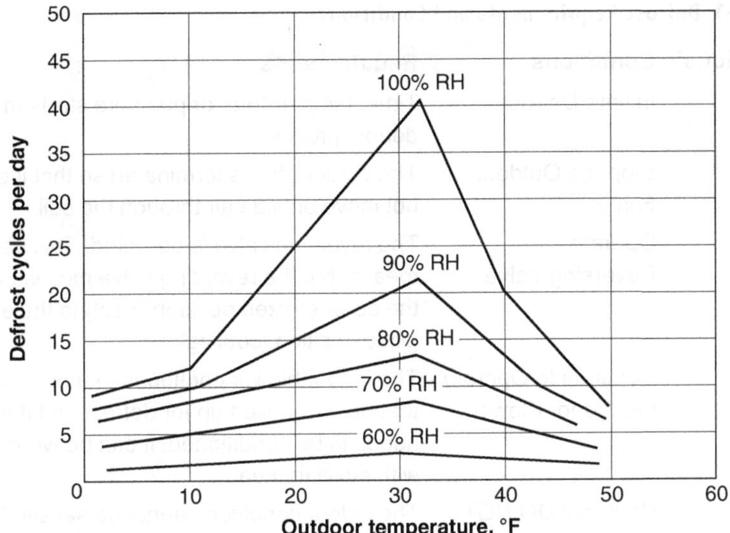

Figure 9–1
The number of defrost cycles needed in one day will differ with the temperature and moisture levels. Temperature and moisture levels may be different from location to location. A heat pump installed in the Northeast will need more defrosting cycles than a heat pump installed in the Southwest. (Courtesy of RSES)

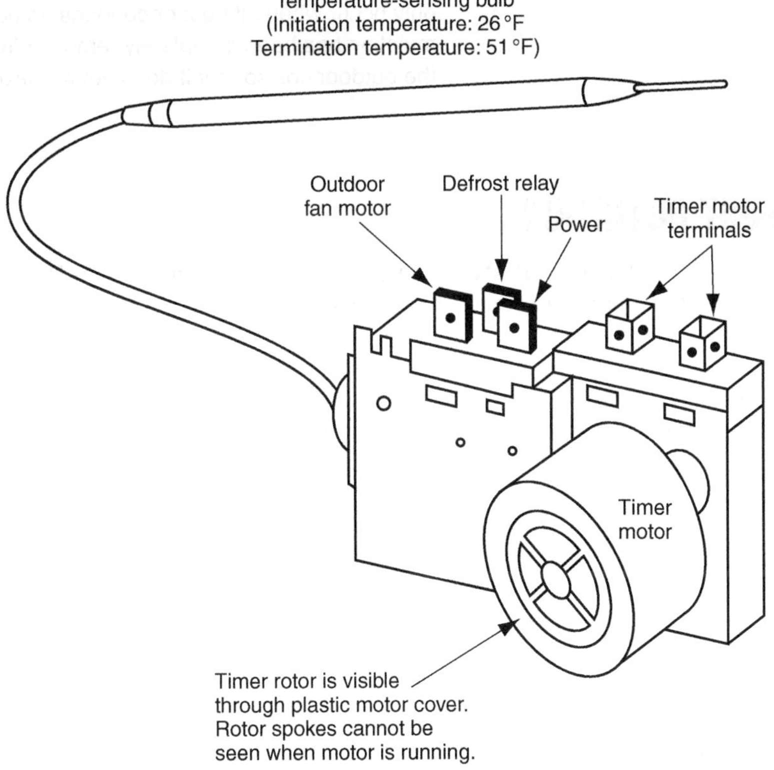

Figure 9–2
Time/temperature defrost control. (Courtesy of RSES)

If any one or more of these operations does not occur, the defrost cycle will be affected; which will ultimately affect system operation and customer comfort. The sequence of operation must occur to maintain energy efficiency and customer comfort conditions.

Table 9-1 Defrost Requirements and Conditions

Operations	Conditions	Requirements
1	Initiate Defrost	Time, temperature, or pressure starts the defrost process.
2	Stop the Outdoor Fan	The outdoor fan is terminated so that it does not move outside air through the coil.
3	Operate the Reversing Valve	The reversing valve is operated. This usually means that the reversing valve moves from the de-energized position (heat) to the energized position (cooling).
4	Continue to Operate the Indoor Blower	The indoor blower continues to move air so that heat is picked up for defrost and if the air needs to be conditioned, it can be warmed with auxiliary heat.
5	Heat, but DO NOT Satisfy the Indoor Temperature Setpoint	The indoor condition cannot be satisfied or the compressor will stop and the defrost cycle will terminate. At the same time, there may be a need to warm indoor air during defrost. Auxiliary heat is brought on to minimally satisfy indoor temperature conditions.
6	Terminate Defrost	Time or temperature will end the defrost cycle. In this part of the process, termination should only occur when all frost or condensate accumulation has been completely removed from the outdoor coil so that it does not refreeze.

DEFROST INITIATION

There are several defrost initiation and termination methods. When the method is named it is usually described as the method of initiation and the method of termination—for example, time initiation/temperature termination, or time/temperature. Many of the defrost methods use the same termination. For this reason the initiation methods will be discussed separately. It should be noted that most controls have a high temperature lock-out that will not allow defrost to occur at temperatures above a certain high temperature setting. Most controls have temperature settings above 40°F, and some are above 50°F.

Time

Timed defrost involves the use of a time clock control and two methods of initiation. In the first case, the clock is used to time the number of defrost cycles that will occur in one day. In the second case, the clock is used to account for the amount of time the compressor runs and is sometimes called an "accumulative timer." The reason it is accumulative is because the clock only operates when the compressor operates. When the compressor starts for a heating cycle, the defrost clock starts running. In both cases the timer control can be adjusted to meet local defrosting needs (Figure 9-3).

The clock control is typically designed or adjusted to provide the longest possible amount of defrost time without compromising indoor temperature too much. The expertise of the technician is needed to adjust the defrost control to minimize owner discomfort and maximize efficiency. Figures 9-4 through 9-6 show the defrost timer.

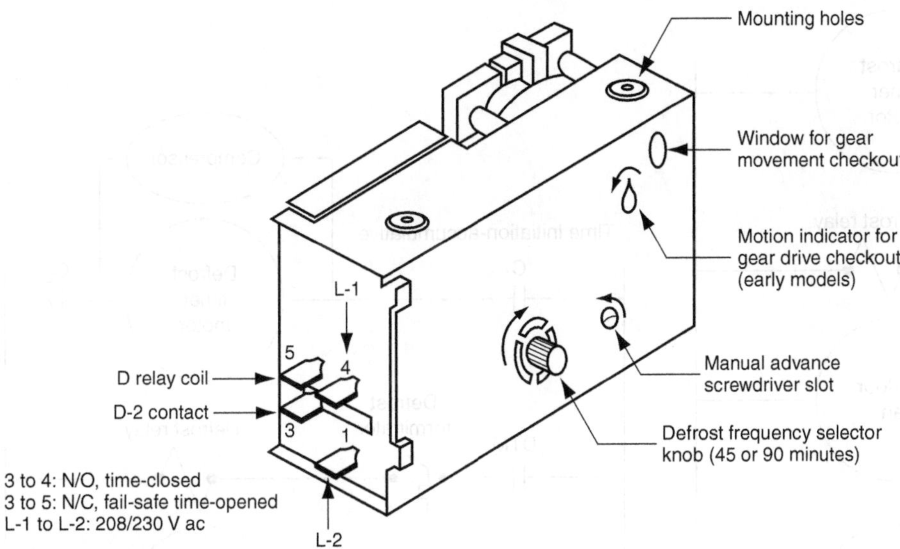

Figure 9–3
The defrost timer has several adjustments that can be made to meet installation location defrost needs. (Courtesy of RSES)

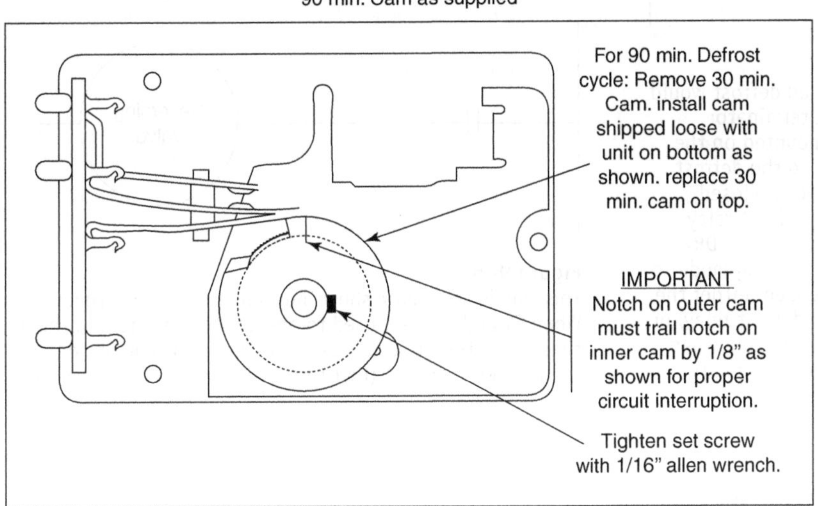

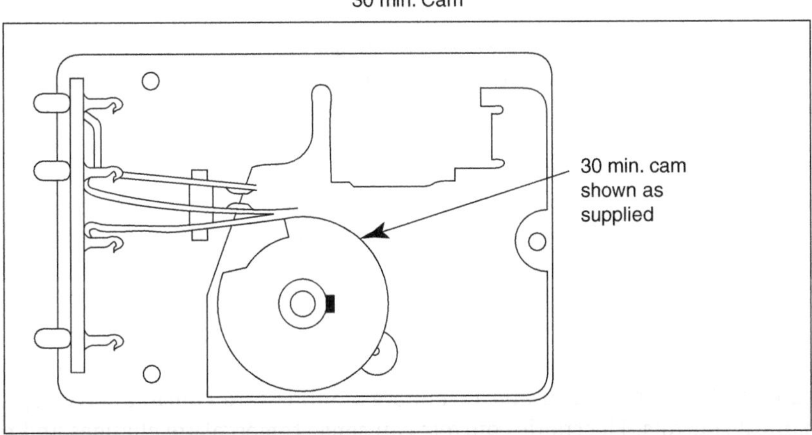

Figure 9–4
The defrost timer can be set up as a 30-, 60-, or 90-minute timer. (Courtesy of Addison Products Co.)

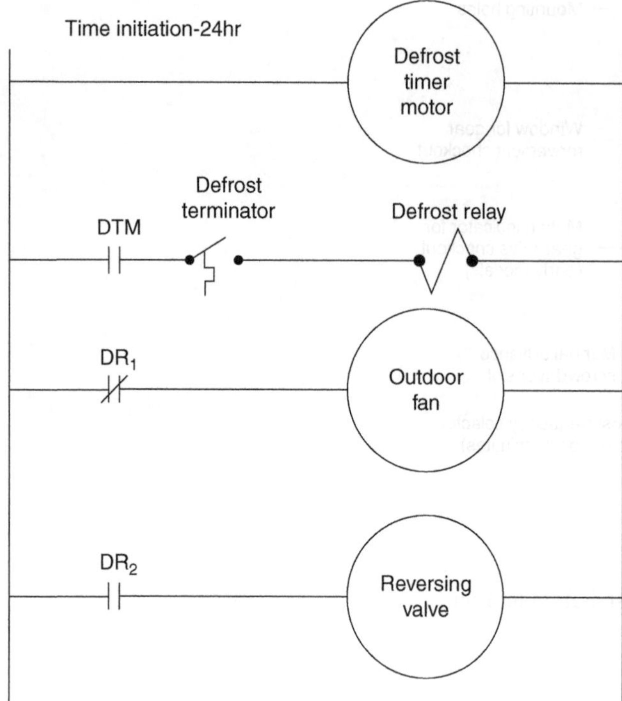

Figure 9–5
This simple schematic shows how a 24-hour timed defrost would work. The timer operates 24 hours. The defrost terminator will cool down during operation because it is mounted on the outdoor coil. When the defrost clock cam turns to the defrost position, DTM will close. If the defrost terminator is closed (indicating frost accumulation possibility), the defrost relay will operate. DR1 will open, turning off the outdoor fan. DR2 will close, moving the reversing valve from the heating mode to the cooling mode. When the defrost cycle has ended, either the defrost terminator will open, de-energizing the defrost relay, or the timer will continue timing and end the defrost cycle when the cam opens DTM. (Courtesy of Delmar/Cengage Learning)

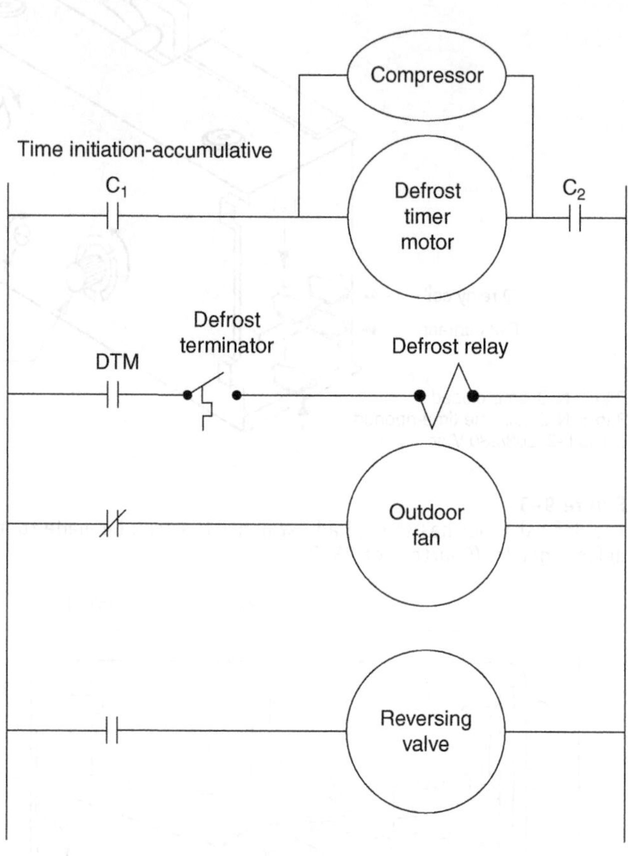

Figure 9–6
This simple schematic shows how an accumulative timer operates. When C1 and C2 are closed to operate the compressor, the defrost timer operates. It accumulates or counts the amount of time the compressor operates. (Courtesy of Delmar/Cengage Learning)

Temperature

As the frost builds, the saturation pressure/temperature in the outdoor coil will drop. A drop in suction pressure is an indicator of how much frost has accumulated and is reducing heat transfer. As the pressure goes lower, a greater amount of ice is building. Typically, if the temperature difference between the outside air and the temperature of the refrigerant in the coil is greater than 10°F, the defrost control will initiate and turn on defrost. This control is a temperature differential control and may also have reset capability. It may be able to adjust the temperature difference to meet outside temperature conditions. This means that the control may operate the defrost with a temperature difference of 15°F when outside conditions are 40°F, or call for defrost with a temperature difference of 10°F when the outside temperature is 12°F. In other words, the control "resets" to optimize the call for defrost as outside conditions change (Figures 9–7 and 9–8).

At the end of the defrost cycle, the temperature control senses higher temperatures at the outside coil and terminates. Solid-state controls have replaced mechanical controls. Electronic controls can sense temperature changes and react to them better than mechanical controls. These controls have reduced problems with cool air coming from indoor diffusers and customer complaints about rushes of cold air.

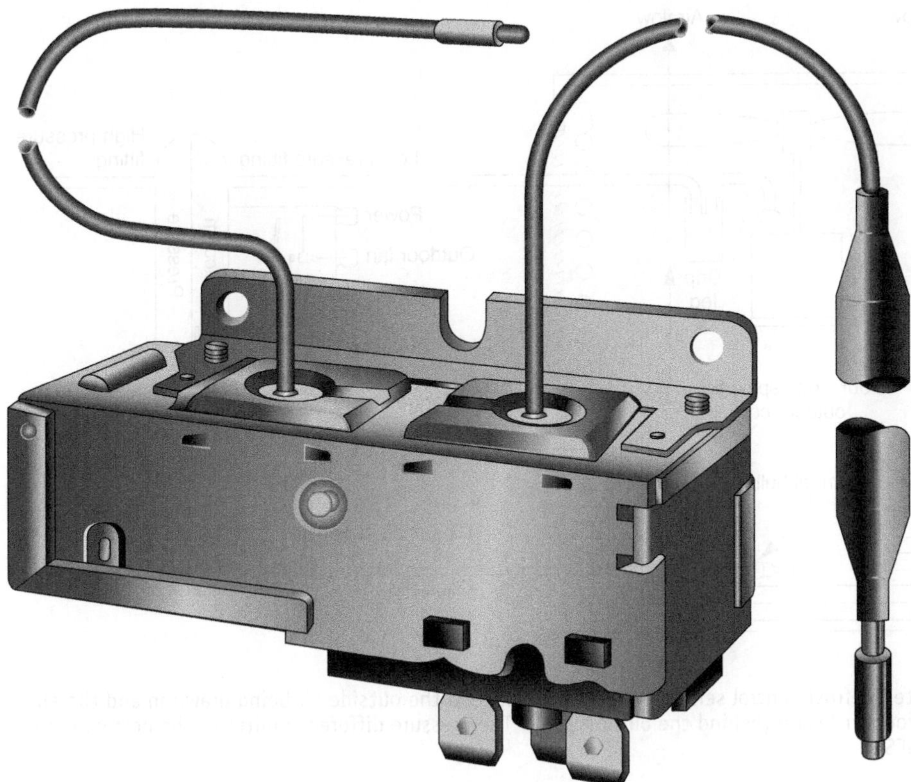

Figure 9–7
The temperature initiation defrost control has two temperature-sensing bulbs. The short element on the top left senses outdoor air temperature. The element on the left with the larger bulb is attached to the outdoor coil to sense coil temperature. (Courtesy of Ranco Inc.)

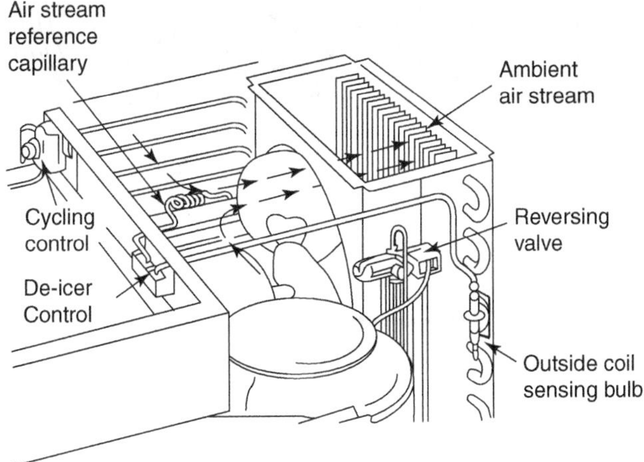

Figure 9–8
This diagram shows the location of the defrost (de-icer) control and sensing elements. (Courtesy of Ranco Inc.)

Pressure

Pressure difference between the outside air being drawn in and the air inside the outdoor enclosure (behind the outdoor coil) is sensed for pressure drop. Pressure drop indicates the amount of frosting that has occurred and is blocking air passage through the coil. When the pressure drop exceeds the pressure difference setting, the control initiates the defrost cycle (Figures 9–9 and 9–10).

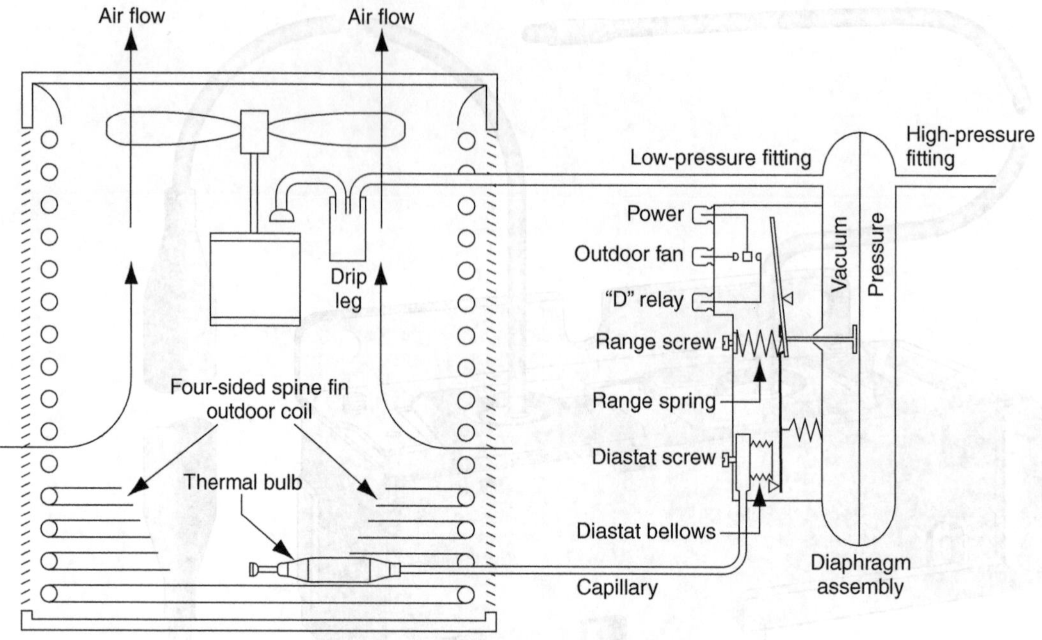

Figure 9–9
Pressure-initiated defrost control senses the pressure between the outside air being drawn in and the air inside the outdoor enclosure (behind the outdoor coil). The pressure difference initiates the defrost cycle. (Courtesy of RSES)

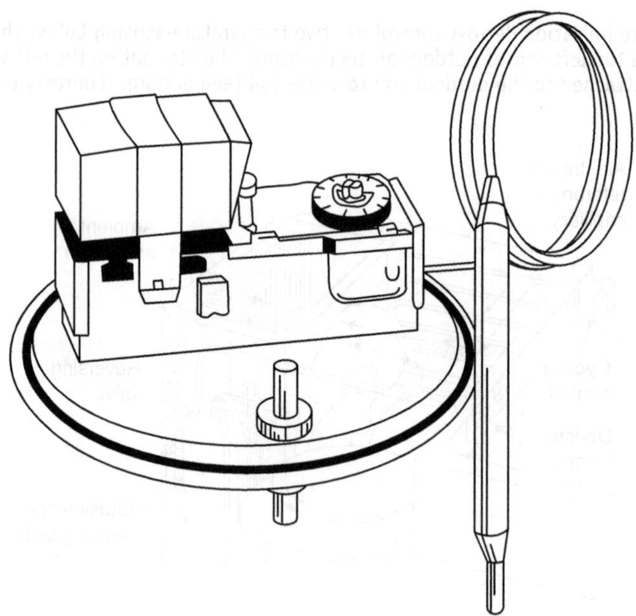

Figure 9–10
Drawing of a pressure-initiated defrost control with temperature termination. (Courtesy of Delmar/Cengage Learning)

The pressure difference can be adjusted and is usually between .4 and .8 inches of water column. This type of control works well when the outdoor coil is clean and the sensing tubes are not blocked. Pressure adjustment is accomplished through sequential steps that are slightly different, depending on the manufacturer of the unit. Always consult the equipment installation and service manual for the appropriate steps and conditions.

Demand

Demand defrost may combine all of the functions of pressure, time, and temperature into one control. Electronically, this control is able to overcome some of the faults of the other systems. If pressure initiated, the control is able to measure the amount of time that the pressure difference exists and requires up to 20 seconds before initiation. Unlike mechanical controls that activate a switch when the set pressure difference is reached, the demand control is able to eliminate the possibility of wind gusts initiating a nuisance defrost. It is also able to store previous defrost information and modify the defrost cycle to suit the installation location. The demand defrost control can store information about compressor run cycles, air temperature, and defrost cycle time.

Demand controls vary in their configuration. Many of the demand defrost systems today use two thermistor-type sensors—one for outdoor air temperature, and one for coil temperature. The ability to measure conditions, store the information, and adjust the defrost initiation, cycle, and termination means that the control can adapt to varying outdoor conditions. The problem with pre-set controls is that as the outside ambient temperature starts to drop, the relative humidity becomes very high and moisture is more likely to deposit on the coil quickly. More frost on the coil means that the coil needs more frequent defrost cycles. As the temperature continues to go down, the frosting condition reduces slightly, requiring less defrost, but it may need more time to shed the moisture from defrosting. The demand control is able to adapt using previous defrost information as a guide (Figure 9–11).

TERMINATION

All defrost systems use temperature or time to terminate a defrost cycle. If the temperature sensed by the temperature sensor is more than 40°F, many systems will terminate the defrost cycle. If the temperature sensor never reaches 40°F, the timer continues the defrost cycle until the cycle times out. Sometimes the time allocated for defrost can be as much as 20 minutes or more.

The length of the defrost cycle is determined by the terminator. Referring to the examples of typical defrost cycles, the first cycle was initiated and

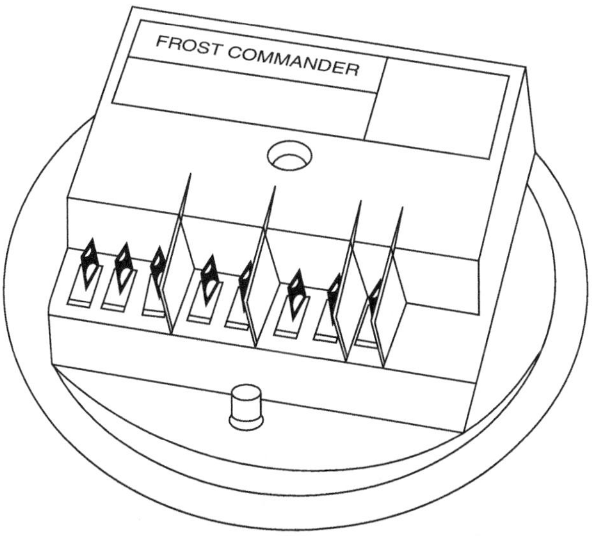

Figure 9–11
Solid-state pressure-initiated demand defrost control. (Courtesy of Ranco Inc.)

completed within 2 minutes because of outdoor conditions. As the outdoor conditions changed, the next cycle time took 8 minutes to complete. The third cycle never reached temperature termination and was finally terminated by time (Figures 9–12 and 9–13).

Examples of typical defrost cycles

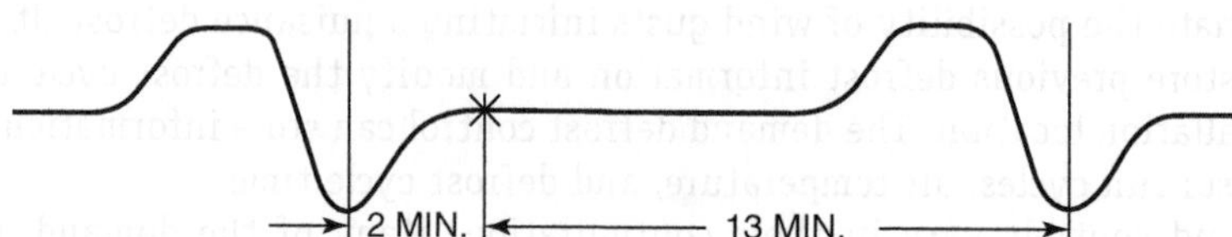

No wind, relatively warm ambient-termination occurs based on coil temp. after only 2 min. of defrosting. Next initiation requires 13 min. of pressure sensor output.

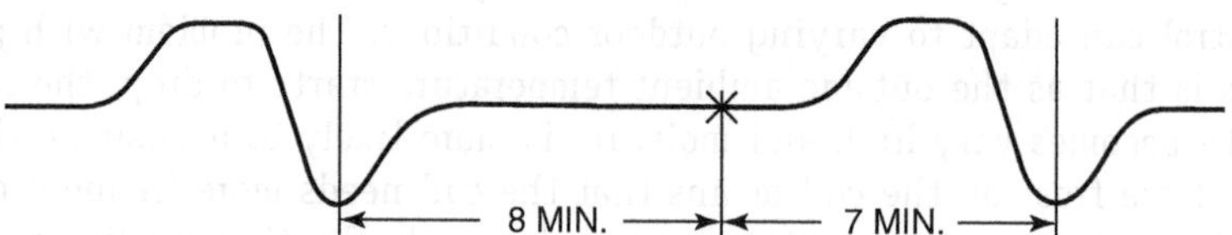

Windy, cold ambient-termination requires 8 min. of defrosting. Next initiation requires 7 min. of pressure sensor output.

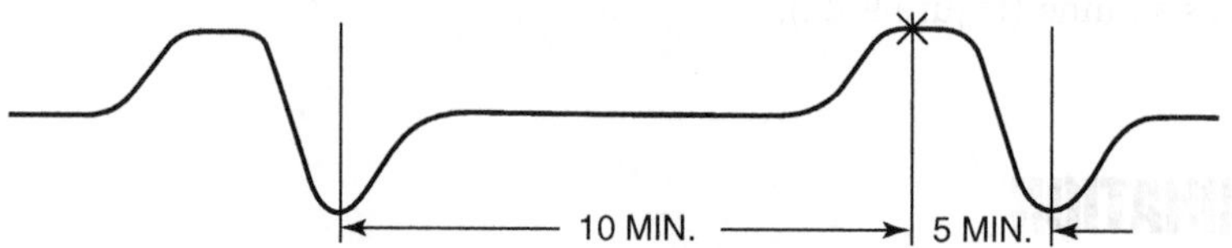

High winds, very cold ambient-termination temp. cannot be reached. Defrost continues for 10 min. before termination based on time. Next initiation requires only 5 min. of pressure sensor output.

Figure 9–12
Examples of defrost cycle time. (Courtesy of Addison Products Co.)

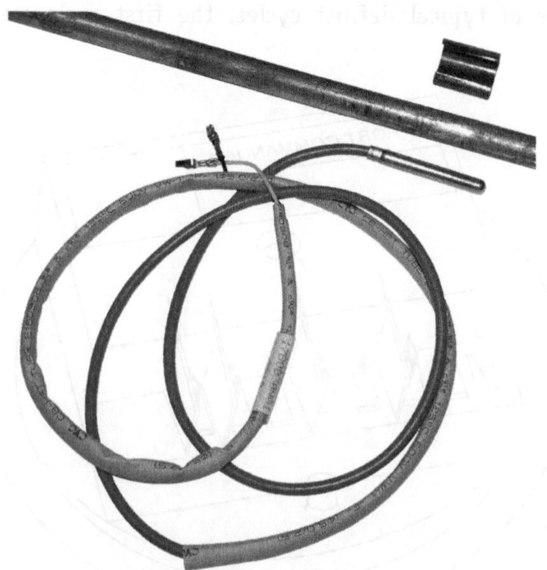

Figure 9–13
A thermistor attached to a refrigeration line. The thermistor is a solid-state device used as a termination sensor for heat pump defrost controls. Similar thermistors can also be used to measure outdoor temperature for demand defrost systems. (Courtesy of Delmar/Cengage Learning)

SUMMARY

All ASHPs require defrost to shed accumulated moisture, in the form of frost, on the outdoor coil. When an ASHP defrosts, several operations must occur to ensure good defrost. If any one of these does not occur, the customer will complain of poor operation.

Defrost control methods are named by their initiation and termination functions. For instance, timed initiation and temperature termination is referred to as time/temperature defrost. Of the varied types of defrost initiation methods that have been used, time, temperature, pressure, and demand defrost systems dominate the industry. Termination is largely done by temperature or time. The length of time for a defrost cycle is determined by termination. Either termination temperature is reached or time runs out for the defrost cycle.

Of all of the control systems, the electronic demand defrost method combines elements of all of the other methods and provides additional abilities to store and modify the defrost cycle in an effort to minimize defrost time and maximize outdoor coil efficiency by keeping it clear of frost.

REVIEW QUESTIONS

1. Describe the reason why air-source heat humps (ASHPs) require defrost systems.
2. Describe how the outdoor coil is warmed to melt frost.
3. Describe the reason why certain operations need to occur to have a good defrost cycle.
4. Describe the operation of a pressure-initiated/temperature-terminated defrost system.
5. Explain how electronic controls have helped improve heat pump defrost.
6. Describe the benefits of demand defrost controls.

CHAPTER 10

Electrical Schematics

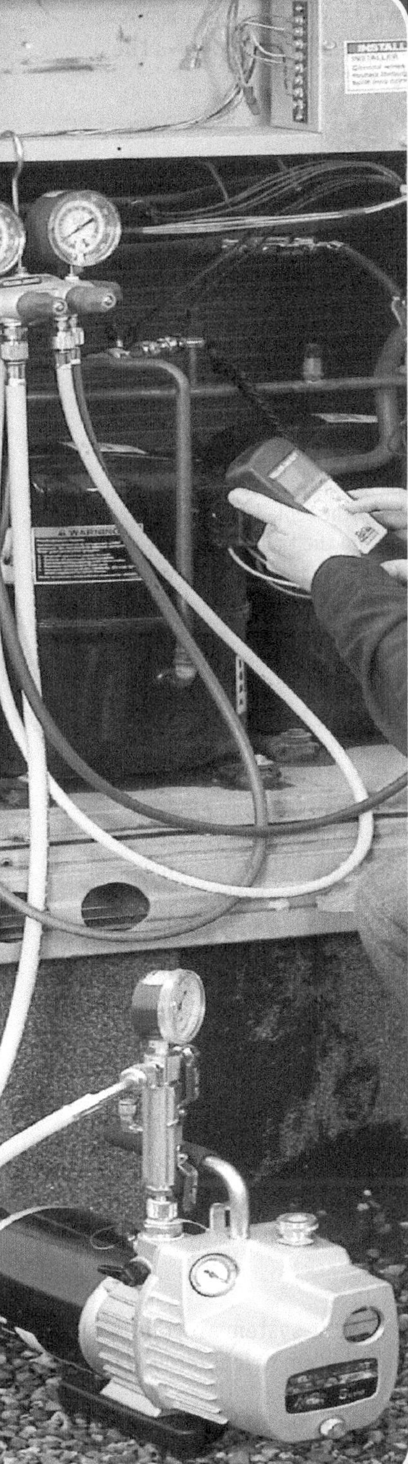

LEARNING OBJECTIVES

The student will:

- Describe what is meant by the term "electromechanical" and identify those types of components
- Describe the use of a wiring diagram legend
- Name the five similar electrical circuits that heat pumps share with air conditioning systems
- Explain how the crankcase heater circuit works
- Describe the four modes of operation for air-source heat pumps (heating/cooling/defrost/emergency heat)
- Relate why a hard-start kit might be used and describe its operation
- Describe the benefits that electronic circuits provide for heat pumps

INTRODUCTION

This chapter reviews the electrical components and wiring of an air-source heat pump (ASHP). It should be mentioned that an ASHP and a ground-source heat pump (GSHP) share many of the same components and that this chapter could be used to understand the fundamental operation of all heat pumps. For instance, the outdoor blower motor of an ASHP could easily be replaced with a loop pump used in a GSHP. The manufacturer's diagram in Figure 10–1 shows the system as it is wired and in the de-energized mode. Different manufacturers represent their electrical drawings in similar ways.

A good understanding of the operations and the sequences of events that occur in the electronic/electromechanical system is required to service heat pumps. This chapter explains, in detail, the electrical system of an ASHP and follows the operation of electrical components in a step-by-step process.

Field Problem

The customer was flustered while talking to the technician. He related that while he was mowing the lawn, a small rock or object was thrown toward the outside unit of the heat pump. It left a hole in the side of the outdoor unit and he was afraid that he damaged something. When asked if the system was running satisfactorily, the customer reported that everything seemed okay. The day was a relatively warm day that required cooling. The control panel was bent, with a hole through it the size of a penny.

Symptoms:
None

Possible Causes:
Damage to the electrical system; damage to the refrigeration system; possibly no damage other than cosmetic.

Opening the control panel, the technician used his senses to find the object that caused the hole and to try to determine if there was anything obviously damaged. The object was a bolt, likely from the riding lawn mower. Putting it under the clip of his board, he made a mental note to suggest that the homeowner get the mower checked as well.

It seemed that the bolt entered the control panel, but there was no other obvious damage. Still, the technician generated a list of things that needed to be checked to ensure there was no damage:

- Check the contactor operation.
- Check capacitors for physical damage and operation.
- Check for loose wiring or broken wires.
- Check for switch settings (dual inline packet [DIP] switches) on control board.
- Check the control board for operation and visually for cracks.
- Check any relay in the control panel for operation.

As the technician proceeded with the checklist, he noticed that if the bolt traveled in a straight line, the control board would have been the direct path, so he started there. Using the manufacturer's directions, he reviewed the microprocessor light-emitting diode (LED) and diagnostics table. First on the table were the DIP switch settings. He checked the settings on the table with those on the board. Two of the switches were in the wrong position. Looking further, a visible crack ran across the DIP switches. Both of these switches had to do with the reversing valve. Checking the valve operation, the technician verified that the valve was working and in the cooling mode.

It was obvious, now, that the bolt had hit the control board and damaged the DIP switches. Even though the DIP switches were in the wrong position, the control was operating properly. The technician refrained from attempting to move the DIP switches to their proper position. If he had, the system might have failed immediately. The decision was made to tell the customer that the board needed to be replaced and that the hole in the control panel needed to be patched (or the panel replaced), and the technician told the customer about where the bolt might have come from.

The customer approved the system repair and thanked the technician for the suggestion to check the mower. The system operated through the night without a problem. The next day the panel door and the control board were replaced and the technician made a mental note to thank himself for not moving those DIP switches, because the switches fell off the board in pieces when he removed the board.

THE ELECTRICAL/MECHANICAL SEQUENCE

A heat pump wiring diagram is very similar to diagram for a split-system air conditioner. Like an air conditioning system, there are some similar electrical circuits:

1. Compressor
2. Outdoor fan

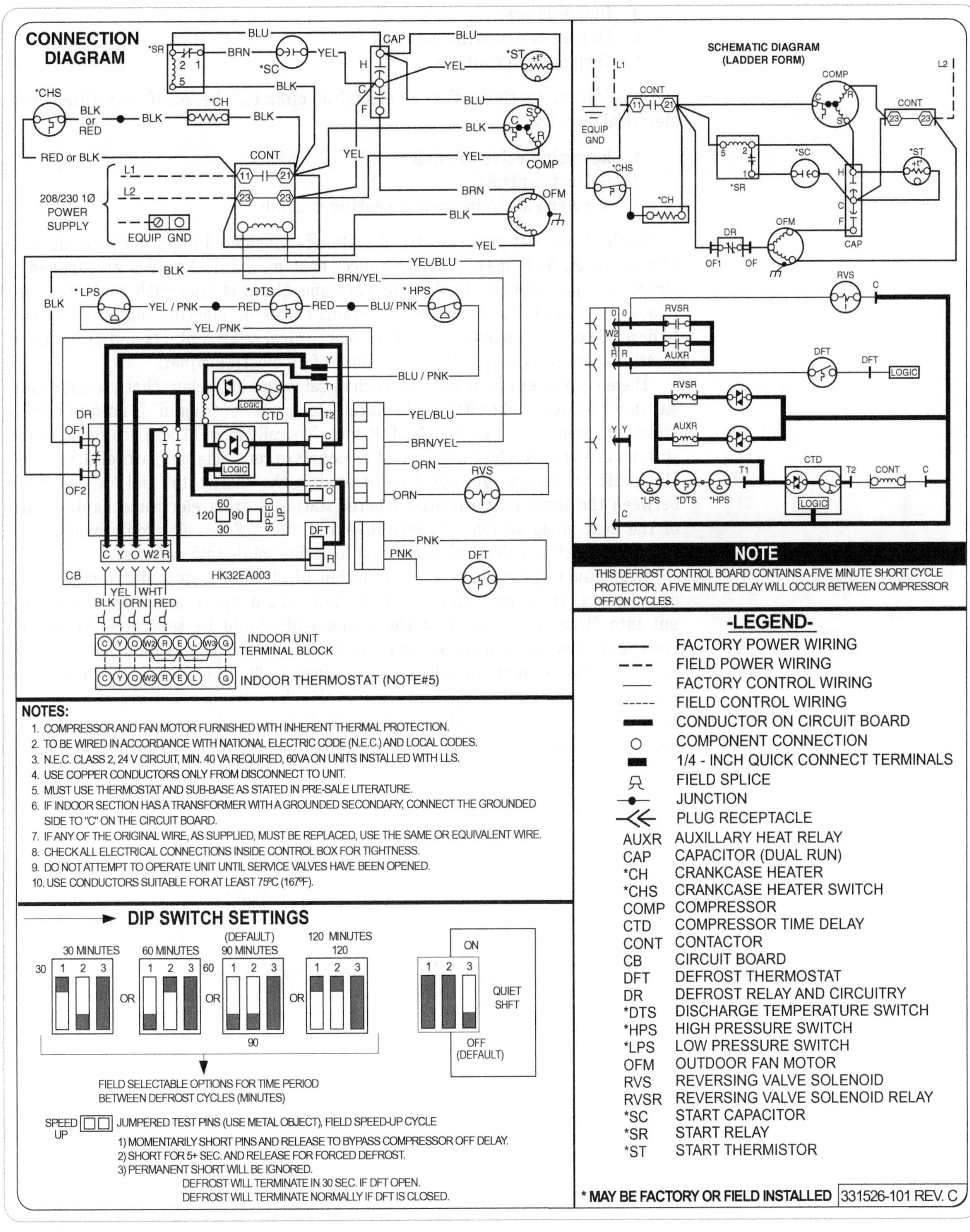

Figure 10–1
This is a diagram from a manufacturer of heat pumps. DIP switch settings can be seen in the lower left of the diagram. The upper-left diagram is the Connection Diagram. Wiring shown in this part of the diagram shows the color coding of each wire and the label of terminals where each wire is connected. The upper-right diagram is the Schematic Diagram (Ladder Form), showing how each circuit of the system operates. The legend in the lower right is used to understand the labels used in the wiring diagram. The "Notes" section, left-center, contains essential information that is supplied by the manufacturer. (Courtesy of Carrier Corporation)

3. Indoor blower
4. Compressor starting controls
5. Temperature controls

In addition to these similar electrical circuits, the ASHP has three more electrical circuits:

1. Reversing valve solenoid
2. Defrost controls
3. Supplemental/emergency heating controls

Much of the heat pump electrical system is considered to be electromechanical (Figure 10–2). This means, for instance, that mechanical valves are opened or closed using an electrical solenoid. The same thing is true with contactors and relays. A solenoid is also an electromagnet. The term "electromechanical" is used to describe physical movement of both mechanical relay contacts and valves using electromotive force, typically in the form of an electromagnet.

There is another way to control mechanical valves and relays that use solid-state electronic sensors and switches. The term "electronic" or "digital" is used to refer to boards or assemblies of circuits that do not physically move. These boards or control systems use small DC electrical current or signals to control larger parts of the system. Electronic boards are typically incorporated in the thermostat and are connected between the temperature control (thermostat) and other electromechanical parts of the heat pump system (Figures 10–3 and 10–4). Some electronic boards have embedded electromechanical relays that look like solid-state switches.

Heat pump systems are a combination of electronic and electromechanical components. If a heat pump system were taken apart and the components put into "like" piles, most of the components could be separated easily into electronic, electromechanical, and mechanical. Some parts would not easily fit the description of just one pile of parts, however. The compressor, for instance, is an electromechanical component. It uses the electromagnetic force of a motor to turn a compressor and produce a pressure difference. It would be considered an electromechanical device, but it is also a mechanical device. Because the electric

Figure 10–2
This relay is an example of an electromechanical component. On the label of the relay are electrical symbols that show where the contacts are connected and how they operate. Note the symbol of a circle that stands for the electromagnetic load and the two switches; one is normally open (NO) and the other normally closed (NC). (Courtesy of Emerson Climate Technologies)

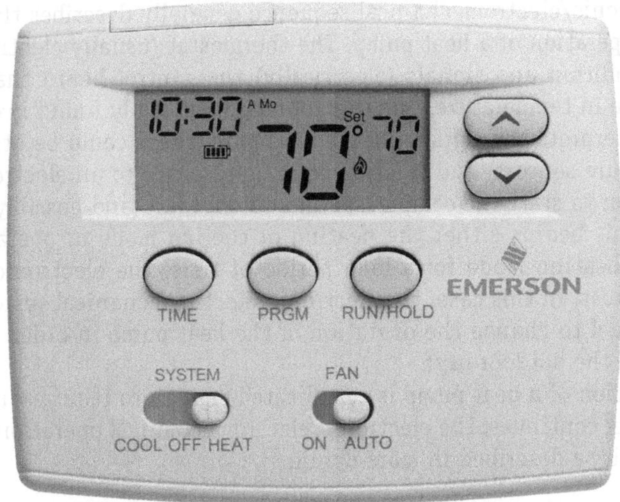

Figure 10–3
An example of an electronic thermostat. (Courtesy of Emerson Climate Technologies)

Figure 10–4
Electronic control boards are made up of many smaller parts. One or more of these smaller parts can fail, rendering the board inoperable. When checking the board, always follow the manufacturer's recommendations for determining input and output values. (Courtesy of Emerson Climate Technologies)

motor and the compressor are both housed within the same enclosure, a hermetic container, the two cannot be separated. In this case the compressor could be placed in the electromechanical pile with the contactor and relay. But, it could just as easily be placed in the mechanical pile along with the metering valve, drier, and system access valves.

Tech Tip

Using a jumper as a diagnostic procedure is NOT recommended or advised. This procedure requires that the system is operated while a jumper wire is used to short out or bypass electrical devices. Arcing, burned and overloaded wiring, and destruction of electrical components are all possible while using this method of troubleshooting. Unless the manufacturer specifically provides and marks a jumper location (as in the lower-left corner of the electrical information sheet), it is advised that technicians not use jumping-out as a diagnostic method or to energize a device.

The electronic/electromechanical sequence generally describes the electrical-mechanical operation of a heat pump. The thermostat (usually electronic) senses the room condition and signals (electrically) the control board that there has been a change in temperature. The control board (typically found inside the outdoor unit) interprets the signal, checks all inputs (these could be other pressure and temperature sensors) and sends a signal (electrical) to an electromechanical relay/contactor to start the compressor and outdoor fan (and possibly the reversing valve). This begins either the heating or cooling mode of operation. If the system is in heating mode for a long period of time, the electronic board may signal for defrost. In this case, the electronic/electromechanical system would be given the signal to change the operation of the heat pump in order for ice to be removed from the outdoor unit.

The operation of a heat pump is complicated enough to require further study. As this chapter continues, the electronic/electromechanical operation of a typical heat pump will be described in more detail.

Field Problem

The customer complaint was that the indoor temperature setpoint was 68°F, but the indoor temperature was 66°F in the heating mode and the thermostat was never satisfied.

Symptoms:
Frosted outdoor coil; compressor is running.

Possible Causes:
Defrost board; defrost sensor; refrigerant charge; bad outdoor fan motor; defrost contacts stuck; outdoor fan runs continuously; stuck reversing valve.

The technician observed that the outdoor coil was badly frosted and the outdoor fan motor was running. Opening the control panel, a voltmeter was used to determine if the defrost thermostat (DFT) was open, indicating a bad thermostat. There was no difference in voltage across the DFT, indicating that it was closed and calling for defrost or there was no voltage in the circuit.

Reading the electrical information sheet on the back of the control panel, the technician found the "speed up" jumper pins provided by the manufacturer to be used to put the system in defrost mode. Using a jumper, the system was manually placed in the defrost mode. The outdoor fan motor continued to operate, but the reversing valve switched and the refrigerant tubing could be felt to change modes from heating to cooling. Using a voltmeter, the technician checked the defrost relay (DR) switch and found no line voltage difference across the contacts, indicating that the DR (across OF1 and OF of Figure 10-1) contacts were stuck closed, preventing the outdoor fan from turning off as required to perform a proper defrost. The defrost control board would need to be replaced.

MANUFACTURER'S ELECTRICAL INFORMATION SHEET

Heat pump electrical diagrams and schematics can be found inside the electrical control box and with the installation manual supplied by the manufacturer. Usually, the electrical information sheet includes more than one diagram and is accompanied by notes from the manufacturer, switch settings, and a legend, so that electrical component symbols can be identified. The diagram may also show the different types of wiring, such as factory wiring and field wiring. One important thing to remember is that every manufacturer may label its own electrical components in a proprietary format—there are no standard electrical symbols or naming conventions. That is why there are differences in how wiring diagrams are drawn and labeled. In this chapter we will be working with one manufacturer's electrical sheet. This manufacturer's electrical sheet has been supplied by a prominent manufacturer of heat pumps, but other heat pump diagrams and schematics will be similar in their style and layout.

LEGEND

——	FACTORY POWER WIRING
- - -	FIELD POWER WIRING
——	FACTORY CONTROL WIRING
-----	FIELD CONTROL WIRING
▬	CONDUCTOR ON CIRCUIT BOARD
○	COMPONENT CONNECTION
■	1/4 - INCH QUICK CONNECT TERMINALS
	FIELD SPLICE
—•—	JUNCTION
—<<	PLUG RECEPTACLE
AUXR	AUXILLARY HEAT RELAY
CAP	CAPACITOR (DUAL RUN)
*CH	CRANKCASE HEATER
*CHS	CRANKCASE HEATER SWITCH
COMP	COMPRESSOR
CTD	COMPRESSOR TIME DELAY
CONT	CONTACTOR
CB	CIRCUIT BOARD
DFT	DEFROST THERMOSTAT
DR	DEFROST RELAY AND CIRCUITRY
*DTS	DISCHARGE TEMPERATURE SWITCH
*HPS	HIGH PRESSURE SWITCH
*LPS	LOW PRESSURE SWITCH
OFM	OUTDOOR FAN MOTOR
RVS	REVERSING VALVE SOLENOID
RVSR	REVERSING VALVE SOLENOID RELAY
*SC	START CAPACITOR
*SR	START RELAY
*ST	START THERMISTOR

***MAY BE FACTORY OR FIELD INSTALLED**

Figure 10–5
The legend is a guide to decoding the symbols of a wiring diagram. For instance, read the legend to find what "CTD" stands for. Try to find it in the diagram or schematic. The CTD is a compressor time delay control—an electronic timer. (Courtesy of Carrier Corporation)

The legend is always very helpful in identifying electronic/electrical components in the wiring diagram. You will need to refer to the legend in Figure 10–5 often in this chapter, just as you will need to refer to similar electrical wiring diagrams in the field when working with wiring diagrams. This legend will be used throughout the chapter.

The "Connection Diagram" used to re-wire or to check the wiring connections is necessary to find the terminal points of connected components (Figure 10–6). When the technician wants to check a component, the wiring diagram needs to be consulted to determine where the component is connected and what the wire color code might be.

Theoretically, if the technician wanted to check the start capacitor (SC; reminder: check the legend), the Connection Diagram would be read to find the SC and how it's connected. By disconnecting the SC from the circuit on one side (the BRN-colored wire at terminal 1 of the start relay [SR]), the technician could check the SC for capacitance (microfarad) from the disconnected BRN wire to the C terminal on the capacitor (CAP). Electrically, it would also be possible for the technician to check the SC for capacitance (microfarad) from the BRN wire to terminal 23 on the contactor (CONT), or at the end of the YEL-colored wire at the start thermistor (ST), or at the R terminal of the compressor (COMP). This check could be done if the wire connections were good, because there is no other component between these four terminals and SC. Look at the diagram and note that

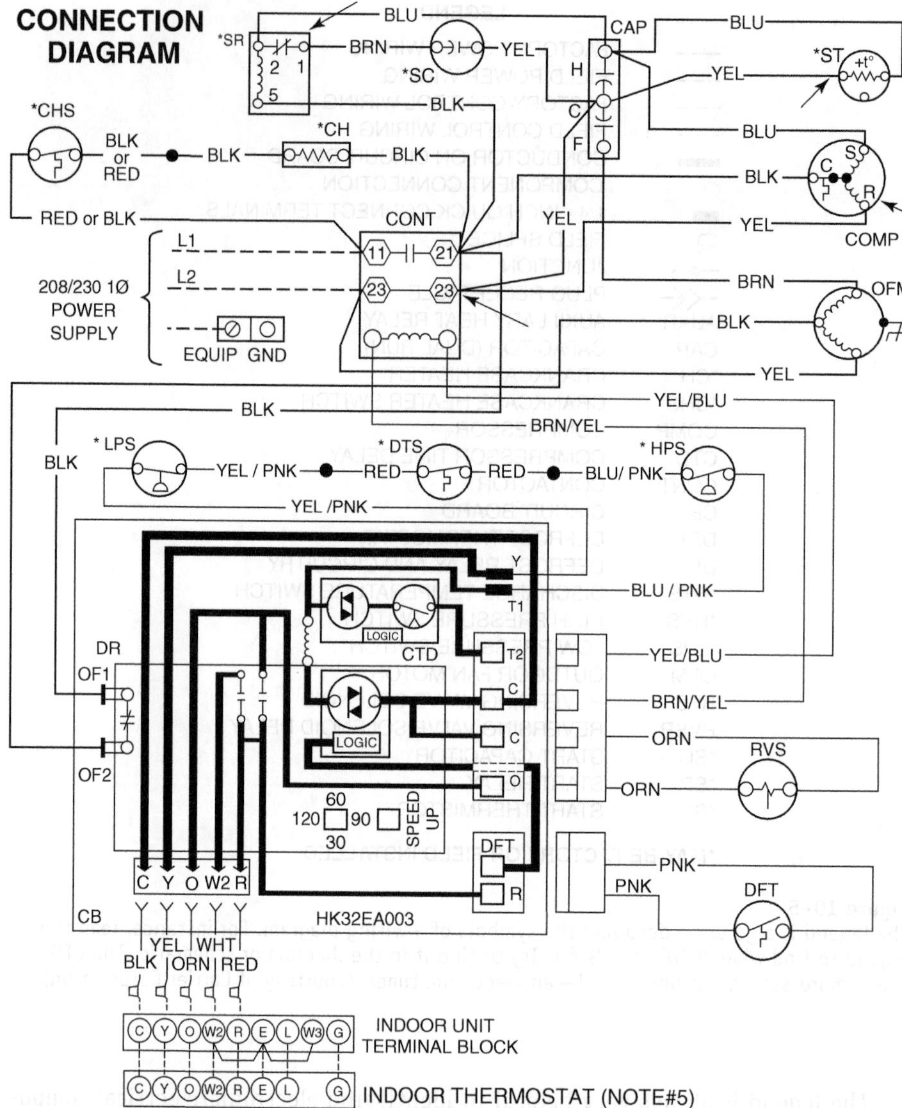

Figure 10–6
The Connection Diagram shows how each electrical component of the heat pump is wired, the color code of each wire, and the terminal designation where wires are connected. SR, SC, and ST are optional components. ST is never used in the same unit with SR and SC. (Courtesy of Carrier Corporation)

there is just wire running from terminal to terminal, no electrical components. The YEL-colored wire connected at SC is the same colored wire that is connected to terminal C at CAP and to the other three terminal points at CONT, ST, and COMP. Table 10–1 shows wire color-code designations.

The "Schematic Diagram (Ladder Form)" is used to read the electrical operation of the heat pump (Figure 10–7) and understand the electrical/mechanical sequence. Each electrical switch that operates a load is placed in the same "rung" of the schematic. In this way the technician can easily read which switch or series of switches will be used to operate each load.

For instance, if the outdoor fan motor is not operating, the technician could look for the OFM (outdoor fan motor) labeled component in this schematic and see the DR switch connected to the left of it. The DR, or defrost relay, switch should be closed; this is shown by a diagonal slash through the vertical lines representing the contacts of the relay. If this switch is open, the outdoor fan will

Table 10–1 Wire Color-Code Designations

Designation	Color
YEL	Yellow
BRN	Brown
BLU	Blue
BLK	Black
PNK	Pink
ORN	Orange
WHT	White
RED	Red

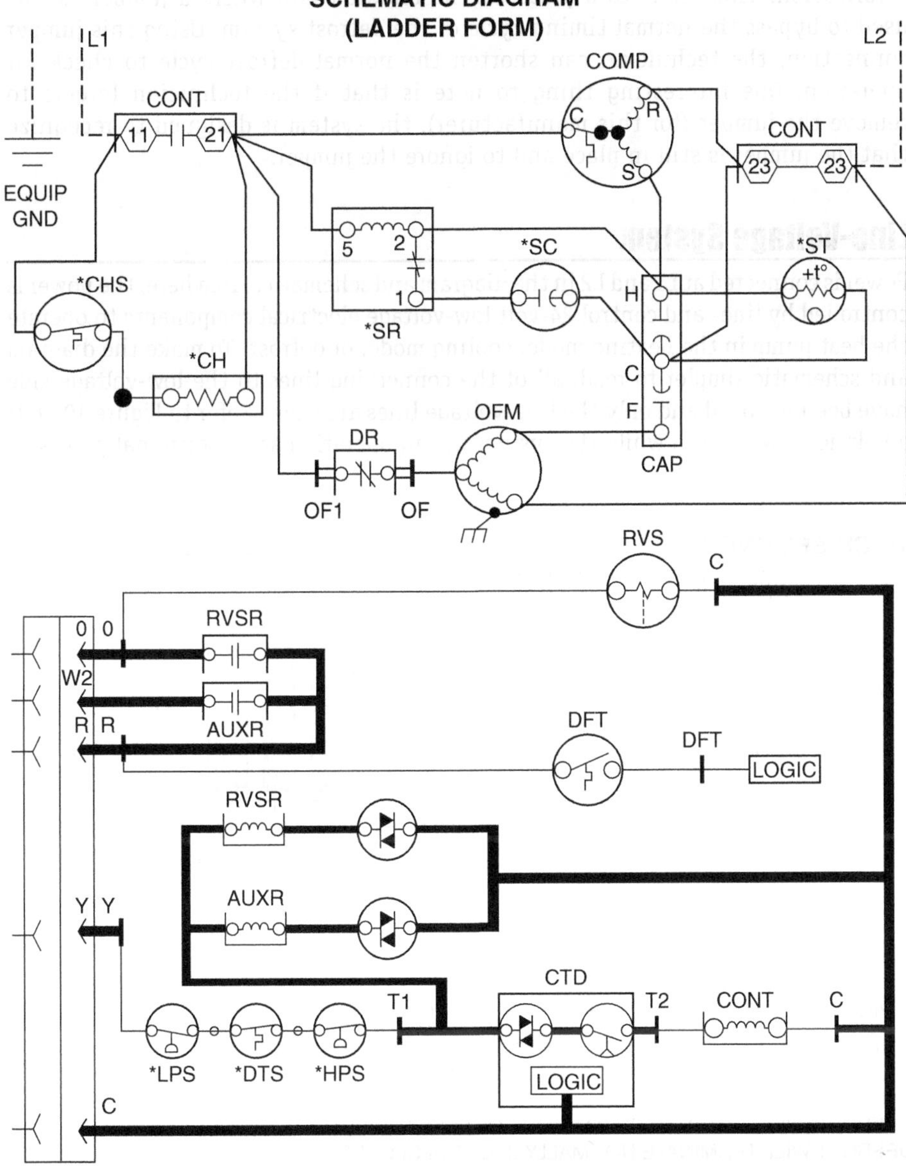

Figure 10–7
This electrical diagram shows how electrical switches are connected to their respective loads. Each switch is shown connected in series with the load that it operates. (Courtesy of Carrier Corporation)

not operate. Following the wire to the left of the DR, it connects to terminal 21 at the CONT (contactor). If the contactor is open (as shown) the outdoor fan motor will not operate either. Another way to read this circuit is that both the CONT (contacts between terminals 11 and 21) *and* the DR (contacts between OF_1 and OM) need to be closed to operate the outdoor fan motor.

Another part of the schematic is a depiction of the "DIP Switch Settings" (Figure 10-8). DIP stands for dual inline package and refers to the side-by-side arrangement of switches. Many manufacturers have switch settings that can be changed to match installation conditions or expected operating conditions. This depiction shows the DIP switch setting for the defrost control, 30 through 120 minutes. The default setting at 90 minutes is marked.

On the right is the setting for "quiet shft" (quiet shift) which is one manufacturer's way of turning off the compressor prior to shifting the reversing valve so that the changeover is made quietly.

Under the DIP switch settings is the "speed up" jumper and test connection information. This refers to a place on the circuit board where a jumper can be used to bypass the normal timing cycle of the defrost system. Using this jumper connection, the technician can shorten the normal defrost cycle to check for operation. One interesting thing to note is that if the technician forgets to remove the jumper (for this manufacturer), the system is designed to recognize that the jumper is still in place and to ignore the jumper.

Line-Voltage System

Power is connected at L1 and L2 in the diagram and schematic. From here, the power is controlled by line- and control 24-volt low-voltage electrical components to operate the heat pump in the heating mode, cooling mode, or defrost. To make the diagram and schematic simpler to read, all of the connecting lines to the low-voltage side have been removed and only the line-voltage lines are show. Refer to Figure 10-9. If a voltmeter were used while the system was in operation at the terminal points 11

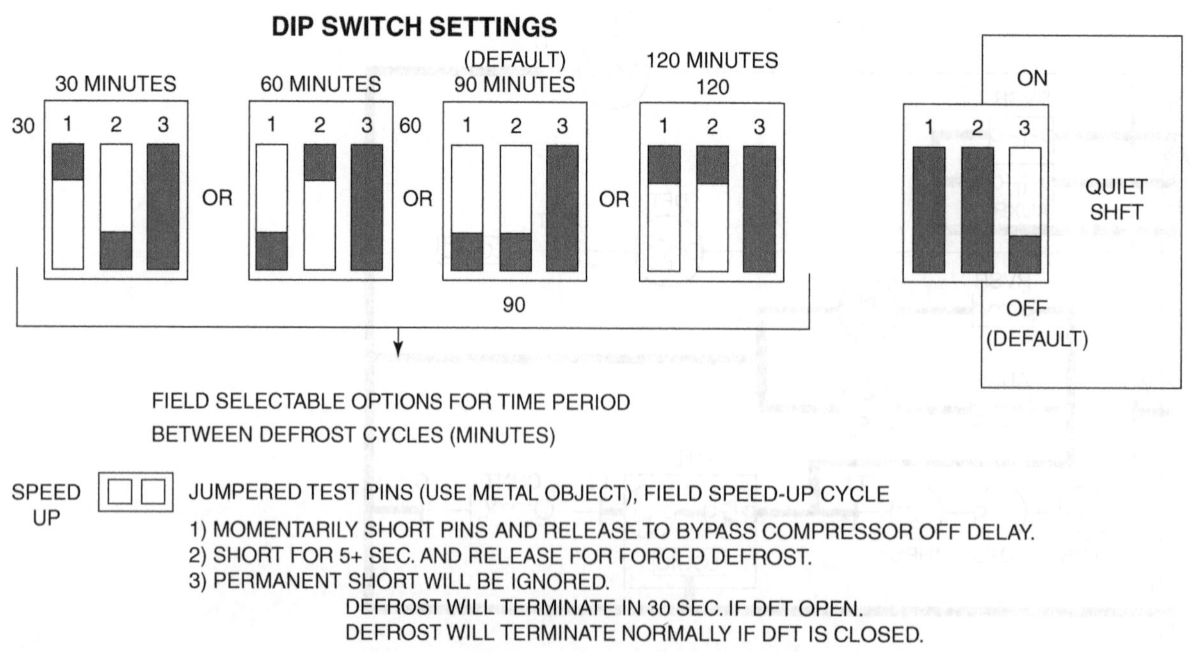

Figure 10-8
This section of the electrical information sheet shows how DIP switches are to be set for various defrost time settings and defrost operation. The "speed up" jumper helps the technician speed up the defrost cycle to check for operation. (Courtesy of Carrier Corporation)

Field Problem

The customer complaint was that the indoor temperature setpoint was 68°F, but the indoor temperature was 66°F and the thermostat was never satisfied.

Symptoms:
Frosted outdoor coil; compressor is running.

Possible Causes:
Defrost board; defrost sensor; refrigerant charge; bad outdoor fan motor; defrost contacts stuck; outdoor fan runs continuously; stuck reversing valve.

The technician observed that the outdoor coil was badly frosted and the outdoor fan motor was running. Opening the control panel, a voltmeter was used to determine if there was a difference in voltage across the DFT, indicating that it was open. The DFT did not register control voltage; it was closed and calling for defrost.

Reading the electrical information sheet on the back of the control panel, the technician found the "speed up" jumper pins used to put the system in defrost mode. Using a jumper, the system was manually placed in the defrost mode. The outdoor fan motor shut down, but the refrigerant tubing did not change temperature, indicating no change in mode from heating to cooling. Using a voltmeter, the technician checked for control voltage to the RVS by placing the probes on O and C at the control valve. Control voltage was being supplied to the RVS during the defrost mode and when the system returned to heating mode, no control voltage was detected at O and C.

De-energizing the heat pump and removing one wire of the RVS, the technician used an ohmmeter to determine if the RVS was open; it showed some resistance indicating continuity. The RVS was electrically okay, but the solenoid valve of the four-way valve was mechanically stuck in the heating mode and did not move the piston to the cooling/defrost mode when energized. The technician conferred with the customer and changed the reversing valve. He tested the operation of the system to confirm that the problem was solved.

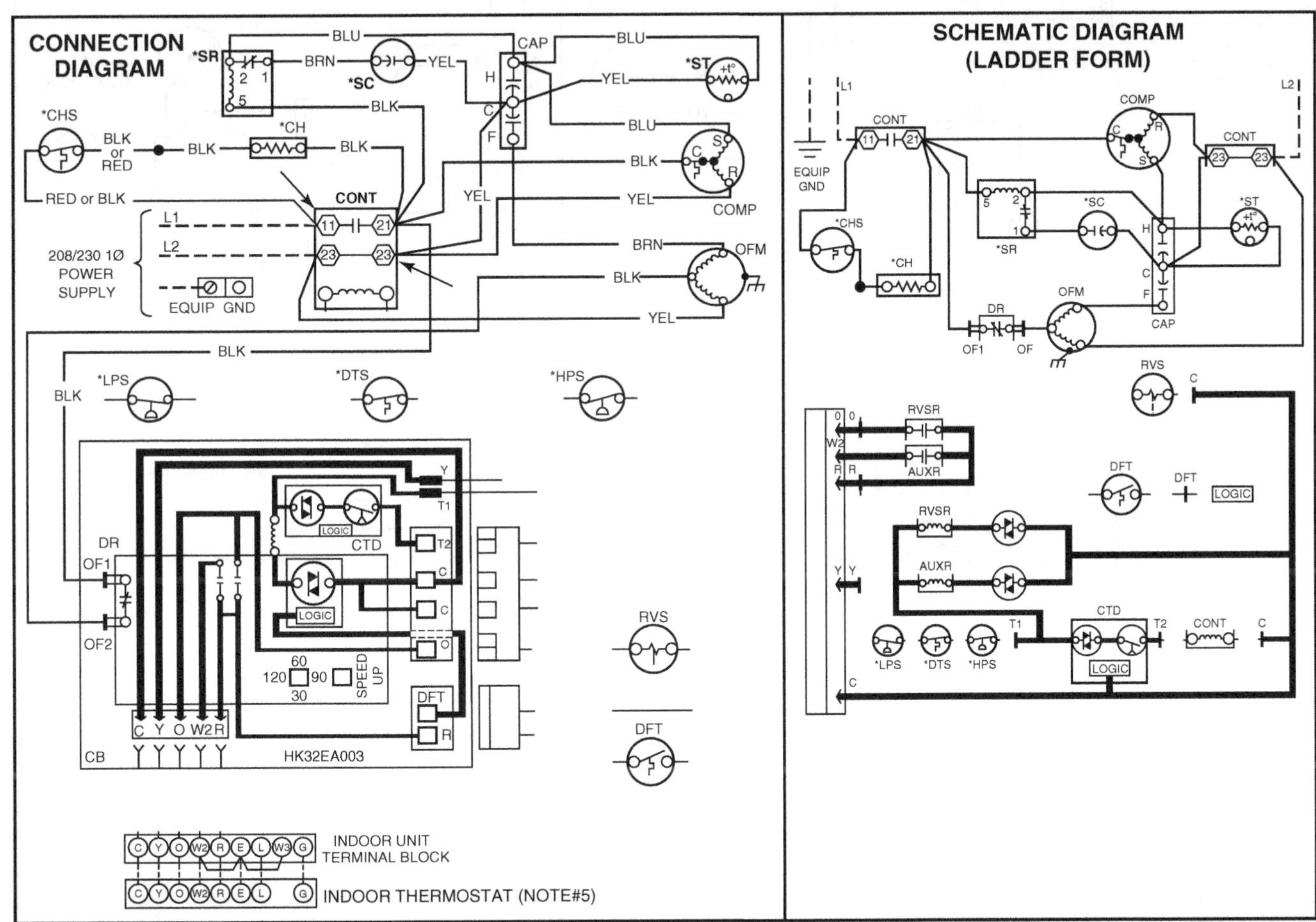

Figure 10–9
The graphic shows all of the line-voltage wiring. The low-voltage wiring has been eliminated in both the diagram and schematic. (Courtesy of Carrier Corporation)

and 23 shown in the diagram or schematic, the test meter would be expected to read 208 or 230 volts, single phase.

The line voltage connected to the heat pump comes from field wiring (look at the symbol and legend) that connects the fused disconnect (not shown) to the compressor contactor (CONT). Field wiring is run by the technician during installation of the heat pump.

Line-voltage components include the compressor, contactor, outdoor fan, crankcase heater and thermostat (CHS), both capacitors (SC and CAP), start relay, and start thermistor—a total of nine line-voltage electrical components.

Low Voltage Control System

The low voltage control system consists of a thermostat, low-voltage relays, safety switches, transformer, and an electronic control board. Look at Figure 10–10, which shows the wiring diagram and schematic with the line-voltage wiring removed. The connected circles that have symbols in them are the switches and loads that are part of the low-voltage control system. Sometimes the manufacturer's schematic and the diagram do not show the transformer or how it connects to provide low voltage to the control circuit.

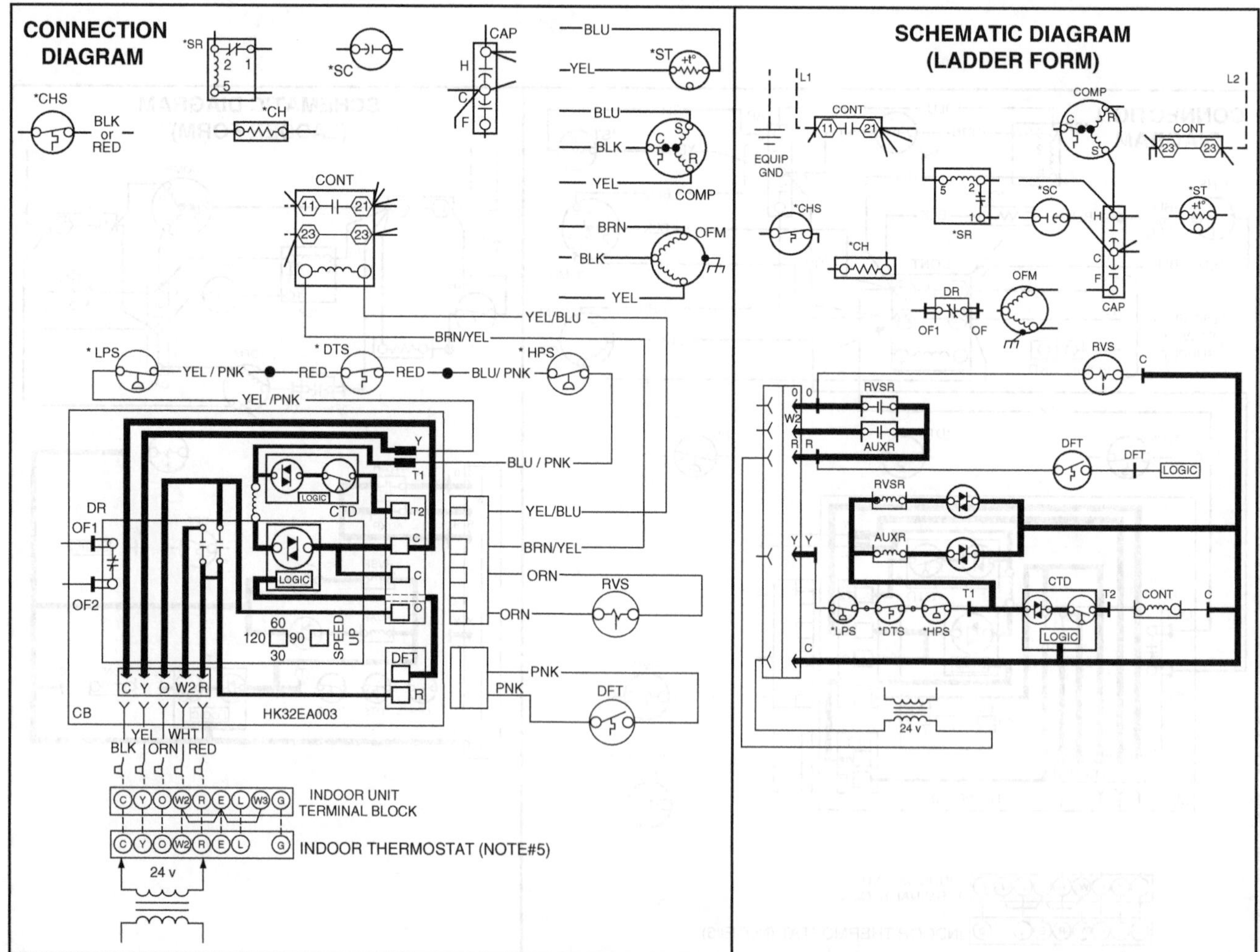

Figure 10–10
The graphic shows only the low-voltage wiring of the system. The line voltage wiring has been eliminated. The control-voltage source (a transformer symbol) has been added to the diagram. (Courtesy of Carrier Corporation)

The low-voltage transformer is usually found on the indoor unit, but may be located on another piece of equipment or it may need to be added, depending on the equipment configuration during installation. For our purposes, we need to keep in mind that the transformer is supplying low voltage at terminals C and R. The symbol for a low-voltage thermostat has been added to the diagram and schematic.

Compressor and Starting System

The manufacturer of this electrical information sheet and the ASHP has supplied a wiring diagram and schematic that includes three possible compressor starting configurations: (1) an ST (start thermistor), (2) an SR and SC (start relay and start capacitor), or (3) with neither, starting instead as a PSC without additional start assist. The ST would be used if the heat pump encountered line-voltage fluctuations from the power supplier. The SR and SC would be used when long refrigerant lines are installed and/or low ambient cooling applications are encountered. The SR and SC would also be used if a hard-shut-off TXV was used on the indoor coil. *Either one set of starting components or the other would be used, but not both.* For this reason, each of these starting systems will be explained separately.

If the ST is installed, power to the compressor is received through the contactor. The compressor start windings receive line voltage through the ST until it heats up, increasing the resistive value, and blocks the current flow. The compressor operates as a permanent split capacitor (PSC) motor; CAP supplies modified power to the start windings. Start windings operate as auxiliary windings while the compressor is energized. When the compressor shuts down, the ST requires time to cool before the compressor is able to start again. This is one reason why there is a 5-minute on/off cycle delay that is built into the control board (see the "NOTE" in the middle of the right column of the Manufacturer's Electrical Information Sheet above the legend in Figure 10–11a).

If the SR and SC are installed, power to the compressor is received through the contactor. The compressor is a capacitor start/capacitor run. To start the compressor, the SR and the SC are used until the compressor comes to within 85% of its rated running speed. When that occurs, the SR opens the NC (normally closed) switch and disconnects the SC from the compressor circuit. The CAP is not being used during start but is part of the circuit after the SR opens. The compressor continues to operate both the run winding and the start winding, as an auxiliary winding, while the compressor is energized (Figure 10–11b).

If a manufacturer does not have a starting relay as part of the original design, a hard-start kit may be applied to help the compressor start under unusually high load conditions or when it encounters low voltage. Notice the similarity of Figure 10–12 and the ST circuit in the previous manufacturer diagram and schematic.

The hard-start kit consists of a start capacitor, a current-limiting resistor that is permanently attached (to limit the current through the start relay contacts so as to not weld them), and a relay. The kit may be packaged so that all components are housed in one container, so that wiring can be reduced to a minimum. The purpose of the hard-start kit is to increase starting torque. Always refer to the manufacturer's specifications and wiring instructions when applying a hard-start kit.

Crankcase Heater Operation

The crankcase heater (CH) is located in, on, or around the bottom of the compressor in a circuit that operates while the compressor is off and when the crankcase heater switch (CHS) is closed. The CHS closes at outdoor temperatures below approximately 65°F. When the compressor starts, the CH circuit turns off. To understand how this happens, we first need to know why we need this kind of operation.

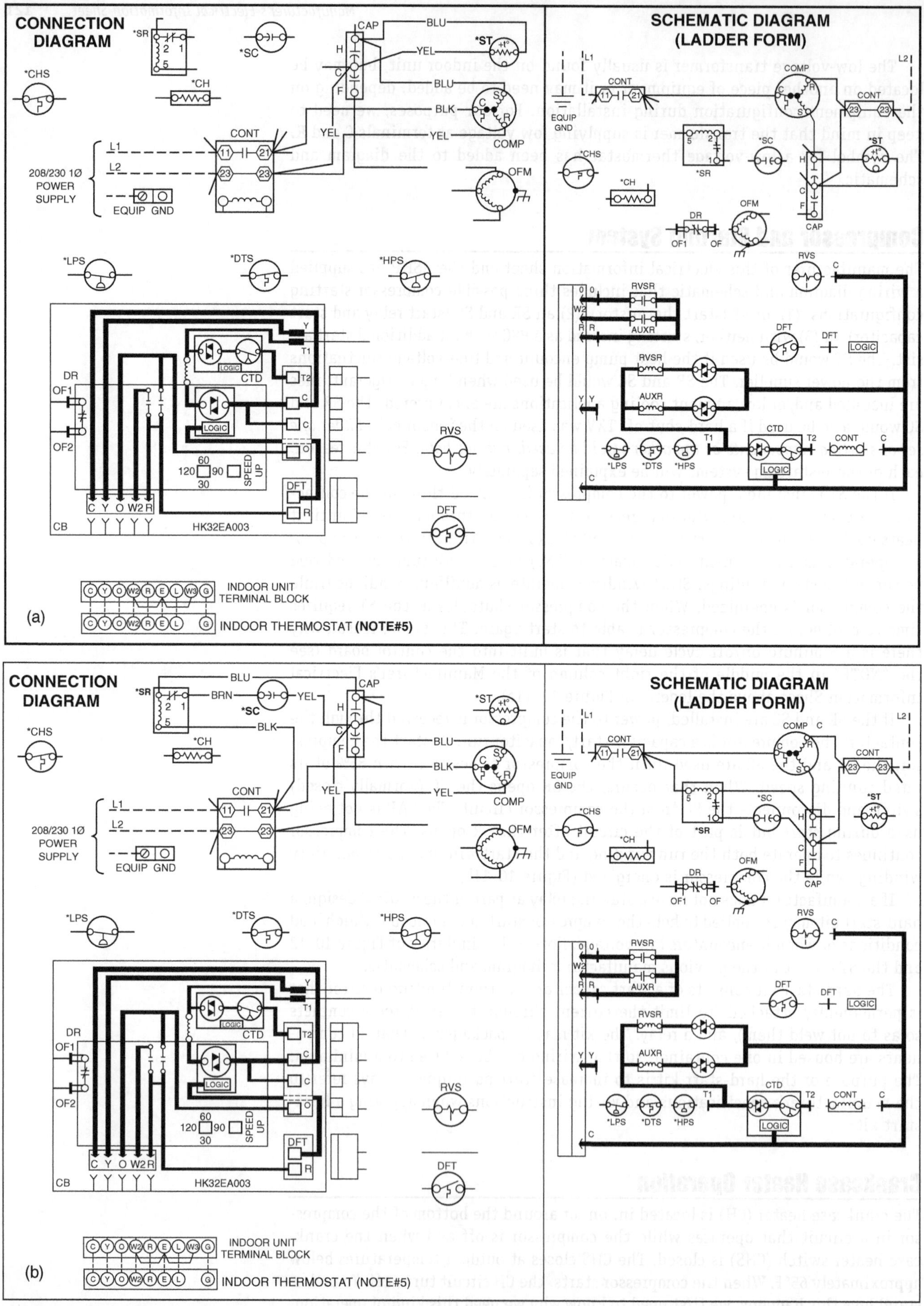

CONNECTION DIAGRAM

SCHEMATIC DIAGRAM (LADDER FORM)

208/230 1Ø POWER SUPPLY

INDOOR UNIT TERMINAL BLOCK

INDOOR THERMOSTAT (NOTE#5)

(a)

CONNECTION DIAGRAM

SCHEMATIC DIAGRAM (LADDER FORM)

208/230 1Ø POWER SUPPLY

INDOOR UNIT TERMINAL BLOCK

INDOOR THERMOSTAT (NOTE#5)

(b)

Figure 10–11 (a), (b)
The graphic shows all of the line-voltage wiring, except those wires needed to start the compressor. (Courtesy of Carrier Corporation)

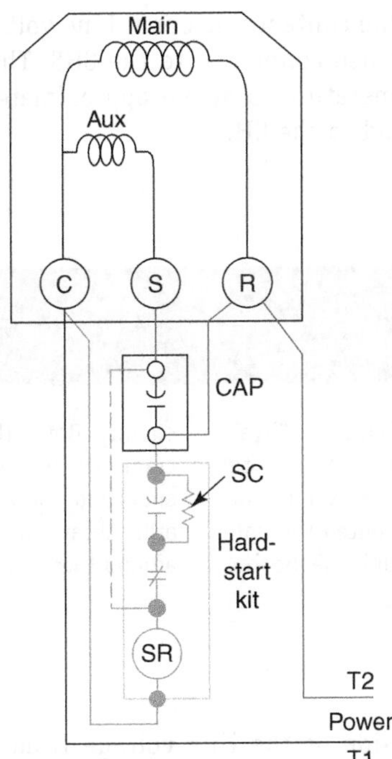

Figure 10–12
The diagram shows how the hard-start kit is wired to the compressor. Inside the hard-start kit is a start capacitor with a resistor and a potential relay operator and contacts or a solid-state relay. Refer to Figure 10-11b for further details. (Courtesy of Delmar/Cengage Learning)

When the compressor is not running, refrigerant will mix with oil in the sump or bottom of the hermetically sealed compressor. In cold weather (or if the compressor gets colder than the rest of the system), refrigerant can migrate and collect as a liquid under the oil in the compressor. When the compressor starts, the refrigerant and oil are subjected to low pressure and the refrigerant expands rapidly, causing the oil to excessively foam. Excessive foaming oil can be pumped with the refrigerant through the compressor and into other parts of the system. If the compressor loses oil, it will lose the lubrication that it needs and could fail.

The crankcase heater circuit is ON to heat the oil in the sump of the compressor, when the compressor is not running and the crankcase heater switch is below its setpoint (approximately 65°F). Heat prevents refrigerant migration and dilution with oil, and minimizing oil foaming when the compressor starts. Oil remains in the sump of the compressor rather than being pumped through the system (Figure 10–13).

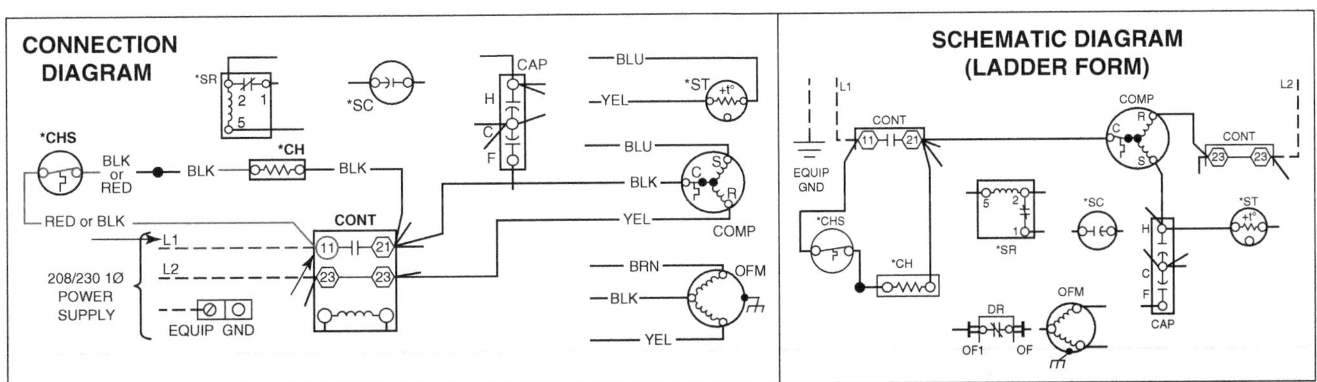

Figure 10–13
All of the other wiring has been removed so that the crankcase heater wiring can be easily seen. Voltage can be traced from L1 to terminal 11 on the CONT and to CHS through the RED-or BLK-colored wire. Because the CHS is closed, voltage continues to be traced on the BLK or RED wire to the junction (the black dot; see the legend) and then through the BLK-colored wire to the CHS. (Courtesy of Carrier Corporation)

During the off cycle, the contactor is open. Line voltage is connected to CONT at terminal 11. Voltage is also connected to the CHS. The CHS is closed and only opens when outdoor temperature is above approximately 65°F. Voltage can be traced through the CHS and to the CH.

Tech Tip

The crankcase heater circuit is "live" when the compressor is shut down. Note that even though the compressor is not operating, power is being routed through the compressor to the crankcase heater circuit. Both the compressor and the crankcase heater circuit are electrically energized and a potential electrical hazard. Always be cautious when working around these circuits.

Now trace the other side of the line voltage from L2 to the CH. From L2, follow the connection from terminal 23 to terminal 23. There is only a wire; there is no switch or contact. Continue on the YEL-colored wire to terminal R on COMP. The compressor run windings are usually the lowest resistance of all of the motor windings, so current will tend to follow the path of lowest resistance—it will go through the windings without making the motor run (Figure 10–14).

The current continues through the run windings and internal overload thermostat within the compressor to terminal C and the compressor thermostat, shown by the thermostatic switch symbol. From the C terminal, voltage can be traced on the BLK-colored wire to terminal 21 on CONT. From terminal 21 the voltage continues on the BLK-colored wire to the CH. If both sides of the potential (voltage) are applied in this way, the CH will operate and become warm.

When CONT is operated and the contacts between 11 and 21 are closed, both sides of the circuit change to the same potential, becoming the same side of the circuit. At this point, the CH is not receiving a potential or voltage difference. The operation of the crankcase heater circuit is the same in both the heating and cooling modes.

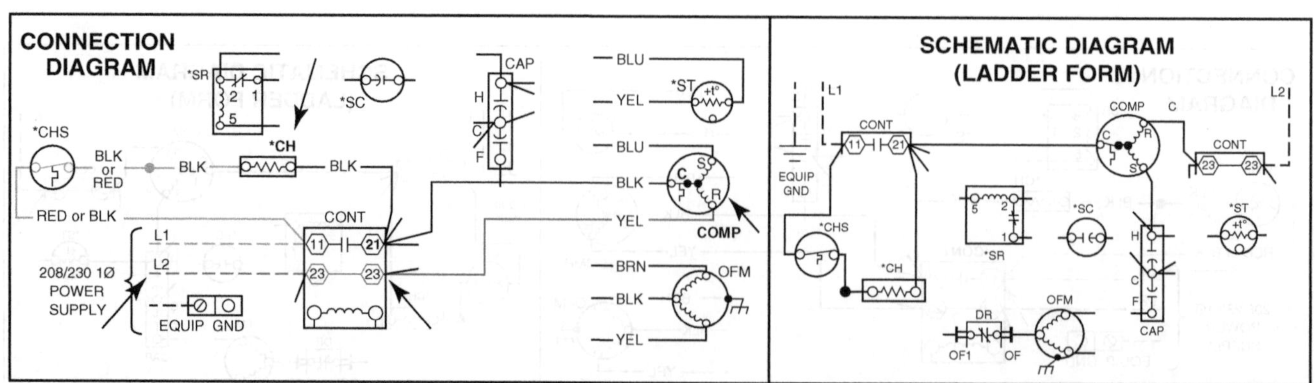

Figure 10–14
L2 voltage is traced through the contactor, because there is no set of contacts. Voltage continues on the YEL-colored wire to terminal R on COMP. (Courtesy of Carrier Corporation)

Case Study

The service technician was dispatched to a commercial account that used an ASHP to condition an office area. The scheduled maintenance was a routine check of the system on a 50°F fall day. One of the checks performed was to check the crankcase heater circuit. When the heater was checked, it was cold.

Symptoms:
Crankcase heater was cold; compressor was not running.

Possible Causes:
Power to the system is de-energized; crankcase heater is burned out; heater thermostat is open; wiring problems with the heater circuit.

The technician checked for power across the contacts (terminals 11 and 21 of the line-voltage schematic) of the contactor and found line voltage. This confirmed that power was being applied and that the heater circuit should be operating while the compressor was off. Putting the right voltage probe on terminal 23 and the left on terminal 11, the technician read line voltage. Moving the left probe to the left side of CHS, the technician read line voltage. Moving the left probe to the right side of CHS, the technician read "0" volts, indicating that the CHS was open when it should have been closed since the outdoor temperature was measured at 50°F. To confirm that the CHS was open, the technician de-energized the circuit, removed a wire from the CHS, and used an ohmmeter for a continuity test; the CHS was open.

The technician knew that the problem was not the heater, but instead, it was the crankcase heater thermostat. The thermostat should have been closed when the compressor was cold and should only open when the compressor is warm. Reporting the problem to the customer and getting permission to make the repair, the technician replaced the thermostat, and the compressor crankcase heater operated when the power was restored to the unit.

Arc Flash

There are five classification categories of personal protective equipment (PPE). Each of these categories requires specific PPE. Table 10–2 shows what is required to wear for each category.

Arc flash hazards are to be identified with signage that warns of the danger, states the category number, and lists the PPE required. In addition to this knowledge, the technician must be able to identify a potential arc flash hazard and reduce or eliminate this potential. For instance, simply de-energizing a circuit will eliminate an arc flash potential. If the circuit must be in operation for testing, tools that reduce a potential arc flash are to be used. Standards that apply to arc flash are:

- OSHA 29 Code of Federal Regulations Part 1910 Subpart S
- NFPA 70 National Electrical Code
- NFPA 70E Standard for Electrical Safety Requirements for Employee Workplaces
- IEEE Standard 1584 Guide for Performing Arc Flash Hazard Calculations
- ANSI Z535.4 Product Safety Signs and Labels

An example of a warning label is shown in Figure 10–15.

Table 10–2 Personal Protective Equipment (PPE)

Category #	Clothing Required
0	Untreated cotton, wool, rayon, silk, or blend (1 layer)
1	Flame-retardant shirt and pants or coverall (1 layer)
2	Cotton underwear AND flame-retardant shirt and pants or coverall (2 layers)
3	Cotton underwear AND flame-retardant shirt and pants PLUS flame-retardant coverall (3 layers)
4	Cotton underwear AND flame-retardant shirt and pants PLUS multi-layer flash suit (3 or more layers)

Safety Tip

Troubleshooting with a voltmeter requires that the heat pump is operating. Both line- and low-voltage circuits are live and the possibility of accident is high. Always use a voltmeter with the correct rating for the voltage being tested. The probes of the voltmeter should meet probe-length standards and the technician should be wearing the right personal protective equipment (PPE). Use one hand to probe electrical connections while keeping one voltmeter probe continuously connected. Using one hand reduces the chance that electrical shock could cause cardiac arrest.

Safety Tip

Arc flash is a rapid release of energy due to an electrical fault (as in a short circuit). The resulting energy discharge can cause eye damage from the light, hearing damage because of the concussion, and electrical burns from the heat. While arcing can occur at any voltage, voltage levels of 120 volts or more can sustain an arc. It is important to know the industry standards to prevent an arc flash and to be in compliance with tools and clothing (PPE) required on the job.

⚠ DANGER

Arc Flash & Shock Hazard
Appropriate PPE Required

8.4	Inch Flash Hazard boundary
0.4	Cal/cm2 Flash Hazard at 18 Inches
#0	PPE Level
	Leather Gloves, Face Shield
5.28	kV Shock Hazard When Cover Is Removed
18.21	kA Bolted Fault Current

Equipment Name: HVAC System Panel – 5KV

Figure 10–15
Example of a warning label. (Courtesy of Delmar/Cengage Learning)

Safety Tip

When checking the crankcase heater, a noncontact thermometer should be used. Sometimes, crankcase heaters can be hot enough to burn. Do not touch a crankcase heater or test it with your fingers.

FOUR MODES OF AN ASHP

The ASHP has four modes: heating mode, cooling mode, defrost mode, and emergency heat mode. Each of these modes or cycles is controlled by temperature sensors and electronic controls. Recall that the thermostat is most likely an electronic unit that senses indoor conditions. On some heat pumps, outdoor temperature sensors are used to determine outdoor ambient conditions. The DFT, in our example, measures the temperature for the defrost control and is located in the outdoor unit. Installations in colder climates may require additional outdoor temperature sensors for operating supplemental heat at the designed thermal balance point for an ASHP.

In the following four sections, the operation of the heat pump will be explained in terms of each mode and electrical function. The electrical operation will be explained in a step-by-step manner. Read the words and follow the electrical operation by tracing or reading the electrical schematic diagrams provided. The diagrams have been simplified to make reading and understanding easier.

Heating Mode

The heating mode for the example heat pump is the default mode (see Figure 10-17). The outdoor coil will be cool and the indoor coil will be hot. In this mode the reversing valve solenoid (RVS) will not be energized. This mode may be interrupted by the defrost mode, to remove frost and ice accumulation on the outdoor coil.

Refer to the Heat Mode section in Figure 10–16 in the schematic top section of the line voltage. Power moves through the contacts on L1 from terminal 11 to terminal 21 at CONT when the thermostat signals the control board for heat mode and the board sends a signal to the coil in the contactor. In the schematic bottom low (control) voltage, three switches must close or continue to remain closed. Those switches are in the low-voltage control circuit and are LPS, DTS, and HPS. All of the wires that are connected at terminal 21 will now measure as line voltage when tested from terminal 21 to terminal 23 (L2). Measuring voltage in this way

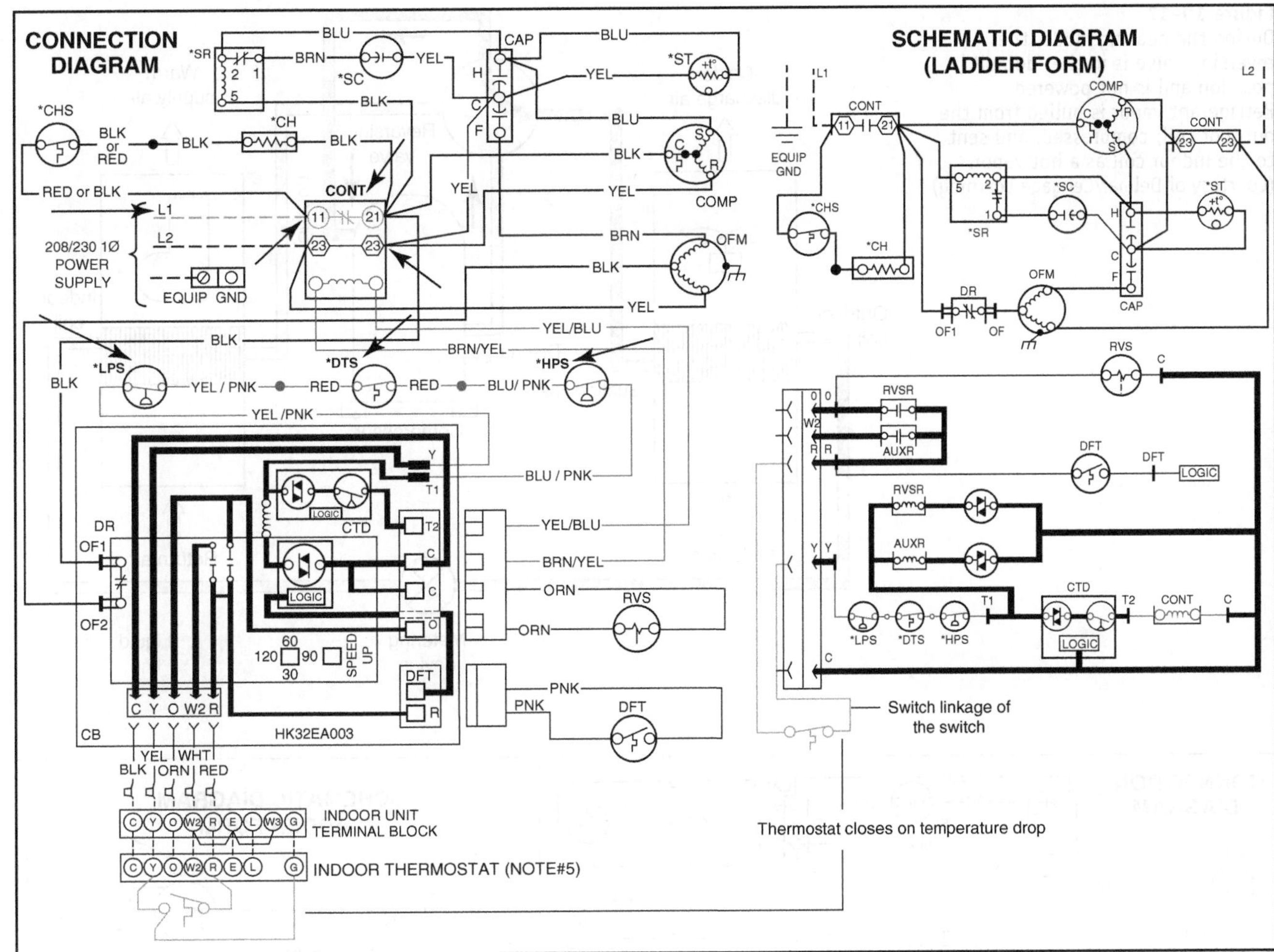

Figure 10–16
CONT contacts close and power travels from L1 and contact 11 through the contacts to terminal 21. (Courtesy of Carrier Corporation)

is called "measuring across the line" (across the load (in parallel) of the compressor and OFM or measuring from L1 to L2).

Line voltage is received at the C terminal at COMP, terminal 5 at SR, and through the DR switch to the common terminal (unmarked) at OFM. Notice that the stopping points for this explanation are each at a load: the compressor, the relay, and the outdoor fan. When tracing the circuit or reading a wiring diagram, always trace from the voltage source to a load and stop. Next, read or trace from L2 to the same loads: compressor, relay, and outdoor fan motor (Figure 10–18).

Start at L2 and trace to the contactor. Notice that both terminals at CONT have the same number designation, 23. This means that there is no switch between these two terminals. Both of these terminals are at the same electrical potential (or voltage). Trace the wire connections at each of the terminal 23 connections. On the left terminal, one wire (YEL) is connected to the OFM, completing the circuit to that load. The outdoor fan would be energized. If the outdoor fan starting circuit works (BRN wire), the fan should be running. On the right terminal, YEL-colored wires run to the R terminal at the compressor. The same thing would be true for the compressor; it is energized. If the compressor start circuit works (BLU wire), the compressor should be running (Figure 10–19).

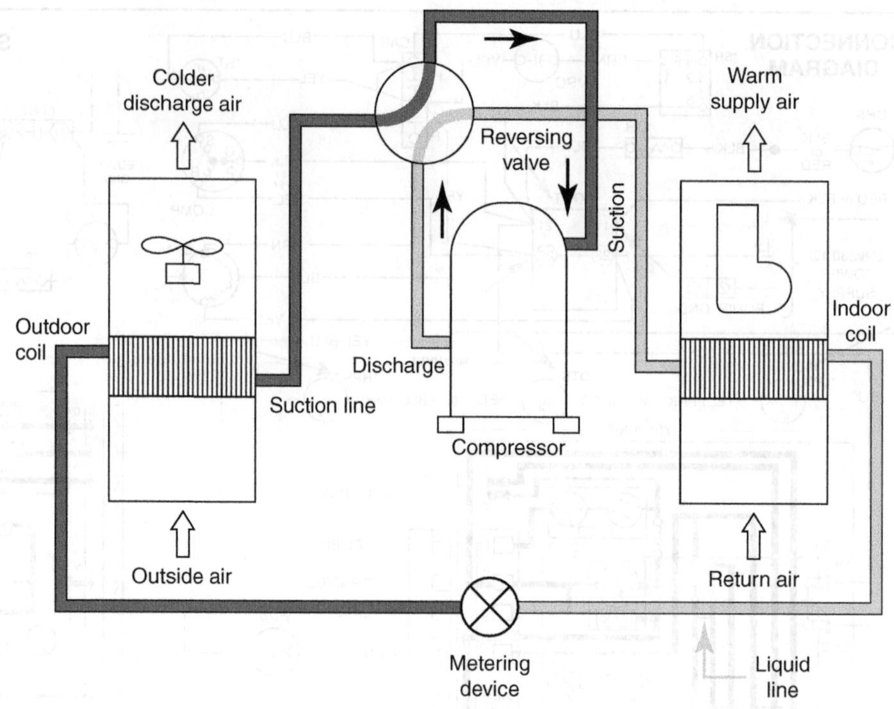

Figure 10–17
During the heating mode, the reversing valve is in the "default" position and is not powered. Refrigerant vapor is pulled from the outdoor coil, compressed, and sent to the indoor coil as a hot vapor. (Courtesy of Delmar/Cengage Learning)

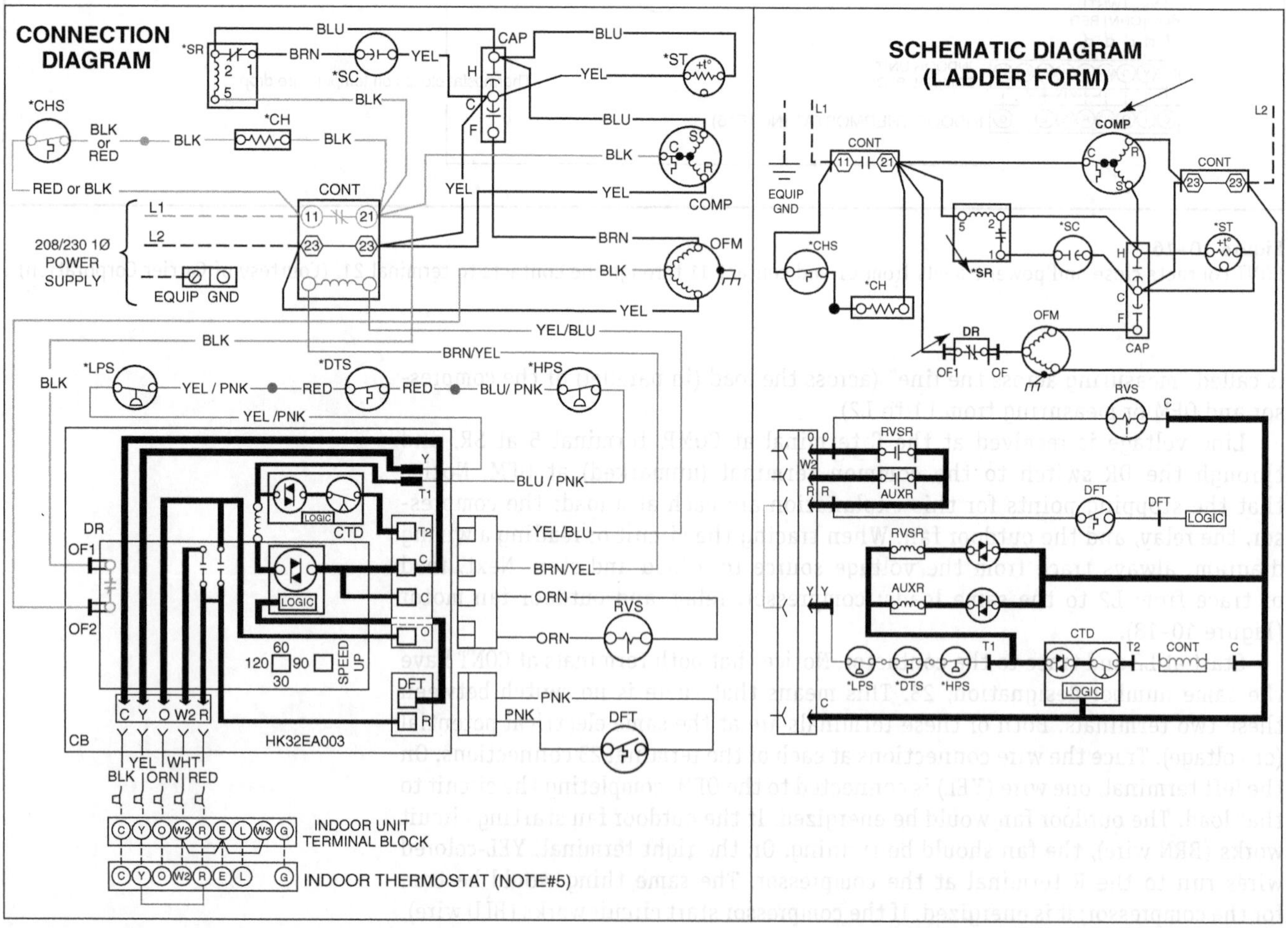

Figure 10–18
Voltage is distributed from terminal 21, CONT, to the COMP, OFM, and SR. (Courtesy of Carrier Corporation)

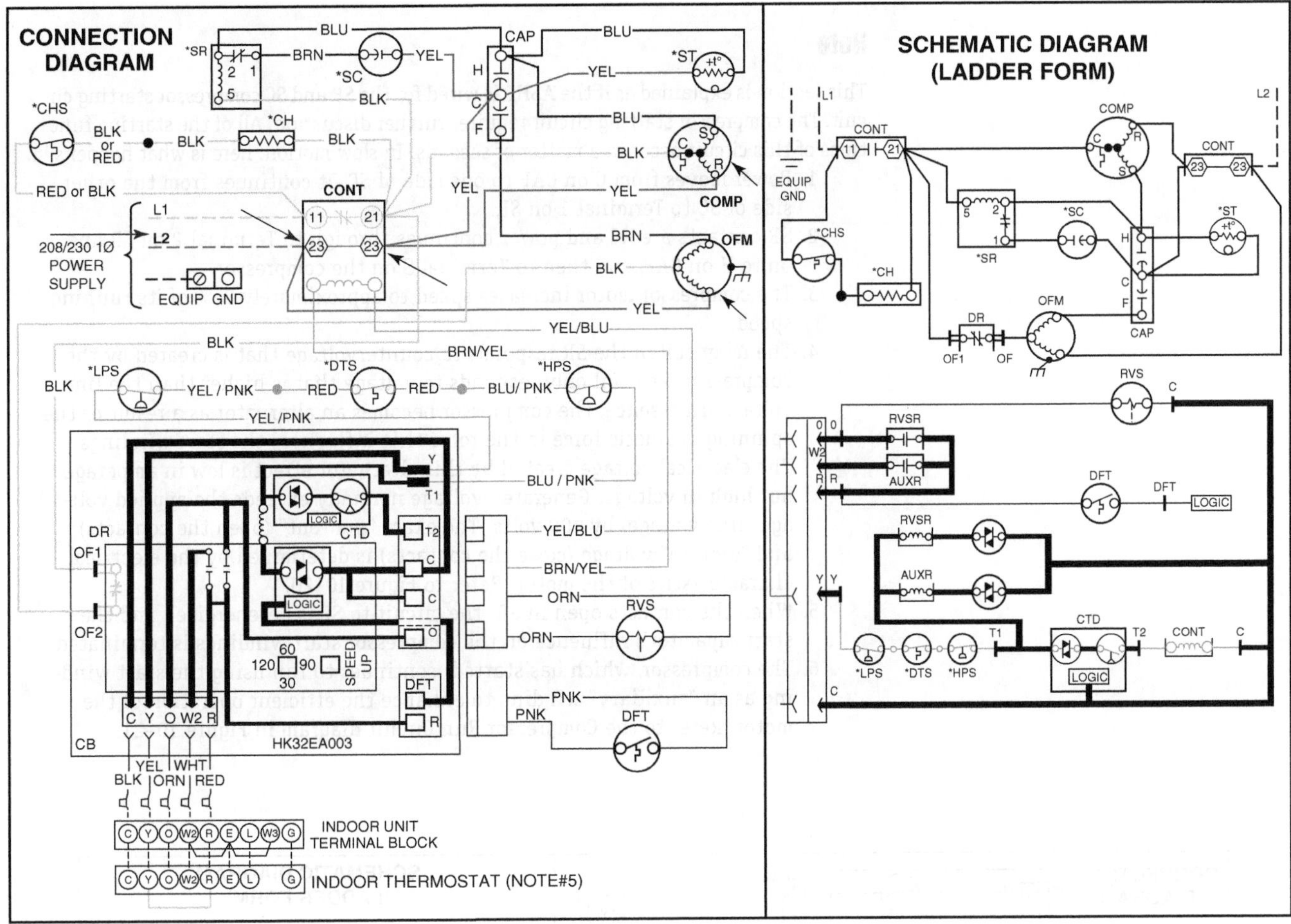

Figure 10-19
The other side of the potential, L2, moves through CONT and is distributed on the YEL-colored wires to each of the loads: COMP, OFM, and SC. (Courtesy of Carrier Corporation)

> **Note**
>
> This section is explained as if the ASHP is wired for the ST (start thermistor) compressor starting circuit (see Figure 10-19). Another YEL-colored wire connects to C terminal at CAP and continues to ST. The capacitor is a dual capacitor used to start the outdoor fan motor and the compressor. If both capacitors inside this single package are good, both fan and compressor will operate. The outdoor fan motor starts, but the compressor starting circuit is not that simple. The YEL-colored wire going to the ST is cool and allows power to move through the ST and through the BLU-colored wire to the H terminal of the CAP.

Voltage is supplied to the compressor start windings and the compressor starts. After that, the ST is warm enough to increase resistance and effectively turn off the voltage to the start windings. The CAP continues to supply modified power to the start windings so that they act as auxiliary windings, assisting the compressor while running as a PSC motor.

Note

This section is explained as if the ASHP is wired for the SR and SC compressor starting circuit. The compressor starting circuit requires further discussion. All of the starting functions of this circuit occur in a matter of seconds. In slow motion, here is what happens:

1. Power moves from C on CAP to one side of SC. It continues from the other side of SC to Terminal 1 on SR.
2. SR contacts are NC and power continues through to Terminal 2 on SR and on to H on CAP, and then to Terminal S on the compressor.
3. The compressor motor increases speed to approximately 85% of its running speed.
4. The relay coil in the SR responds to countervoltage that is created by the compressor. The coil only responds to voltage that is higher than the line voltage. In essence, the compressor becomes an alternator as a result of the spinning magnetic force in the rotor that influences the start windings. The electrical voltage created by this alternator effect is low in amperage, but high in voltage. Generated voltage normally exceeds the applied voltage (line voltage) by 100 volts. The exact "pull-out" (open the contacts) and "drop-in" voltage (close the contacts) is determined by the electrical characteristics of the motor. Refer to Figure 10–20.
5. When the contacts open in SR, the circuit to SC is de-energized and the start capacitor's influence on the compressor start windings is terminated.
6. The compressor, which has started, continues to run using the start winding as an "auxiliary" winding to enhance the efficient operation of the motor. Refer to the Compressor Run Circuit diagram in Figure 10–21.

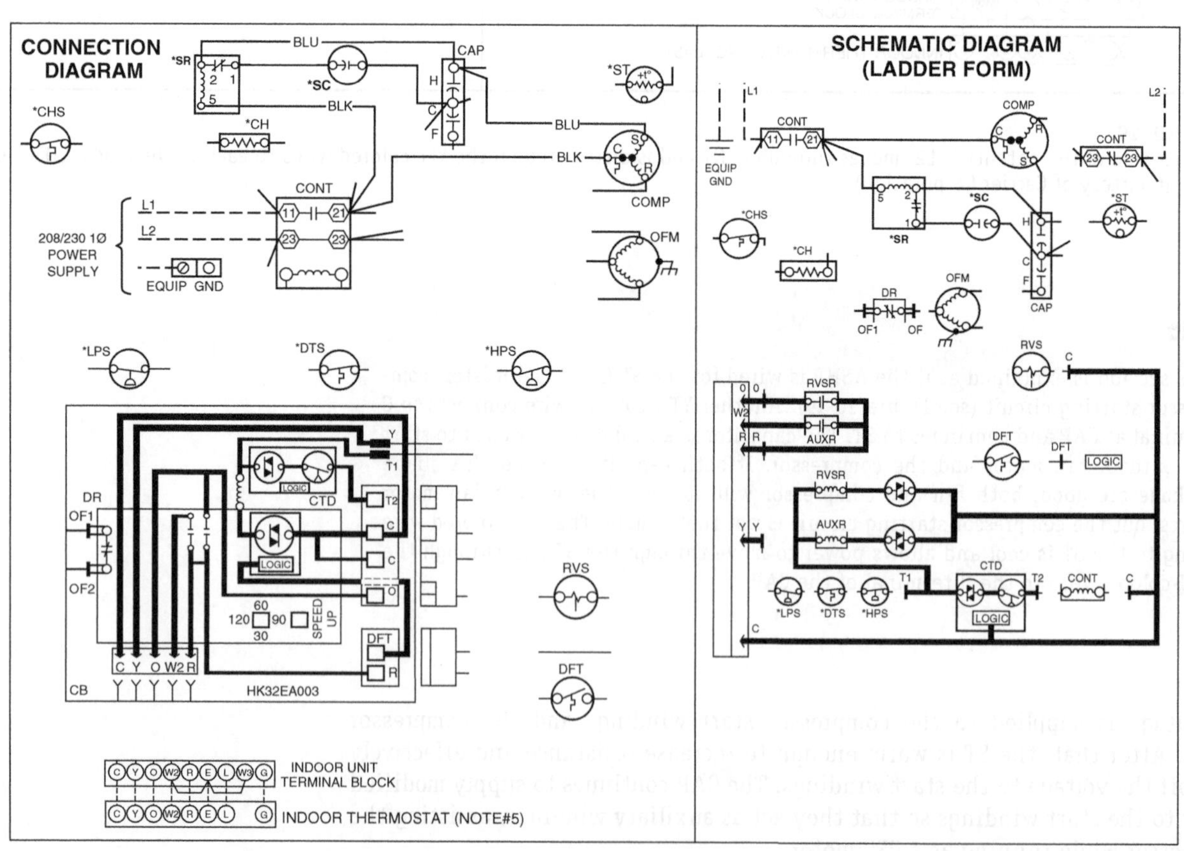

Figure 10–20
The coil in the start relay (SR) and the start winding in the compressor (COMP) remain wired in series during the start and run operation of the compressor. The compressor produces a countervoltage that is higher than the applied voltage to pull the contacts open after the compressor starts. If the compressor should lug, or reduce RPM, the amount of countervoltage drops and the SR contacts close to engage the start circuit. (Courtesy of Carrier Corporation)

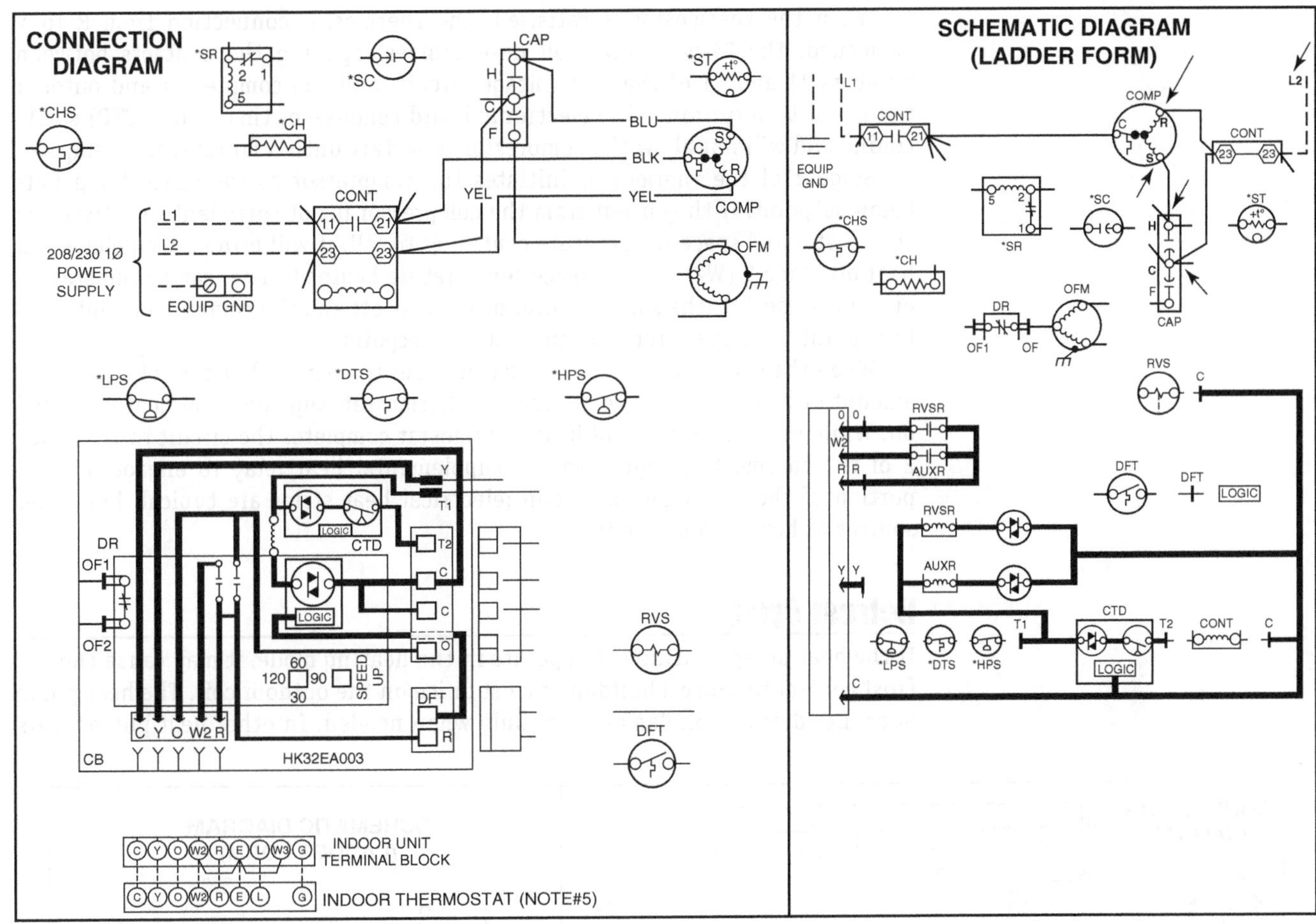

Figure 10–21
The compressor continues to run, after it starts, using the start windings as "auxiliary" windings to enhance the running characteristics. Notice the capacitor connected from L2 to "C" and then from "H" to terminal S on the compressor. The capacitor modifies the sine wave to the start windings. Both the R and S terminals are electrically connected to L2. (Courtesy of Carrier Corporation)

Field Problem

The customer's complaint was insufficient heat. The heat pump was running, but the indoor temperature was lower than the thermostat setpoint. The customer reported that the system was operating intermittently.

Symptoms:
Outdoor unit not running; indoor unit running on electric heat, indoor fan is running; discharge air is too hot; outdoor unit is cycling on and off.

Possible Causes:
Plugged indoor coil, dirty filter, fan problems; insulation loose; collapsed duct; blocked return; too many diffusers closed; dampers closed; dirty blower wheel.

The technician inspected the outdoor unit and observed that the outdoor coil was free of frost buildup and that the unit ran for several minutes and then shut down. After several minutes of shutdown, the outdoor unit started again and the operation repeated itself. Using a voltmeter, the technician removed the electrical access panel and probed the contactor coil to determine if the coil was de-energizing when the unit shut down (YEL/BLU and BRN/YEL wire connection at the contactor); he found that it was de-energizing. See Figure 10-22.

Next, the technician placed the voltmeter probes on the Y and T1 connections on the circuit board where the LPS, DTS, and HPS series of switches were connected. He observed that there was control voltage when the system shut down, indicating that one of these switches was opening. Taking voltage readings across each of these electrical switches, the technician found that there was no voltage across LPS and DTS, but there was voltage across HPS, indicating that the HPS was open. The heat pump was shutting down on the high-pressure switch.

Checking the indoor unit, the technician looked for anything that could cause high pressure. The indoor filter was extremely dirty, restricting airflow and causing low air volume across the indoor coil in the heating mode. Replacing the filter eliminated the HPS switch opening and increased the airflow to the building.

When the thermostat is satisfied, the thermostat connection from R to Y is opened, the 24-volt CONT coil de-energizes, opening the contacts between numbers 11 and 21 of the line-voltage circuit, and the compressor and outdoor fan motor stop running. The electronic board compressor time delay (CTD) starts timing and will not allow the compressor to restart until 5 minutes have elapsed.

Stage 1 of the thermostat initiates the compressor as the space temp falls below setpoint. If the output from the compressor is not sufficient to satisfy the thermostat and space temperature continues to fall, it will bring on supplemental heat off stage 2 (W2). As the space temperature begins to rise, satisfying stage 2 of the thermostat, the supplemental heat turns off and the compressor continues to run until the space temperature reaches setpoint.

When the compressor of the ASHP continues to operate, but cannot supply the amount of heat necessary to satisfy the thermostat, supplemental heat is turned on. While in the first stage of heat, thermostat completes the circuit to W2 (stage 2 of the thermostat), signaling the supplemental heat relay to engage all or a portion of the electrical heat strip (electrical heat strips are typical, but other sources of heat may be used).

Defrost Cycle

If the heat pump continues to operate in the heating mode, it may cause the defrost control to sense a buildup of frost or ice on the outdoor coil. The heat pump goes into defrost "on demand," or only when needed. In other heat pumps, this

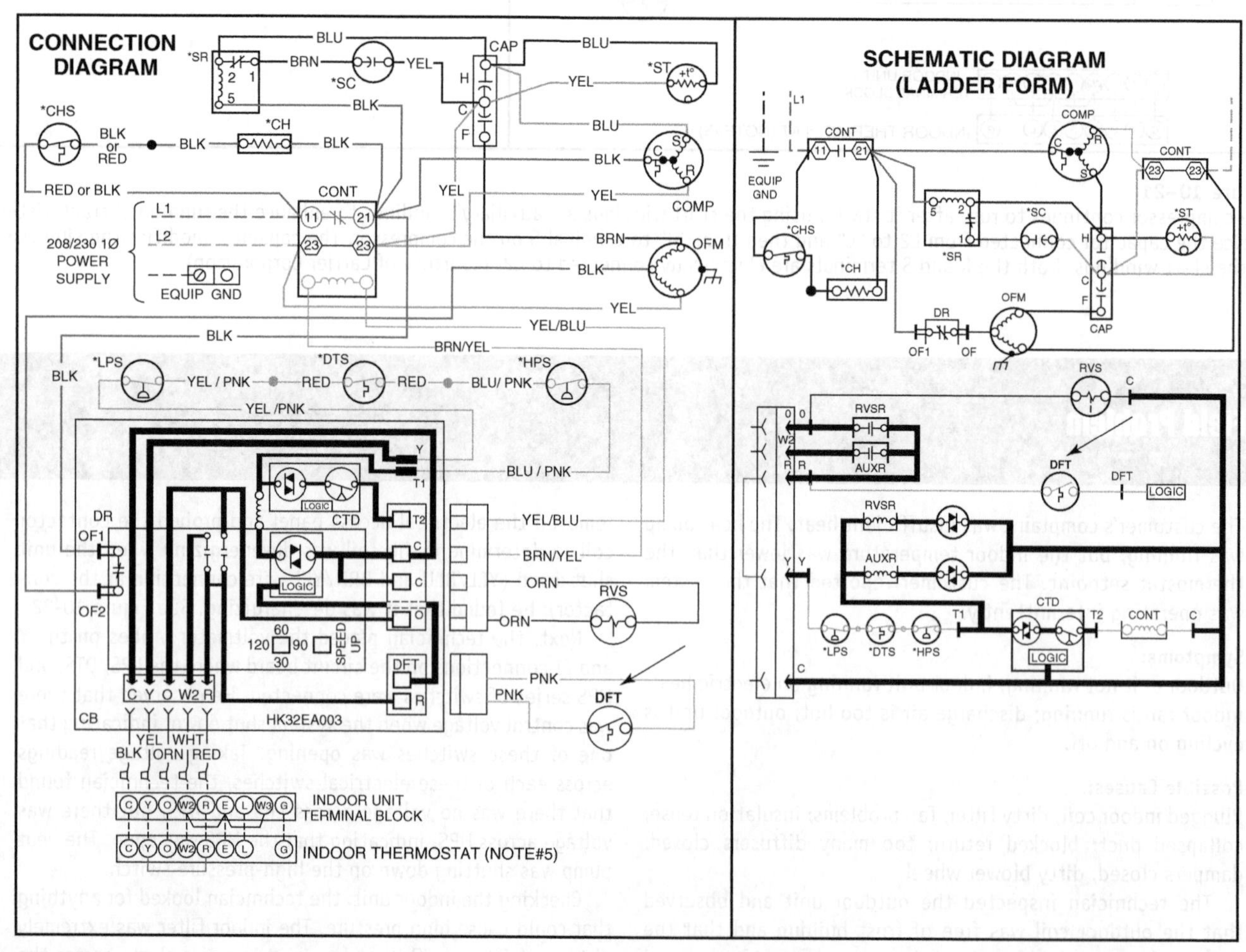

Figure 10–22
This diagram shows the unit in "ready to defrost" mode. (Courtesy of Carrier Corporation)

operation is timed and the heat pump goes into defrost at regular intervals. The system being described in this chapter has selectable defrost times of 30, 60, 90, or 120 minutes. The default time is 90 minutes, meaning that the system will operate in the heating mode for 90 minutes before the defrost cycle is initiated.

When in the defrost mode, three things happen:

1. The reversing valve shifts to cooling mode, causing hot vapor to go to the outdoor coil to melt the ice.
2. The outdoor fan turns off to keep from dissipating heat being sent to the outdoor coil for defrost.
3. The indoor electric heating elements (auxiliary heat strips) are enabled to counteract the cold indoor coil (cooling effect of the system being in the "cooling mode").

The DFT closes after sensing the temperature of the outdoor coil as being approximately 32°F. The DFT closes for a predetermined time, and the defrost cycle begins. The electronic board controls the defrost operation. The outdoor fan is de-energized when the DR switch opens (Figure 10–22).

Auxiliary/supplemental heaters may also be brought on (depending on indoor temperature) through the AUXR switch, which closes. The reversing valve is also energized (the RVS switch closes), while the compressor continues to operate. Hot refrigerant vapor is pumped to the outdoor coil to melt the frost or ice. As the outdoor coil frost or ice melts, the DFT begins to sense warmer conditions. The defrost cycle is terminated when the DFT opens or if the defrost times out. If the defrost is not terminated by the DFT in 10 minutes, the electronic board will take the system out of defrost and return it to the heating mode. At the end of the defrost cycle, the reversing valve is de-energized as well as auxiliary heaters (if they are on) and the system returns to the heating mode.

Cooling Mode

Operation of the heat pump in the cooling mode is similar to operation in the heating mode, with one important difference: the reversing valve is energized for cooling. See Figure 10-24. With that difference in mind, refer to Figure 10–23. The thermostat closes on temperature rise and closes switches from terminal R to G, for the indoor blower; Y, for the compressor; and O, for the reversing valve. Low-voltage power is sent to the contactor coil, which closes the contacts between terminals 11 and 21. Power moves through the contacts on L1 from terminal 11 to terminal 21 at CONT. LPS, DTS, and HPS all need to close or remain closed in the low-voltage safety control circuit. With the contactor closed, all of the wires that are connected at terminal 21 will now measure as line voltage when tested from terminal 21 to terminal 23 (L2). The crankcase heater circuit is energized when the compressor contactor is open and the CHS is closed.

Tech Tip

In some cases the manufacturer has a "lock-out" circuit that requires it to be reset before the compressor can attempt to start again. If any of the safety switches or sensors is activated, the system will lock-out, rather than attempt to start over and over again. This feature is incorporated into the control circuit to protect the equipment from damage as well as to cause someone to take notice of the problem. Check the manufacturer's installation or troubleshooting manual for the sequence of operation to determine if the heat pump has this feature.

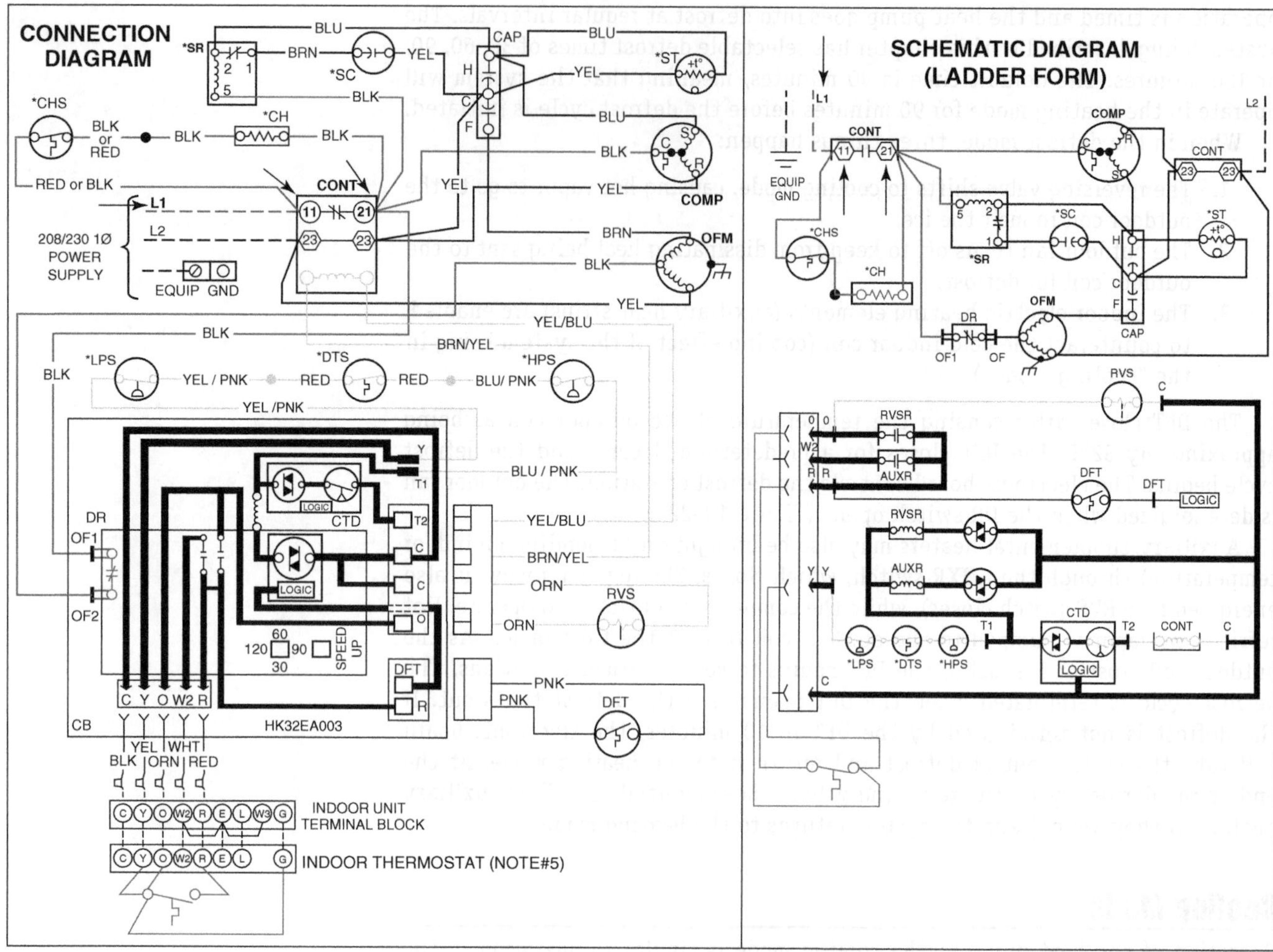

Figure 10–23
An external room thermostat closes on temperature rise when in the cooling mode. CONT contacts close and power travels from L1 and contact 11 through the contacts to terminal 21. Voltage is distributed from terminal 21, CONT, to the COMP, OFM, and SR. (Courtesy of Carrier Corporation)

Tech Tip

Some ASHP systems have manual or electrical lock-outs. This means that the technician needs to reset the system by physically resetting a switch or de-energizing the system for a few seconds before reapplying power. Cycling the thermostat or turning off the power is typically how manual reset is conducted. The ASHP system shown in this chapter is an automatic reset. The system has an internal timing circuit that will not allow the compressor to restart for 5 minutes (see the manufacturer's "NOTE" in the middle of the right column on the electrical information sheet).

Trace L2 of the power supply to terminal 23 on CONT. All YEL-colored wires at both terminals marked 23 will be at the L2 potential. All of the wires connected at these two points are YEL-colored wires. All YEL-colored wires are also at L2 potential (Figure 10–25).

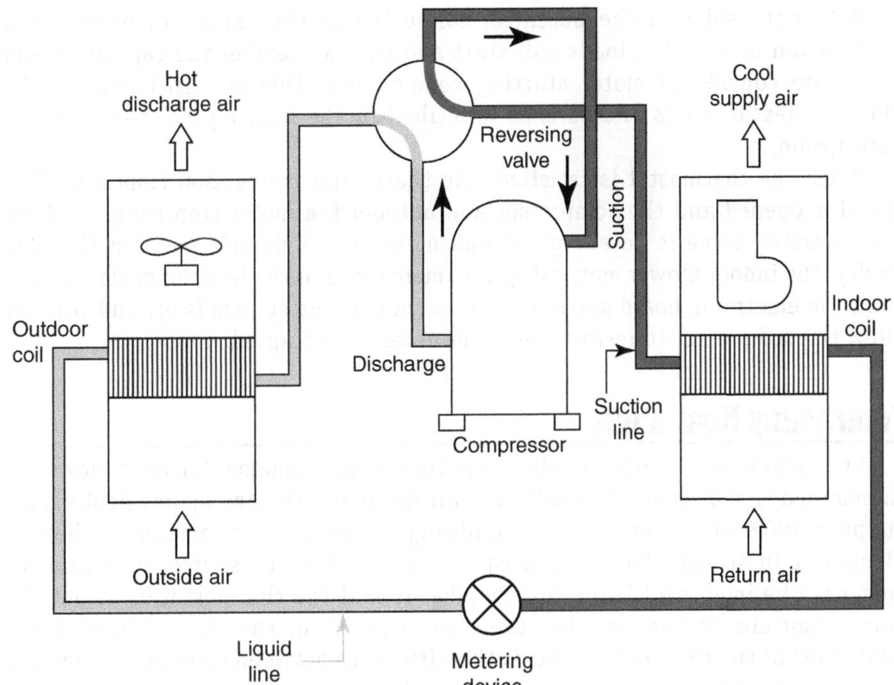

Figure 10–24
During the cooling mode, the reversing valve is powered and moves to the cooling position. Refrigerant vapor is pulled from the indoor coil, compressed, and sent to the outdoor coil as a hot vapor. (Courtesy of Delmar/Cengage Learning)

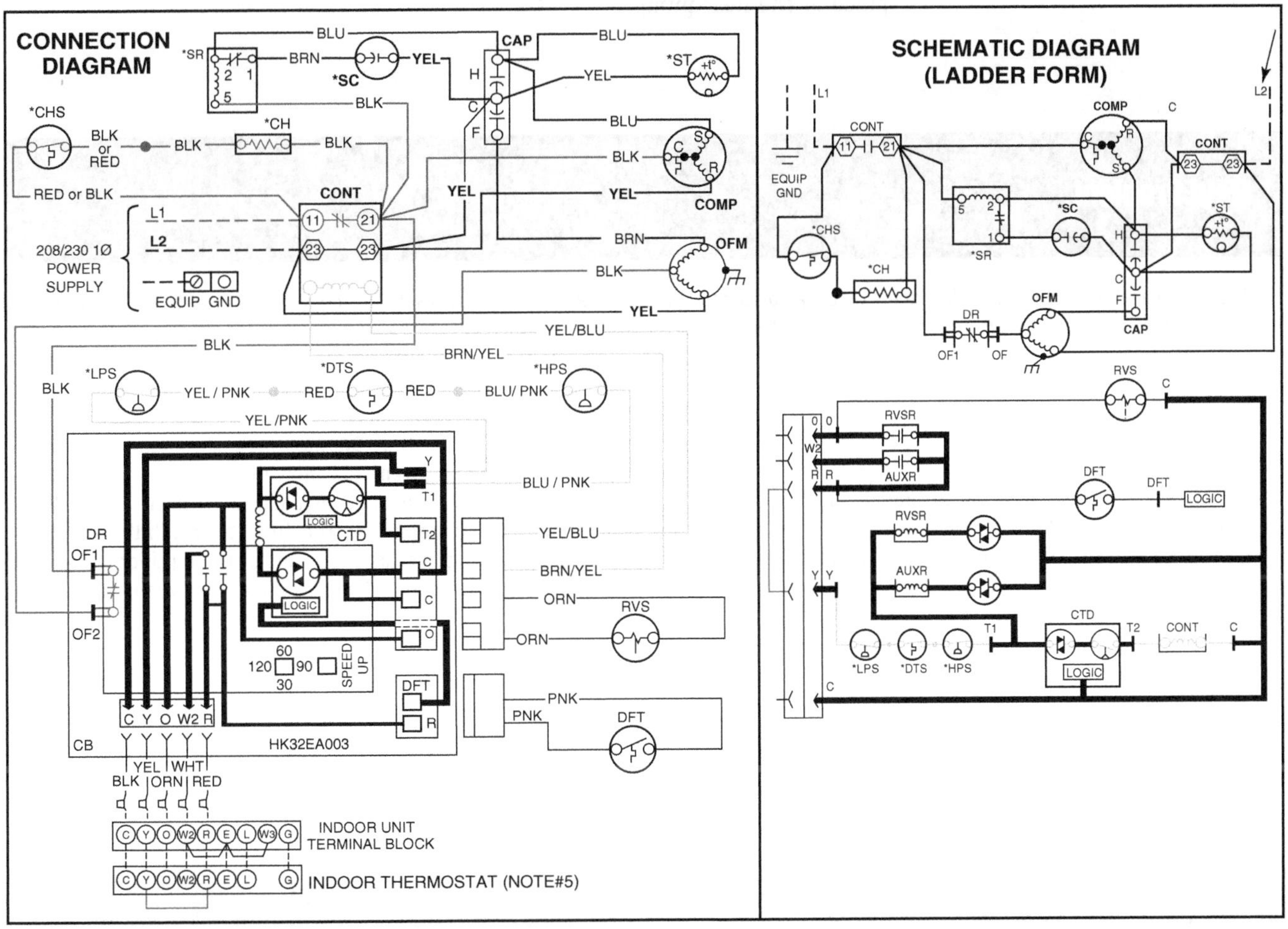

CONNECTION DIAGRAM

SCHEMATIC DIAGRAM (LADDER FORM)

Figure 10–25
The other side of the potential, L2, moves through CONT and is distributed on the YEL-colored wires to each of the components: COMP, OFM, SC, CAP, and ST. (Courtesy of Carrier Corporation)

With both sides of the potential connected to the loads (compressor and outdoor fan motor), the loads will start and run. Remember the capacitors and relays are considered motor starting components. This also includes the ST. The compressor starts and runs as described in the heating mode (see earlier description).

When the thermostat is satisfied, the thermostat connection from R to Y, O, and G is opened and the compressor and outdoor fan motor stop running. Also, the reversing valve is de-energized and moves to its default position (heating mode). The indoor blower motor stops. As mentioned in the heating mode description, the electronic board starts timing as soon as the system is off and will not allow the compressor to restart until 5 minutes have elapsed.

Emergency Heat Mode

Emergency heat is usually the same source as supplemental heat; however, in emergency heat mode, the ASHP is shut down and the emergency heat source supplies 100% of the heat for the building. In most cases, emergency heat is electric strip heaters, but in some cases it is another fuel source, such as a gas furnace. The important thing to remember is that the thermostat locks out the compressor circuit and switches the control point for the electric heat to the first stage of the thermostat (this is the difference between emergency heat and supplemental heat modes). The emergency heat source comes on as a function of the first stage of the thermostat and maintains the indoor temperature in place of the heat pump.

Field Problem

The customer complaint was that the electrical bills for the past month were extremely high, but the indoor temperature was fine.

Symptoms:
System runs continuously

Possible Causes:
Refrigeration system problems; emergency heat was stuck on; electrical system problems.

The technician observed that the compressor was running but the outdoor fan was not. The outdoor coil was clear of frost buildup. Opening the control panel, the technician checked for power at the outdoor fan terminals, 23 and 21, on the contactor. Line voltage was being supplied to the OFM. Reading the electrical diagram, the technician noted that the OFM was also controlled during defrost at the circuit board, at the DR switch. Placing the probes of the voltmeter on terminals OF1 and OF2, line voltage was measured, indicating that DR was open. De-energizing and re-energizing the system did not dislodge the stuck DR switch and the technician concluded the DR circuit board would need to be replaced. The high energy bill was explained to the customer as the result of the outdoor fan and supplemental heat strips being brought on when the outdoor unit could not pick up and transfer enough heat to the building.

SUMMARY

In this chapter we have discussed the electrical component operation of a typical ASHP and related the operation to the electrical information provided by one manufacturer. All other manufacturers have similar electrical information and provide similar diagrams. Both diagrams and component symbols can be different for each manufacturer, because there is no electrical standard for drawing diagrams or symbols for electrical components. However, if the technician is very familiar with one manufacturer's electrical information sheet, it can generally be said that the technician will understand other manufacturers' information.

It was also noted that ASHPs are similar to split air conditioning systems. Both share five standard electrical components:

1. Compressor
2. Outdoor fan
3. Indoor blower
4. Compressor starting controls
5. Temperature controls

The difference between the split air conditioning system and the ASHP is that three more electrical components are required:

1. Reversing valve solenoid
2. Defrost controls
3. Supplemental heat controls

The electrical diagram and schematic was described and shown as two separate systems: (1) line voltage and (2) low or control voltage. Separating these two electrical systems eliminates some confusion. This is the same separation of systems that occurs when troubleshooting. The compressor and motor-starting components are found in the line-voltage system. The thermostat and related devices that initiate either the heating or cooling mode are found in the low- or control-voltage system. The electrical devices that connect the two systems and have electrical parts in both voltages are the contactor and the circuit board.

The crankcase heater circuit was described. This circuit is in operation while the compressor is off. As soon as the compressor starts, the crankcase heater is turned off.

The starting circuit of the compressor was also an interesting electrical study. The contacts of the starting relay are normally closed (NC) and the coil that electromagnetically opens the contacts requires a higher voltage (backward EMF) than the line voltage supplied to the heat pump. Generally, the coil requires 100 volts above the line voltage to operate.

It was noted that the default position of the reversing valve is heating mode. To get defrost to occur, the reversing valve must be moved to the cooling mode position. Likewise, the reversing valve is energized during the cooling mode.

Four modes of the heat pump were described. Each of these modes was accompanied by electrical diagrams that followed the electrical operation being described. Tracing the electrical circuit started with the voltage connection at L1. The electrical connections were traced to each load (motor or other device). Next, L2 was traced to each load. This method of tracing or reading of electrical diagrams and schematics is important to remember.

This chapter is a very good primer for understanding basic heat pump electrical systems. For that reason, this chapter should be read and re-read until there is a good understanding of the operation of the electrical components and a good understanding of how to read electrical diagrams and schematics.

REVIEW QUESTIONS

1. Describe what is meant by the term "electromechanical" and identify those types of components.
2. Describe the use of a wiring diagram legend.
3. Name the five similar electrical circuits that heat pumps share with split air conditioning systems.
4. Explain the function of the crankcase heater.
5. List the four modes of operation for air-source heat pumps.
6. Relate why a hard-start kit might be used and describe its operation.
7. Describe three benefits that electronic circuits provide for heat pumps.

CHAPTER 11

ASHP Installation

LEARNING OBJECTIVES

The student will:

- Describe the reason for pre-planning the heat pump installation
- Describe the importance of the manufacturer's clearance recommendations
- Relate how duct leakage can affect heat pump performance
- Describe how sound is transmitted through metal ductwork
- Explain how to measure airflow from a diffuser
- Describe how the ACCA Quality Installation (QI) standards can be used

INTRODUCTION

Installation of heat pump equipment involves pre-planning, review of the structure, building science, and forethought. Pre-planning involves customer-specific requirements, building load calculations, system design, and equipment selection. Review of the structure is required to determine if there are any unforeseen obstructions, obstacles, or changes that have been made that will prevent or change the installation. Forethought is needed to try to plan for the future. Future service, repair, and regular maintenance needs to occur to extend the life of the equipment—is there enough room to service the equipment?

This chapter will explore some of the more important installation considerations that affect service. The chapter will end with the Air Conditioning Contractors of America (ACCA) Quality Installation (QI) standards that help to define HVAC installations.

Field Problem

During a scheduled maintenance call, the customer complained that the heat pump was costing him more to run during this summer as compared to the previous summer. It was September, and the technician asked if the customer had noticed any difference in operation last winter. The customer was quick to report that it worked as well as the previous winter.

Symptoms:
High operating costs

Possible Causes:

1. System developed a leak.
2. Indoor air filters are dirty.
3. Airflow is being blocked at the outdoor unit.
4. Outdoor fan motor has bad bearings.
5. Heat exchangers are dirty and need to be cleaned.
6. Outdoor fan blade is damaged or loose.
7. Inefficient compressor.
8. Leaking reversing valve.
9. Metering device or refrigerant flow problems.

With this list of checks, the technician began the scheduled maintenance procedures, looking for any evidence of the complaint. The indoor unit checked out. The controls functioned properly. Air filters were clean, and so forth. It was time to go to the outside unit to continue the maintenance call.

As the technician rounded the corner, he couldn't see the outside unit and thought he was on the wrong side of the house. Looking further, he discovered the unit hidden among the shrubs that had exploded with growth during the summer. The branches of evergreen shrubs were blocking most of the outside unit. Summoning the customer, the technician asked for the hedge clippers and the customer provided some input as the technician cut a service opening and clearance for the outdoor unit. At the conclusion of the maintenance call, the technician followed up with an explanation of clearances and why they are important to maintaining proper airflow through the outdoor unit. Most customers will appreciate you taking the time to educate them about their systems.

Field Problem

A newly installed system stopped working one day after being placed into service. The customer was concerned and shared that with the technician assigned to investigate the problem. The customer had already reset the heat pump, adjusted the thermostat, and made sure the breaker was closed. The technician rechecked all of that, but the system did not respond.

Symptoms:
Inoperable system, no cooling or heating, no blower.

Possible Causes:

1. Burned-out control transformer
2. Disconnected control wires
3. Failed thermostat
4. Bad breaker (indoor unit)

The technician had already checked for blower operation at the thermostat and did not get a response. Going to the system control panel, he checked for low voltage on the control side of the transformer and found power. By finding power, he ruled out

Field Problem (Continued)

the breaker and a burned-out transformer. Checking the control wiring in the control panel, he made sure the thermostat wires were connected correctly and tight. Holding in the door interlock switch and jumping R to G, the blower responded. Going back to the thermostat, he removed the thermostat from the wall and found the red wire broken. Turning the thermostat off, the technician connected the red wire to the R terminal. Moving the blower switch to manual caused the blower to respond. Ensuring that all of the other wires were connected correctly and tight on the thermostat, the technician placed the system into the cooling mode and verified correct operation. He explained to the customer what had happened.

GENERAL INSTALLATION REQUIREMENTS

A quality heat pump installation not only works well but looks good, but a poor installation can look good and not function well. It is not enough to install the equipment according to the manufacturer's specification and local code; it is important to install the equipment using the best knowledge and test equipment to ensure that it functions within the parameters that the manufacturer has designed.

All of the preparation prior to installation must have already been done. Heat load/gain calculations, duct design, equipment selection, and control selection precede the installation. When the equipment is delivered to the site, it has to be inspected for shipping damage. Suspicious dents and scratches can uncover underlying damage that would prevent the system from functioning. Positioning of the equipment means checking equipment components to maintain recommended clearances. Forethought is needed to envision how the equipment is to be used and maintained. With this forethought, the equipment may need to be repositioned or extra clearance allowed. Electrical and piping runs need to be checked for clear routes and adjusted for obstructions. These are just a few of the considerations that have to be made to get ready to perform a heat pump installation. Equipment installation manuals should be reviewed before beginning any installation.

Indoor Unit Installation

The indoor unit is placed in a location where service can be performed. The indoor unit can be placed in several locations, depending on the type, style, and system design (Figure 11–1). These locations could be:

1. Attic
2. Mechanical room on the living level
3. Basement
4. Horizontal ceiling or under floor mount
5. Crawl space
6. Outside (package unit)
7. Garage

Six pre-installation checks should be made prior to starting any work.

1. Clearances and Service Space: The manufacturer's clearances must be maintained. In some cases, specified clearances may not be enough for comfortable work and the unit will need to be adjusted to provide those clearances. National/local codes dictate accessibility to units located in attics and crawlspaces.
2. Power Wiring: Identify where power wiring is located and review any requirements for grounding, disconnect wiring, and conduit sizing (correct for the unit installed) by the manufacturer and the *National Electric Code* (NEC).

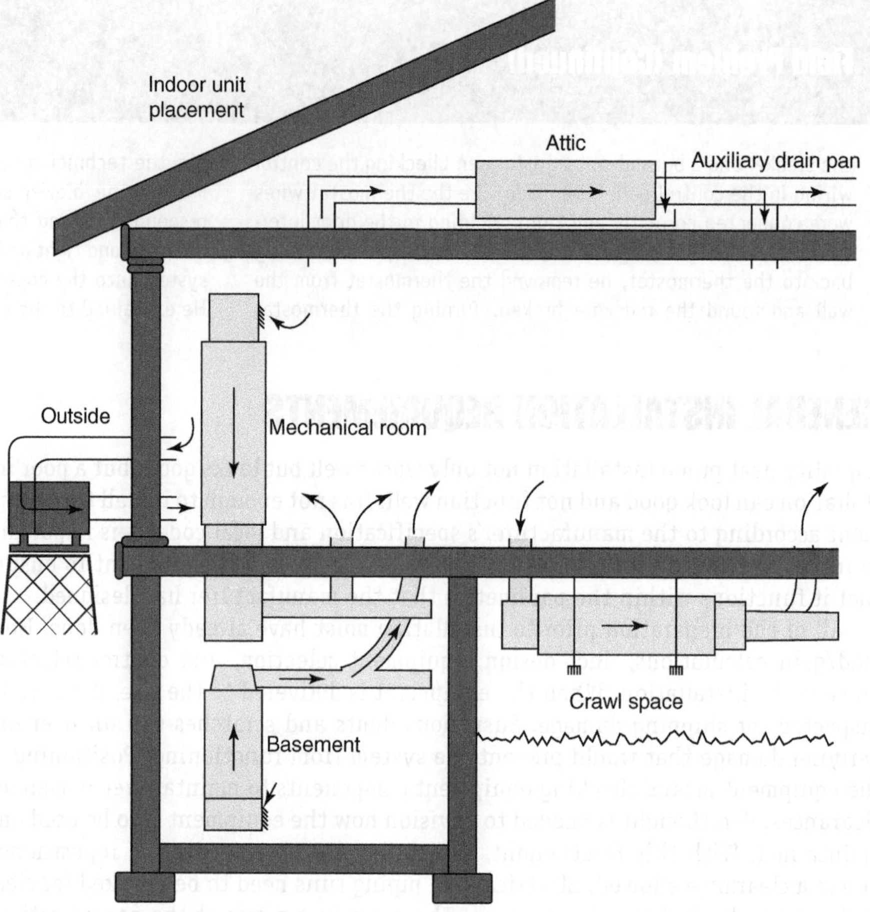

Figure 11–1
This figure shows several possible locations for the indoor unit. (Courtesy of Delmar/Cengage Learning)

3. Drains: Locate the drain and review requirements by the manufacturer regarding traps, drain length, termination, and pitch of drain line.
4. Structural Members: Determine if structural members will support the equipment or if they need to be reinforced before installation. Ensure that the weight of the unit will be carried by any hangers, ties, or clamps used to secure the unit to the ceiling or floor joists (Figure 11–2). Ensure that system parts will pass by structural members and that these members are not damaged, cut, or removed during installation. Vibration isolation may be considered as well.
5. Ductwork: Review the path that the duct will follow and ensure that the ductwork will pass through the structure as designed without removing or damaging structural members. Review the duct size and ensure that headroom and other clearances will be maintained after installation. Ensure that if metal ductwork is installed, flexible connectors are used for both the supply and return connections to the unit.
6. Refrigerant Line Routing: Review the path that the refrigerant lines will follow and ensure that the lines will be able to pass through structural members without damaging the structure. Refrigerant lines should not need to make tight bends or be placed in a location where they can be physically damaged by normal activity (lawn mowers, egress, etc.).

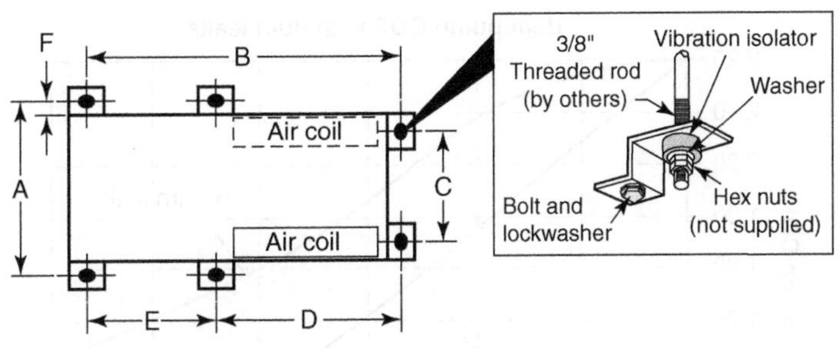

Model	Hanger kit part number	A	B	C	D	E	F
009-012*	995500A04	25.1	44.7	21.4	N/A	N/A	1.3
015-018*	995500A04	25.1	53.7	21.4	N/A	N/A	1.3
022-030*	995500A04	24.8	63.4	21.1	N/A	N/A	1.1
036-038	995500A03	27.8	72.4	24.1	43.1	29.3	1.1
042-049	995500A03	27.8	77.4	24.1	48.1	29.3	1.1
060-072	995500A03	27.8	82.4	24.1	53.1	29.3	1.1

Note: *only the four corner brackets are needed on sizes 009-030.

Figure 11–2
Hanger location points are identified by the manufacturer. These locations must be used as well as the correct size of hanger material. (Courtesy of WaterFurnace International)

Ductwork

As the installation proceeds, the ductwork is connected to the unit and branching runs are made to the conditioned space. The duct connections must be sealed and air-tight. Duct mastic and metal duct tape are typically used to seal duct connections; UL 181 materials only are allowed by code. If return air uses floor joists and other structure spaces, attention should be made to sealing any structural member that could allow uncontrolled air to enter the return air. In some cases, this air could cause moisture problems, leading to mold and structural damage. Energy use can be reduced by sealing the duct system and comfort can be increased (Figure 11–3). Before connecting ductwork to the unit, shipping material is removed from the blower housing and the blower is spun by hand to ensure smooth operation.

Vibration and noise reduction is of primary interest when installing the ductwork. Metal duct is prone to transmitting noise (Figure 11–4). If metal duct is used, most manufacturers recommend that the duct be lined to reduce conduction (losses or gains) and acoustical noise transmission (flexible duct connectors). Both the supply and return plenums are also recommended to be lined with 1 inch of insulation to reduce blower noise transmission while operating. The lining can also insulate. Duct insulation is also required when ductwork passes through an unconditioned space. Insulation can also be applied to the outside of a metal duct. Department of Energy (DOE) "Energy Star"–rated homes require a minimum of R-8 duct insulation for regions 1 and 2 and a minimum of R-6 for regions 3 and 4. Local code may require other minimum amounts of insulation to be applied to ductwork in unheated locations.

Table 11–1 shows the advantages of fiberglass and metal ductwork.

Duct board may have the advantages of increased system performance by reducing duct leakage, eliminating sound conduction, and reducing noise by the blower. When duct board is used as a plenum, the flexible duct connector may be eliminated. (Note: The flexible duct connector may still be required in some

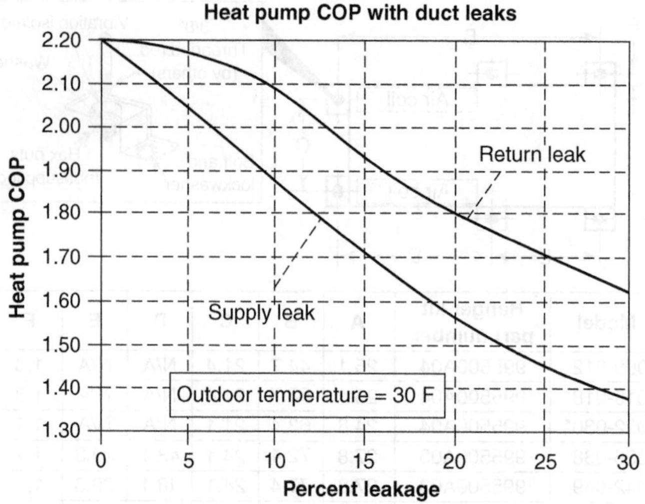

Figure 11–3
This graph shows the effect of duct leakage on the performance of the heat pump. Efficiency decreases as duct leakage increases. (Courtesy of ACCA)

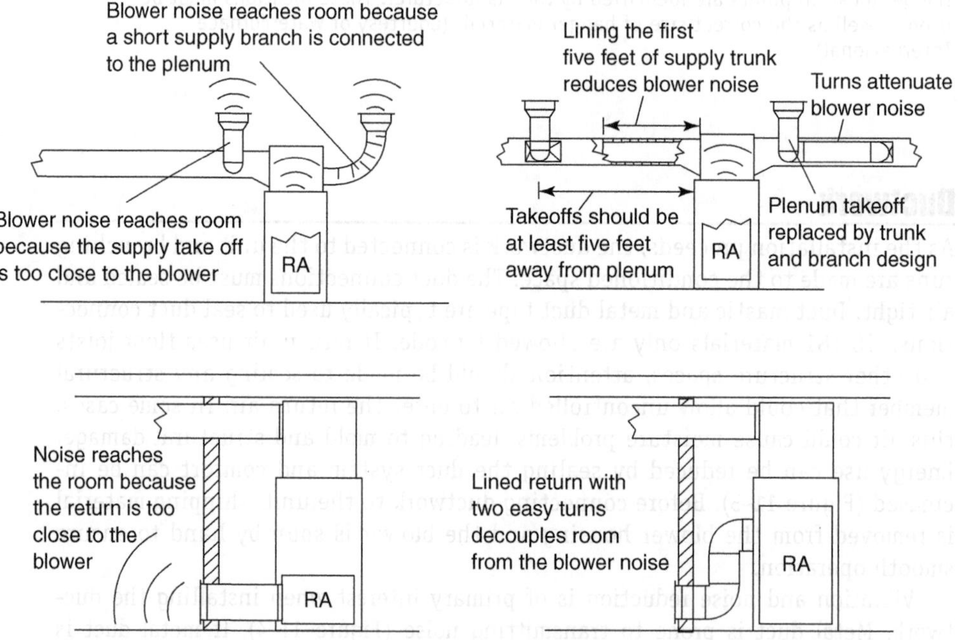

Figure 11–4
Sound transmission can be caused by many factors. Shown are a few causes of sound and one solution. (Courtesy of ACCA)

local regulations.) This material has also been used as a pad for the unit to sit on as part of the unit mounting. Studies of duct board use have shown that:

- It does not erode over time when properly installed.
- Workers can shield themselves from glass fiber during installation.
- The International Agency for Research on Cancer (IARC) reclassified fiberglass as a Group 3, noncarcinogenic, product. With proper installation and blown-down (clean-out) techniques, it leaves too little material to be a cause of human discomfort.

Outside make-up air should not be brought into the return air ductwork any closer than 3 duct diameters to the unit. In the heating mode, outside air may drop the temperature of the air being returned to the air handler. If return air temperature is too low, the air temperature may cause low discharge pressures, which could affect the suction pressure. The discharge air may feel drafty. Lower

Table 11–1 Ductwork Types and Advantages

Attributes	Fiberglass Duct Board	Metal Ductwork
Sealed, when fabricated and installed properly	X	X
Self-insulating	X	
Lightweight	X	
Material is sound absorbing	X	
Easy onsite fabrication	X	
Easy custom fabrication onsite	X	
Less mass to heat and cool	X	
No expansion or contraction	X	
Less labor required to fabricate and install	X	
Less cost for fabrication equipment	X	
Lower static resistance (if not lined)		X
Not easily damaged		X
Structurally strong		X
Easily cleaned (if not lined)		X
Lower cost if insulation is not needed		X
Weather resistant		X

Note: Both of these ducting materials are fire resistant and have life expectancies over 35 years. Often, it is a contractor or a customer's personal feeling that determines whether one type of product or the other is installed.

suction pressure will cause more frost to form on the outside coil. More frost accumulation can cause additional defrost cycles. There is also a possibility, depending on conditions, of a false defrost signal to be generated, signaling defrost to occur when it is not necessary.

One of the greatest concerns for ductwork is the size. Heat pumps need to supply from 350 to 450 cubic feet per minute (CFM) of air per ton. A 3-ton (or 36,000-BTUH) unit may need to supply 1,350 CFM (3 tons × 450 CMF = 1,350 CFM), which translates into larger-size ductwork than a fossil-fuel furnace, which may only require 476 CFM to transfer the same amount of energy (36,000 BTU/70°F temperature rise/1.08 air factor = 476 CFM). To get all of that air to flow through the ductwork also requires consideration for friction loss because of the type of ductwork, transitions, takeoffs, boots, filters, and diffusers or grilles. The length of the duct will also add to the friction loss. All of this is taken into account by using a good duct design computer program or manual calculation. A good installer can use a manual calculation on the job to ensure that no additional friction loss is built into the duct system as the installer encounters unexpected obstructions and changes. ACCA, Manual D, provides a method for designing duct systems that perform well. Installers and technicians may not be called upon to design a duct system, but the manual is a good resource for understanding how to reduce and eliminate duct system problems and possible problems during installation.

Airflow and Face Velocity

Airflow and face velocity have to do with how the air moves through the various duct system components and exits the diffuser. Air moving at high rates can cause noise. In order to control noise, air velocity should stay under the maximum recommended by ACCA, as shown in Table 11–2.

Table 11–2 Air Velocity for Noise Control[1, 2 and 8] (Courtesy of ACCA)

	Supply Side (Fpm)				Return Side (Fpm)			
	Conservative		Maximum		Conservative		Maximum	
Component	Rigid	Flex	Rigid	Flex	Rigid	Flex	Rigid	Flex
Trunk Ducts	700	700	900	900	600	600	700	700
Branch Ducts	600	700	900	900	500	600	700	700
Supply Outlet Face Velocity	Size for Throw		700[7]		—		—	
Return Grille Face Velocity	—		—		—		500	
Filter Grille Face Velocity	—		—		—		300	

1) The design friction rate is affected if air velocity exceeds 900 Fpm (fitting equivalent lengths are for 900 Fpm or less).
2) System resistance considerations supercede velocity considerations (minimum acceptable airway size shall be based on the local Cfm value and the design friction rate). Air way size shall be increased if the local air velocity exceeds the maximum limit.
3) This table applies to metal duct with transverse seams and metal fittings (duct runs and fittings not lined or wrapped with insulating material).
4) This table applies to flexible wire helix duct with duct board junction box fittings.
5) Maximum velocities may be exceeded when construction has less surface irregularities (no transverse seams or less irregularity at transverse seams, and very efficient fittings); and has a sound absorbing attribute (duct board or duct liner).
6) Authoritative guidance concerning velocity limits for aerodynamically efficient and/or sound absorbing designs is not available at this time.
7) The velocity limit for a supply outlet may be ignored if the noise criteria (NC) value for a grille, register or diffuser is 30 or less over the range of Cfm values that will flow through the device (or combination of devices, if a damper is involved), during any mode of system operation.
8) Air velocity limits are superceded by measured noise criteria (NC) values for low rise dwellings (Notes 1 and 2 still apply).
   - NC values measured by sound meter in middle of the room when normal human ear perceives maximum HVAC system noise.
   - Measured NC equals or exceeds 30 with comfort system off; measured NC shall not increase by more than 3 with comfort system on.
   - Measured NC less than 30 with comfort system off; measured NC shall not exceed 33 with comfort system on.

It should also be noted that a larger duct provides a greater area for air to move with less friction. If there is less friction, blower motors work less to provide the same amount of air. For reasons of economy, duct systems are sized to use the minimum amount of material in an effort to reduce the size of the duct, while maintaining reasonable velocity and static pressure. The blower is part of the calculation so that blower amperage is kept low.

Face velocity refers to the speed of air moving through the control louvers of a diffuser as it moves into a conditioned space. It is also a reference to the speed of air that moves through a return grille and back into the return duct. Air in the occupied space is most influenced by the velocity and pattern of air moving from the supply diffuser. It is not significantly influenced by the return grille, the placement of the return grille (high or low), or the velocity of the air in the duct system. The correct diffuser and pattern must be selected and matched to the amount of air need to offset heat loss or heat gain. Using Table 11–2, a room may need 220 CFM. Two floor diffusers that will provide 114 CFM each at a pressure loss of .016 inches of water column may be chosen. This diffuser has a throw of 16 feet (the distance from the diffuser to a terminal velocity of 20 feet per minute [fpm]), with a face velocity of 600 fpm. The designer may have chosen the diffuser for its pattern and throw to move the air in the occupied space without causing a draft.

Tech Tip

Three terms are used for ductwork terminations. Terminations are the end points of a duct system. Registers, diffusers, or grilles can be found at terminations. All of these terms are sometimes used to mean the same thing. However, the proper use of each term is based in usage over time. A grille is a terminator that has no controlling attributes, such as dampers, stationary louvers, or adjustable louvers. A register was formerly known as a terminator for a gravity air furnace system. The register had a grille that prevented large objects from falling into the duct opening, but also had a damper to adjust airflow. A diffuser is a terminator that controls the air pattern and airflow. A diffuser has a damper to adjust the amount of airflow. It also has either stationary or movable louvers that create or modify the air pattern. Air pattern is determined by terminal velocity (lowest air speed) and throw (distance from the diffuser; Figure 11–5). Diffusers can be placed on the floor, wall, or ceiling of a building. Table 11-3 provides technical specifications for a 4 × 12 floor diffuser.

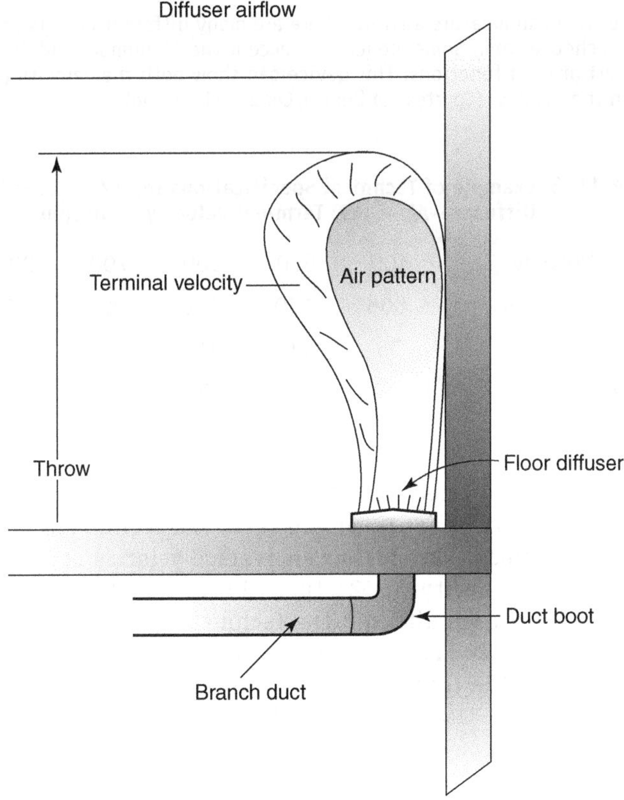

Figure 11–5
Air within the occupied space is influenced by the diffuser pattern and throw. Both the pattern and the throw are defined as the terminal velocity—the speed of the air at the end of the test. Most manufacturers state 50 fpm as the terminal velocity. (Courtesy of Delmar/Cengage Learning)

The technician may use the same manufacturer data to determine the amount of air that is actually coming from the diffuser. Using an anemometer (flow meter), the technician takes an average reading across the diffuser opening and multiplies the average velocity reading by the Ak number (Figure 11–6). This number is the free area in square feet that the manufacturer has determined

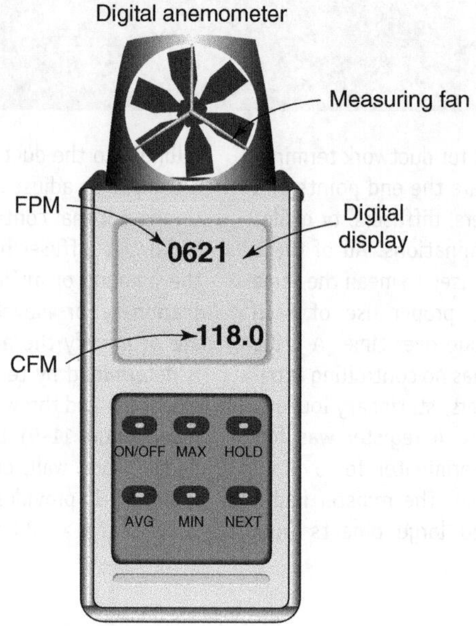

Figure 11–6
This hand-held device can measure airflow. There are many different models and manufacturers to choose from. Some models can accept the Ak number and do the calculation as part of their functions. This device can show both the velocity (fpm) and the volume (CFM) on the display. (Courtesy of Delmar/Cengage Learning)

Table 11–3 Example of Technical Specifications for a 4 × 12 Floor Diffuser—Ak = .19; Terminal Velocity = 50 fpm

Face Velocity	400	500	600	700	800
Pressure Loss	.006	.010	.016	.022	.031
CFM	76	95	114	133	152
Throw	9	12	16	18	20

through testing and is referred to as the "area constant" (k for constant). The Ak number for this diffuser is .19 square feet. The technician measures each of the diffusers and finds that one has an average velocity of 526 fpm and the other has an average velocity of 621 fpm. Multiplying the Ak number times the average velocity of each diffuser, the technician gets 100 CFM (526 fpm × .19 Ak = 100 CFM) and 120 CFM (621 fpm × .19 Ak = 118 CFM). The room needs 220 CFM by design, so the technician adds the two CFM amounts together and gets 218 CFM (100 CFM + 118 CFM = 218 CFM), which is very close to the design condition.

The same measurements and calculation can be done for return air grilles. It doesn't matter which way the air is flowing. If the grille face is large, the technician should use a grid pattern to take measurements. Divide the grille surface into 9 quadrants and take average readings of each of the 9. Add those measurements together and divide the sum by 9 to get the composite average (Figure 11–7).

OUTDOOR UNIT INSTALLATION

Equipment installation starts with well-thought-out equipment placement. This is also true for the outside unit. This unit needs to be as close to the structure as possible to minimize piping and electrical connection lengths. It also needs

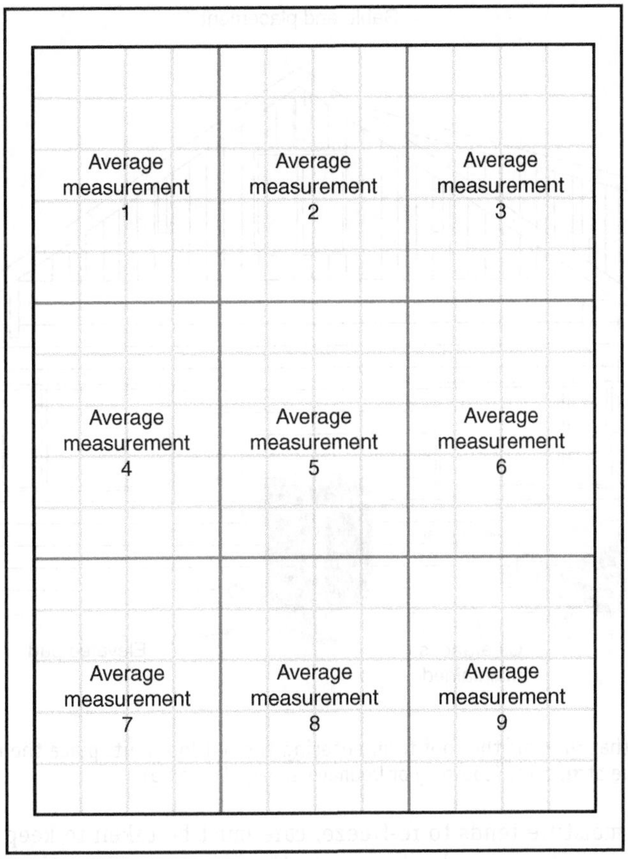

Figure 11–7
If the grille surface is large, divide the grille into several segments and take average velocity measurements of each segment. Add all of the average velocities together and divide by 9 to get the total average velocity of the grille. (Courtesy of Delmar/Cengage Learning)

to be far enough away from the structure to allow for maintenance and airflow. The manufacture provides clearance distances to maintain adequate airflow. All obstructions to airflow should be minimized. Plants, shrubs, and fences need to be at the recommended distances from the unit. NEC requires 3 feet of clearance to the service panels.

Placement should also be in relationship to the structure roof. Water shed from the roof could pour into the outdoor unit if placed too close to the structure. In the winter, this could cause ice jamming and accumulation of ice, resulting in damage to the outdoor unit. If at all possible, the outdoor unit should be placed at the gable end of the structure (Figure 11–8). Additionally, roof overhangs can be a problem by causing the air to be recirculated to the outdoor coil.

The unit must be on a sturdy foundation. The foundation could be a structure, such as a concrete pad, or a platform made of various materials, but not connected to the structure. In some installations, a lightweight pad is used. Because a heat pump is used during the winter months, snowfall is of particular concern in many parts of the country. The outdoor unit must be higher than the average snow depth for the location. The weather service can be consulted to determine the snow pack conditions that might be encountered. If the average snow depth is 1 foot, the unit should be mounted on a sturdy structure that elevates the unit 12 inches or more. Local codes often specify location, such as not on the street side of the house, minimum distances to neighbors, etc.

Drainage is also of concern for heat pump installations. During the heating season, the outdoor unit goes through many defrost cycles. The outdoor unit must be positioned to allow for the defrost water to drain away from the outdoor coil.

Safety Tip

Whenever equipment is in poor condition, it presents an electrical hazard. Be very cautious when encountering equipment in these conditions. It is also advised that the owner be made aware of this condition, with strong suggestions to have it repaired or replaced.

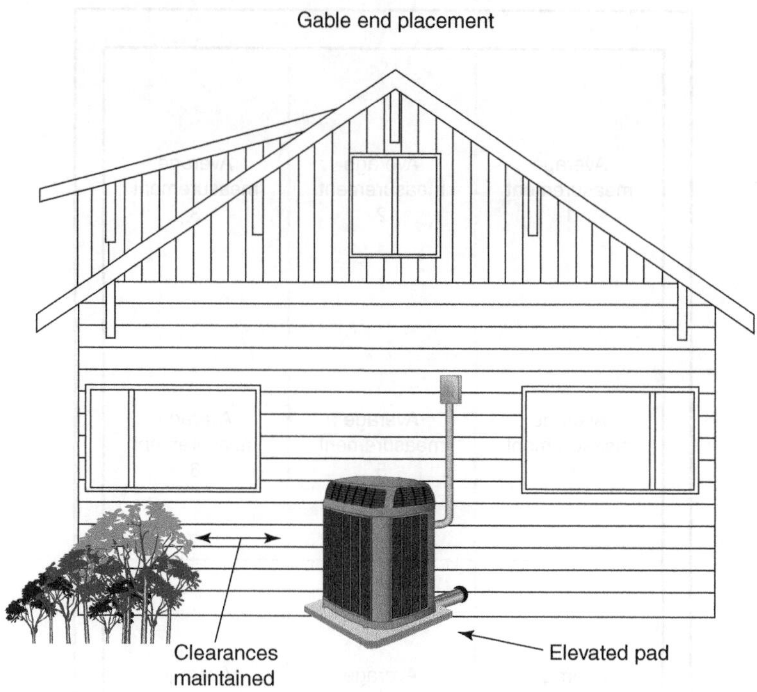

Figure 11–8
To keep water that runs off the roof from entering the outdoor unit, place the unit at the gable end of the structure. (Courtesy of Delmar/Cengage Learning)

Because this moisture tends to re-freeze, care must be taken to keep the outdoor unit away from walkways and entrances to the structure. Water that is allowed to puddle or run across walkways will become a slip hazard during the winter months.

Electrical connections to the outdoor unit must also conform to the electrical code. In most cases this means that electrical conduit and boxes must be rated for outdoor and moisture locations. Electrical equipment that does not meet these requirements will rust, corrode, and become an electrical hazard to the owner and technician. The NEC specifies the length and support requirements for this wiring.

RECOVERY, EVACUATION, AND CHARGING

During the installation process, line sets and other refrigerant piping may be installed. Sometimes this means that the system charge needs to be adjusted for the length of the line set. Additionally, some installations have a desuperheater (domestic water heater feature) that requires an adjustment of the charge. Before the contractor finishes the installation, the charge needs to be verified with the manufacturer's data sheet. Several methods could be used, such as:

1. Weigh in charge
2. Superheat for fixed orifice and cap tube
3. Subcooling for TXV

Whichever method is used, the method must conform to the original equipment manufacturer (OEM) specifications and recommendations. The results should be recorded on a form and kept as a permanent record. A copy of this form with all of the information needs to be placed in the customer's packet and kept in a convenient location next to the indoor system control panel.

Evacuation requires the removal of moisture and vapor. Evacuation to a sustained level of 500 microns or less is required for system longevity. Some manufacturers require lower micron levels. Long evacuation periods are the result of

Safety Tip

If system connections need to be opened for service or repair, the technician should know that in certain conditions, the system pressures may not equalize from low to high side. If lines are not de-pressurized and the refrigerant recovered, lines could burst with heat from a torch and the refrigerant could become a severe hazard. Always ensure that pressure on both sides has been reduced to atmospheric pressure. A good practice would be to cut the system open with a tubing cutter rather than unbrazing.

refrigerant diluted in the oil, moisture, or leaks. Good two-stage deep-evacuation pumps with new evacuation pump oil, evacuation hoses, and accurate micron instrumentation are a must for proper evacuation. Every time an evacuation pump is started, or if it is not pumping to specifications because the oil is contaminated, the oil needs to be changed. Evacuation is complete and no moisture or leaks are present when the vacuum pump is isolated and the system will hold a vacuum of 500 microns or less.

Tech Tip

Polyolester oils (POEs) easily absorb moisture, so they are known as hygroscopic oils. This means that the oil will absorb moisture from the air if it comes in contact with the air. Manufacturers of this oil provide an inert gas cover of nitrogen to prevent the oil and air from coming in contact. When oil is used from the manufacturer's container, nitrogen should be reintroduced or added to the top of the oil to push the air out of the container. When adding this oil to a refrigeration system, care should be taken to minimize or eliminate the oil from coming in contact with the air.

Tech Tip

HFC R-410A is a blend of refrigerants. The "10" designation indicates that it was the tenth that was produced commercially. The capital letter "A" means that the percentage of refrigerants used for the blend differs from R-410B and R-410C, which are all near-azeotropic blends. Near-azeotropic refrigerants exhibit a single evaporating/condensing point, but can also fractionate, or the individual refrigerants can break apart from the blend and leak out of a system at different rates. This changes the percentages of refrigerants remaining behind as a blend and alters their temperature/pressure relationship in a system. This is the reason why the preferred charging method for HFC R-410A is to charge the system by weighing in the charge as a liquid.

ANSI/ACCA QUALITY INSTALLATION STANDARDS

ACCA has developed ANSI sanctioned Quality Installation (QI) standards for HVAC equipment. Excerpted here from those standards are sections that pertain to heat pump installations. Under section 2 of the standard, Scope, heat pumps are covered under three headers:

1. Unitary air conditioners and air-source/water-source heat pumps up to 65,000 BTUH
2. Unitary equipment (packaged and split) greater than 65,000 BTUH
3. Residential equipment used in commercial three-phase applications

Each of the QI standards regarding heat pumps has been taken verbatim from the standard and shows the section number for reference. It is suggested that the complete set of standards be obtained electronically from the ACCA website.

EQUIPMENT INSTALLATION ASPECTS

This section focuses on the HVAC system installation.

Airflow Across Indoor Heat Exchangers

The contractor shall verify that the airflow across the indoor heat exchanger is within acceptable ranges.

Requirements

The contractor shall provide evidence of the following for the measured airflow across the indoor heat exchanger for installed systems (with all accessories and system components in place):[1]

a. For cooling coil (e.g., refrigerant, water) and heat pump applications
   i. Airflow across the coil, at fan design speed and full operating load, is within 15% of the airflow required per the system design; and
   ii. Airflow across the coil is within the range recommended by the OEM product data.[2]

Acceptable Procedures

The contractor shall test using one or all of the following acceptable devices for fulfilling the desired criteria:

a. Pressure matching method[3]
b. An anemometer (e.g., hot wired, rotary style) or other methods (e.g., transverse pitot tubes) for measuring total static and velocity pressures to determine airflow velocity in several traversing locations per AABC, NEBB, or ASHRAE procedures
c. Flow grid measurement method
d. A manometer to determine the pressure drop across a clean cooling coil or fan coil unit and compare with values from the OEM CFM/pressure drop coil tables
e. The temperature rise method (for heating equipment only—gas or oil furnace, electric heat) to verify proper airflow across the heat exchanger or heater elements. [Note: It is not acceptable to use the temperature rise method for cooling (i.e., airflow over the indoor coil).]

Acceptable Documentation

a. Documented field data and calculations recorded on start-up sheet
b. Documented field data and calculations recorded on service records
c. Written job documentation or checklist in job file

Refrigerant Charge

The contractor shall ensure that the HVAC system has the proper refrigerant charge.

Requirements

The contractor shall provide evidence of the following for charging installed systems:[4]

a. For the SUPERHEAT method, system refrigerant charging per OEM charging data/instructions and within ± 5°F of the OEM-recommended optimal refrigerant charge

---

[1]When verifying airflow at full design fan speed, there is little distinction between a permanent split capacitor (PSC) fan motor and a variable-speed fan motor (e.g., electronically commutated motor, or ECM). See "Fan Airflow" in Appendix B. Note: ECM fan motors are designed to modify their RPMs in order to provide a prescribed (programmed) air volume in response to static pressure conditions (actually torque on the output shaft). Hence, an ECM may use more or less power than a comparable PSC motor in the same application.

[2]Airflow across the coil is typically between 350 and 450 CFM per ton.

[3]Use a calibrated fan to match the supply plenum pressure and measure the system airflow through an active fan.

[4]Refrigerant charge tolerances noted (i.e., ± 5°F and/or ± 3°F of the OEM-recommended optimal refrigerant charge) are not additive to any OEM-specified tolerances.

b. For SUBCOOLING method, system refrigerant charging per OEM charging data/instructions and within ± 3°F of the OEM-recommended optimal refrigerant charge
c. Any method approved and specifically stated by the OEM that will ensure proper refrigerant charging of the system

Acceptable Procedures

The system shall be charged according to an approved/acceptable charging method. The charging method used should be documented, including:

- system conditions
- calculations conducted
- results obtained

If ambient conditions require a follow-up visit to finalize the charging process, this should be recorded both at the initial visit and the follow-up visit. The contractor shall use one or all of the following acceptable procedures for completing the desired measurements after confirmation of required airflow over the indoor coil per §4.1:

a. Superheat test done under outdoor ambient conditions, as specified by the OEM instructions (typically, 55°F dry bulb temperature or higher)
b. Subcooling test done under outdoor ambient conditions, as specified by the OEM instructions (typically, 60°F or higher). [charges are not additive to any OEM-specified tolerances.]

Acceptable Documentation

a. Documented field data AND operating conditions recorded on start-up sheet
b. Documented field data AND operating conditions recorded on service records
c. Written job documentation or checklist in job file

Electrical Requirements

The contractor shall ensure all electrical requirements are met as related to the installed equipment.

Requirements

The contractor shall provide evidence of the following:

a. LINE and LOW VOLTAGES per equipment (single and three-phase) rating plate
   - the percentage (or amount) below or above nameplate values are within OEM specifications and/or code requirements
b. AMPERAGES per equipment (single and three-phase) rating plate—the percentage (or amount) below or above nameplate values are within OEM specifications and/or code requirements
c. LINE-and LOW-VOLTAGE wiring sizes per NEC (*National Electric Code*) or equivalent
d. GROUNDING/BONDING per NEC or equivalent

Acceptable Procedures

The contractor shall test using the following acceptable procedures for fulfilling the design criteria:

a. Volt meter to measure the voltage
b. Amp meter to measure the amperage
c. Verify measurements with nameplate and over current protection criteria

Acceptable Documentation

a. Documents showing that selections are in compliance with OEM specifications
b. Written job documentation or checklist in job file

System Controls

The contractor shall ensure proper selection and functioning of system operational and safety controls.

Requirements

The contractor shall provide evidence of the following:

a. Operating controls and safety controls are compatible with the system type and application, and the selected controls are consistent with OEM recommendations and industry practices, and
b. Operating controls and safety controls lead to proper sequencing of equipment functions, with all controls and safeties functioning per OEM or customer design specifications. NOTE: Examples of operating controls include: thermostats, humidistats, economizer controls, etc. Examples of safety controls include: temperature limit switch, airflow switch, condensate overflow switch, furnace limit switch, boiler limit switch, etc.

Acceptable Procedures

The contractor shall use the following acceptable procedures for fulfilling the desired design criteria:

a. Confirmation of the control/safety selections made
b. Supporting OEM literature related to the selections made
c. Verification of correct cycling/operational sequences of controls and safety devices/systems per OEM specifications

SUMMARY

In this chapter we have seen that installation is more than just connecting equipment together. It requires pre-planning, review, and forethought to create a quality installation. Discussed in this chapter were the six pre-installation checks. These are important to do in preparation for installation. Ductwork that includes insulation, sealing, and size is a very important part of the system. Without the right size of ductwork, the heat pump might not be able to move the quantities of air necessary to be an efficient heating and cooling source. Airflow through the duct system, diffusers, and return grilles delivers conditioned air to the occupied space in the right quantities and at the right face velocities to create air movement without drafts.

The outdoor unit has its own special requirements. Where it is placed and how it is positioned must meet the manufacturing specifications and conform to recommended distances. Outdoor weather conditions, especially moisture, can affect the outdoor unit if placed under an eave where water from the roof can fall into the unit. Electrical components must also conform to outdoor use standards.

System charge must be verified with the manufacturer's specifications. The verification procedures must also conform to the procedures recommended by the manufacturer. The results of the test need to be recorded and kept as a record of the installation along with the other customer materials. Recovery, evacuation, and recharge methods must also conform to R-410A standards, as most heat pumps use this refrigerant.

The chapter ended with excerpts from the ACCA QI standards. Those that apply to heat pumps were included. These form a formal guide for the installation of heat pumps and other HVAC equipment.

REVIEW QUESTIONS

1. Describe the reason for pre-planning the heat pump installation.
2. Describe the importance of the manufacturer's clearance recommendations.
3. Relate how duct leakage can affect heat pump performance.
4. Describe how sound is transmitted through metal ductwork.
5. Explain how to measure airflow from a diffuser.
6. Describe how the ACCA Quality Installation (QI) standards can be used.

CHAPTER 12

Scheduled Maintenance

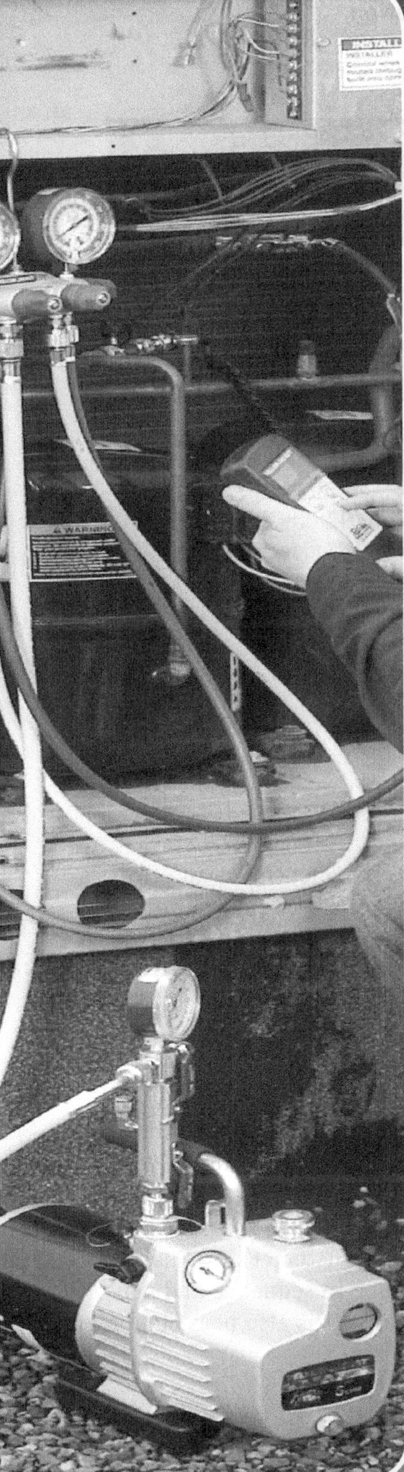

LEARNING OBJECTIVES

The student will:

- Describe the need for standards
- Explain how standards relate to heat pump maintenance
- Discuss how the general inspection tasks relate to heat pump outdoor units
- Describe the difference between Quality Maintenance (QM) and Quality Installation (QI) standards
- Describe what might happen if system controls were neglected in the QM standards

INTRODUCTION

Quality maintenance and quality system installation go hand-in-hand. One is required to have the other. That is what this chapter will discuss. The Air Conditioning Contractors of America (ACCA) has developed Quality Maintenance (QM) and Quality Installation (QI) standards that are designed to guide technicians and maintenance companies. Through the use of standards, several things can occur in the industry. First, the customer is served through quality installation and maintenance activities. The system has better performance and lower energy bills, which are the customer benefit. Second, the technician is provided with a guide to quality practice. And third, dialog is generated in the industry to improve both quality maintenance and quality installation.

Field Problem

The heating system had been recently upgraded from a fossil-fuel heating system to an air-source heat pump (ASHP). Even though the system was new, the owner was complaining that the system didn't seem to heat as well as he had been promised or had expected. The technician was dispatched to determine the cause.

This was a typical heat pump installation. It was a 2-ton split system and the technician had been provided with all of the system specifications by his company. He was wondering where he should start when he remembered that he had just acquired the ACCA QM standards. He remembered that it provided a guide to checking and maintaining heating and cooling systems and that there were some specific checks for heat pump systems. He had a copy in the truck and decided that he would use the standards and check off each item in an effort to determine the problem of low heat for this heat pump.

He found that using the standards was relatively simple as he checked off each task that applied to the heat pump system. He hadn't proceeded very far with the standard outdoor checks before he discovered a possible problem. There was oil residue on one of the field brazed joints. The rest was easy. He confirmed that there was a low charge and a leak. He recovered the refrigerant, fixed the leak, checked the leak, verified with nitrogen, evacuated to 500 microns, and recharged to the manufacturer's specifications. Finishing the maintenance call, he rechecked the operation to see that there was an improvement and reported back to the customer.

QUALITY MAINTENANCE STANDARDS

Heat pump scheduled maintenance has been an imprecise science. In an effort to remedy that shortcoming and to provide guidance for maintenance activities, ACCA has produced standards that are recognized by the American National Standards Institute (ANSI). The following Checklists (Tables 12-1 to 12-9) are standards taken directly from ACCA QM standards.

Outdoor Unit

Outdoor unit maintenance encompasses all things that should be checked or done to the outside coil and outdoor unit (Figures 12-1 and 12-2). You should read through these inspection tasks and recommended corrective actions to familiarize yourself. Note that there is no sequence for inspection tasks. The sequence may be dictated by what is found on the jobsite or by the customer's complaint. A service company might find that some tasks can be done in sequence and may institute a company policy to guide all service technicians. It is also possible for an individual technician to do the same thing.

Table 12–1 Outdoor Unit Checklist

Inspection Task	*Recommended Corrective Actions
Cabinet	
Shall inspect cabinet, cabinet fasteners, and cabinet panels.	Repair or replace insulation to ensure proper operation. Replace lost fasteners as needed to ensure proper integrity and fit/finish of equipment (as applicable). Seal air leaks.
Shall inspect the required clearance (e.g., service) around cabinet.	Record and report instances where the cabinet does not meet the requirements.
Electrical	
Shall inspect electrical disconnect box.	Ensure electrical connections are clean and tight. Ensure fused disconnects use the proper fuse size and are not bypassed. Ensure the case is intact and complete. Replace as necessary.
Shall ensure proper equipment grounding.	Tighten, correct, and repair as necessary.
Shall measure and record line voltage.	Compare to OEM specifications or equipment nameplate data. Notify homeowner and/or utility.
Shall inspect and test contactors and relays.	Look for pitting or other signs of damage. Replace contactors and relays demonstrating evidence of excessive contact arcing and pitting.
Shall inspect electrical connections and wire.	Ensure wire size and type match the load conditions. Tighten all loose connections, replace heat discolored connections, and repair or replace any damaged electrical wiring.
Shall inspect all stand alone capacitors.	Replace those that are bulged, split, incorrectly sized, or do not meet OEM specifications.
Shall measure and record amperage draw to motor/nameplate data (FLA) as available.	If outside OEM rating or specification, inspect for cause and repair as necessary.
Refrigeration	
Shall inspect accessible refrigerant lines, joints, components, and coils for oil leaks.	Test all oil-stained joints for leaks, clean or repair as necessary.
If indoor airflow is within OEM specifications but TD is not, shall measure and record system refrigeration charge, in COOLING mode.	Evaluate metering device for proper installation and operation then add or remove refrigerant as necessary.
Shall inspect refrigerant line insulation.	Repair or replace refrigerant line insulation.
Outdoor Fan Motor	
Shall confirm the fan blade has a tight connection to the fan motor shaft. Shall inspect fan for free rotation and minimal endplay. Measure and record amp draw.	Lubricate bearings as needed, only if recommended by OEM. If amp draw exceeds OEM specifications then adjust motor speed or otherwise remedy the cause. If due to motor failure recommend replacement of fan motor.
Outdoor Coil	
Shall inspect coil fins.	Ensure fins are clean, straight, and open. Clean, straighten, and repair as required.

*All corrective actions should be performed in accordance with the applicable original equipment manufacturer's (OEM's) instructions. Corrective actions that involve health and safety should follow the applicable building codes.

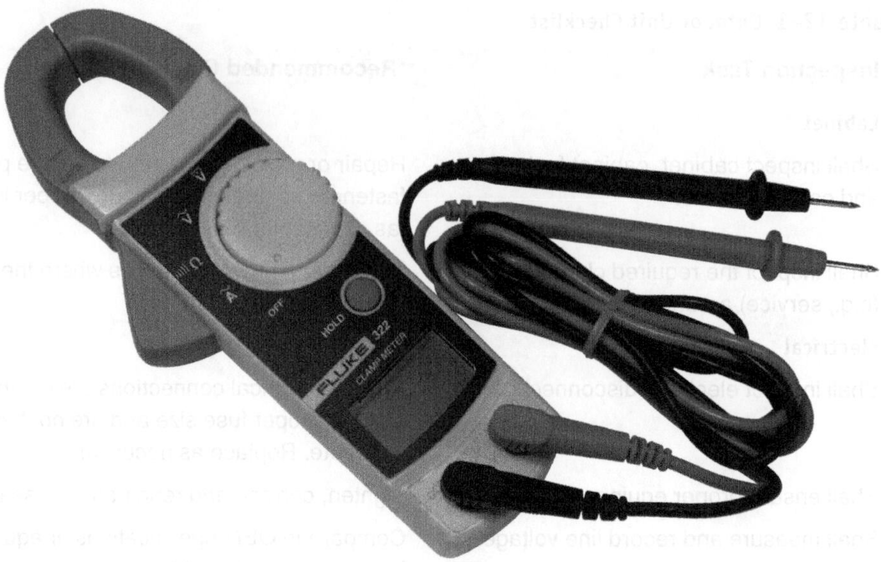

Figure 12–1
Measure and record amperage draw to and compare to the motor/nameplate data (full-load amperage [FLA]). (Courtesy of Delmar/Cengage Learning)

Figure 12–2
Inspect the heat pump outdoor unit. (Courtesy of iStock Photo)

Heat Pump Outdoor Units

Heat pumps have a few additional checks that should be done in addition to the checks that a typical outdoor air conditioning unit might require. These checks have to do with the function and components exclusive to heat pumps. Note that these checks are for air-source heat pump (ASHP) systems.

Table 12–2 Additional Tasks for Heat Pump Outdoors Checklist

Inspection Task	*Recommended Corrective Actions
Shall test reversing valve operation.	Record findings, repair or replace as necessary.
If indoor airflow is within OEM specifications but temperature difference (TD) is not, shall measure and record system refrigeration charge, in HEATING mode.	Evaluate metering device for proper installation and operation then trim charge to manufacturer's specifications.
Shall test defrost cycle controls.	Repair, replace, or adjust controls as needed.
Shall inspect outdoor unit condensate drain ports.	Ensure condensate drain ports are open and the unit is elevated above obstructions to allow free flow of condensate or per local code for seasonal obstructions like snow.

*All corrective actions should be performed in accordance with the applicable OEM's instructions. Corrective actions that involve health and safety should follow the applicable building codes.

Tech Tip

When the charge needs to be trimmed in the heating mode, it should be weighed in or the unit should be put in the cooling mode and charged in per the manufacturer's instructions.

Indoor Coil

The indoor coil of a heat pump is an evaporator for the cooling mode and a condenser coil for the heating mode. The following evaporator coil checks are directed to the indoor coil.

Table 12–3 Indoor Coil Checklist

Inspection Task	*Recommended Corrective Actions
Cabinet	
Shall inspect cabinet, cabinet fasteners, and cabinet panels.	Repair or replace insulation to ensure proper operation. Replace lost fasteners as needed to ensure proper integrity and fit/finish of equipment (as applicable). Seal air leaks.
Shall inspect the required clearance (e.g., service) around cabinet.	Record and report instances where the cabinet does not meet the requirements.
Condensate Removal	
Shall inspect condensate drain piping (and traps) for proper operation.	Clean, insulate, repair, or replace as necessary.
Shall inspect for condensate blowing from coil into cabinet or air distribution system (ADS).	Adjust fan speed, clean coil fins, ensure OEM-supplied deflectors are in place, or replace coil as necessary to eliminate water carryover.
Shall inspect drain pan and accessible drain line for biological growth.	Clean as needed to remove bio growth and ensure proper operation, add algaecide tablets or strips as necessary. Ensure algae tablets and cleaning agent are compatible with the fin and tube material.

(continued)

Table 12–3 Indoor Coil Checklist *(Continued)*

Inspection Task	*Recommended Corrective Actions
Refrigeration	
Shall measure and record TD across indoor coil.	Evaluate this measurement with airflow, refrigerant charge, and operating conditions.
Shall inspect coil fins.	Ensure fins are visibly clean, straight, and open. Clean and straighten as required.
Shall inspect accessible refrigerant lines, joints, components, and coils for oil leaks.	Test all oil-stained joints for leaks, clean or repair as necessary.
Shall inspect refrigerant line insulation.	Repair or replace refrigerant line insulation.
Measure pressure drop across the coil.	Adjust, clean, replace, and repair as necessary to ensure to proper airflow.

*All corrective actions should be performed in accordance with the applicable OEM's instructions. Corrective actions that involve health and safety should follow the applicable building codes.

Controls

Control operation of the heat pump system should be checked in the same fashion as other heating systems. ASHP systems require defrost. Defrost should be checked each time maintenance is done on the heat pump system.

Table 12–4 Controls and Safeties Checklist

Inspection Task	*Recommended Corrective Actions
Shall test modes of operation and control sequences. Shall test system control devices to ensure they are maintaining their expected range.	Repair or replace controls as needed to ensure proper operation.
Shall test zoning control's modes of operation, zone control to ensure proper damper/valve operation, and test bypass dampers for proper function.	Repair or replace components as needed to ensure proper operation.
Shall test remote control thermostat in all modes of operation.	Replace battery annually, check for corrosion on the battery contact points.
Shall initiate a test of the defrost control board's mode of operation, for those with that capability.	Repair, replace, or adjust controls as needed.
Shall test drain pan safety switch(es) for proper operation.	Repair wiring or replace safety switch as needed.
Shall test unit safety switch.	Repair wiring or replace safety switch as needed.

Air Distribution Systems

The air delivery system is a crucial part of a heat pump installation. Adequate airflow across the indoor coil is very important to maintain energy efficiency and adequate heat, and must be verified to ensure proper operation. Filters are of high importance because they are the cause of a significant number of unnecessary service calls.

Table 12–5 Air Distribution System Checklist

Inspection Task	*Recommended Corrective Actions
Shall inspect for particulate accumulation on filters.	Clean or replace filters if accumulation results in PD higher than design or if airflow is outside of established operating limits.
Shall inspect air filter housing integrity and air seal.	Correct as needed.
Shall inspect grilles, registers, and diffusers for dirt accumulation.	Clean as needed.
Shall inspect all accessible ductwork for areas of moisture accumulation or biological growth.	Install access doors as needed. Clean or replace as needed.
Shall inspect integrity of all accessible ductwork insulation.	Observe for proper alterations, rips, tears, or improper duct adhesives. Repair, seal, replace as necessary. Install access doors as needed.
Shall inspect the integrity of all accessible ductwork including: duct strapping, hangers, sections, joints, and seams.	Note improper alterations, straps, air leaks, and improper duct adhesives. Repair, seal, replace as necessary.

*All corrective actions should be performed in accordance with the applicable OEM's instructions. Corrective actions that involve health and safety should follow the applicable building codes.

Packaged Heat Pumps

Some heat pumps are packaged units rather than split systems. The following inspection tasks and recommended corrective actions can apply to heat pumps and general packaged units (Figure 12–3). Those tasks that are specific to packaged heat pumps follow the general list.

Figure 12–3
Inspect the ASHP packaged system. (Courtesy of Carrier Corporation)

Table 12–6 Package Units Checklist

Inspection Task	*Recommended Corrective Actions
Cabinet	
Shall inspect cabinet, cabinet fasteners, and cabinet panels.	Repair or replace insulation to ensure proper operation. Replace lost fasteners as needed to ensure proper integrity and fit/finish of equipment (as applicable). Seal air leaks on indoor air processing sections.
Shall inspect the required clearance (e.g., combustion and service) around cabinet.	Record and report instances where the cabinet does not meet the requirements.
Electrical	
Shall inspect electrical disconnect box.	Ensure electrical connections are clean and tight. Ensure fused disconnects use the proper fuse size and are not bypassed. Ensure case is intact and complete. Replace as necessary.
Shall ensure proper equipment grounding.	Tighten, correct and repair as necessary.
Shall measure and record line voltage.	Compare to OEM specifications or equipment nameplate data. Notify homeowner and/or utility.
Shall inspect and test contactors and relays.	Look for pitting or other signs of damage. Replace contactors and relays demonstrating evidence of excessive contact arcing and pitting.
Shall inspect electrical connections and wire.	Ensure wire size and type match the load conditions. Tighten all loose connections, replace heat-discolored connections, and repair or replace any damaged electrical wiring.
Shall inspect all stand-alone capacitors.	Replace those that are bulged, split, incorrectly sized, or do not meet OEM specifications.
Shall measure and record amperage draw to motor/nameplate data (FLA) as available.	If outside OEM rating or specification, inspect for cause and repair as necessary.
Indoor Blower Motor	
Shall determine and record airflow across heat exchanger/coil.	Verify all grilles and registers are open and free of obstruction. Adjust, clean, replace, and repair as necessary to ensure to proper airflow.
Shall test variable frequency drive (e.g., ECM) for proper operation.	Replace if necessary to ensure proper operation.
Shall inspect fan belt tension. Inspect belt and pulleys for wear and tear.	Repair or replace as necessary to ensure proper operation (if applicable).
Shall confirm the fan blade or blower wheel has a tight connection to the blower motor shaft. Shall inspect fan for free rotation and minimal endplay. Measure and record amp draw.	Lubricate bearings as needed, only if recommended by OEM. If amp draw exceeds OEM specifications then adjust motor speed or otherwise remedy the cause. If due to motor failure recommend replacement of blower motor.
Evaporator Coil Section	
Shall inspect coil fins.	Ensure fins are clean, straight, and open. Clean and straighten as required.
Shall inspect for condensate blowing from coil into cabinet or ADS.	Adjust fan speed, clean coil fins, or replace coil as necessary to eliminate water carryover.
Shall inspect accessible refrigerant connecting lines, joints, and coils for oil leaks.	Test all oil-stained joints for leaks; clean or repair as necessary.
Shall measure and record TD across evaporator coil.	Evaluate this measurement with airflow, refrigerant charge, and operating conditions.

Table 12–6 Package Units Checklist *(Continued)*

Inspection Task	*Recommended Corrective Actions
Condensate Removal	
Shall inspect for condensate blowing from coil into cabinet or ADS.	Adjust fan speed, clean coil fins, ensure OEM-supplied deflectors are in place, or replace coil as necessary to eliminate water carryover.
Shall inspect condensate drains (and traps) for proper operation.	Clean, insulate, repair, or replace as necessary.
Shall inspect drain pan and accessible drain line for biological growth.	Clean as needed to remove bio growth and ensure proper operation, add algae tablets or strips as necessary. Ensure algae tablets and cleaning agent are compatible with the fin and tube material.
Outdoor Blower Motor	
Shall confirm the fan blade or blower wheel has a tight connection to the blower motor shaft. Shall inspect fan for free rotation and minimal endplay. Measure and record amp draw.	Lubricate bearings as needed, only if recommended by OEM. If amp draw exceeds OEM specifications then adjust motor speed or otherwise remedy the cause. If due to motor failure recommend replacement of blower motor.
Outdoor Coil Section	
Shall inspect coil fins.	Ensure fins are clean, straight, and open. Clean and straighten as required.
Shall inspect accessible refrigerant connecting lines, joints, and coils for oil leaks.	Test all oil-stained joints for leaks; clean or repair as necessary.
Refrigeration	
Shall inspect accessible refrigerant connecting lines, joints, and coils for oil leaks.	Test all oil stains for leaks; clean or repair as necessary.
If indoor airflow is within OEM specifications but TD is not, shall measure and record system refrigeration charge, in COOLING mode.	Evaluate metering device for proper installation and operation then add or recover refrigerant as necessary.
Supplemental Electric Heaters	
Shall test electric heater's capacity and sequence of operation.	If outside OEM rating or sequencer specification, inspect for cause and repair as necessary.

*All corrective actions should be performed in accordance with the applicable OEM's instructions. Corrective actions that involve health and safety should follow the applicable building codes.

Table 12–7 Additional Tasks for Package Heat Pumps Checklist

Inspection Task	*Recommended Corrective Actions
Shall test reversing valve operation.	Record findings, repair or replace as necessary.
If indoor airflow is within OEM specifications but TD is not, shall measure and record system refrigeration charge, in HEATING mode.	Evaluate metering device for proper installation and operation then add or remove refrigerant as necessary.
Shall test defrost cycle controls.	Repair, replace, or adjust controls as needed.
Shall inspect outdoor section condensate drain ports.	Ensure condensate drain ports are open and elevated above obstructions to allow free flow of condensate or per local code for seasonal obstructions like snow.

Geothermal Heat Pumps

Geothermal heat pumps incorporate some of the same checks as other systems. However, ACCA saw fit to include them in one complete list of tasks. Some geothermal heat pumps have the ability to preheat water using a desuperheater. Additional tasks for hot water recovery follow the Geothermal list (Table 12–8).

Table 12–8 Geothermal Checklist

Inspection Task	*Recommended Corrective Actions
Cabinet	
Shall inspect cabinet, cabinet fasteners, and cabinet panels.	Repair or replace insulation to ensure proper operation. Replace lost fasteners as needed to ensure proper integrity and fit/finish of equipment (as applicable). Seal air leaks.
Shall inspect the required clearance (e.g., service) around cabinet.	Record and report instances where the cabinet does not meet the requirements.
Electrical	
Shall inspect electrical disconnect box.	Ensure electrical connections are clean and tight. Ensure fused disconnects use the proper fuse size and are not bypassed. Ensure case is intact and complete. Replace as necessary.
Shall ensure proper equipment grounding.	Tighten, correct and repair as necessary.
Shall measure and record line voltage.	Compare to OEM specifications or equipment nameplate data. Notify homeowner and/or utility.
Shall inspect and test contactors and relays.	Look for pitting or other signs of damage. Replace contactors and relays demonstrating evidence of excessive contact arcing and pitting.
Shall inspect electrical connections and wire.	Ensure wire size and type match the load conditions. Tighten all loose connections, replace heat discolored connections, and repair or replace any damaged electrical wiring.
Shall inspect all stand-alone capacitors.	Replace those that are bulged, split, incorrectly sized, or do not meet OEM specifications.
Shall measure and record amperage draw to motor/nameplate data (FLA) as available.	If outside OEM rating or specification, inspect for cause and repair as necessary.
Indoor Blower Motor	
Shall determine and record airflow across heat exchanger/coil.	Verify all grilles and registers are open and free of obstruction. Adjust, clean, replace, and repair as necessary to ensure to proper airflow.
Shall test variable-frequency drive (e.g., ECM) for proper operation.	Replace if necessary to ensure proper operation.
Shall inspect fan belt tension. Inspect belt and pulleys for wear and tear.	Repair or replace as necessary to ensure proper operation (if applicable).
Shall confirm the fan blade or blower wheel has a tight connection to the blower motor shaft. Shall inspect fan for free rotation and minimal endplay. Measure and record amp draw.	Lubricate bearings as needed, only if recommended by OEM. If amp draw exceeds OEM specifications then adjust motor speed or otherwise remedy the cause. If due to motor failure recommend replacement of blower motor.

Table 12–8: Geothermal Checklist *(Continued)*

Inspection Task	*Recommended Corrective Actions
Condensate Removal	
Shall inspect for condensate blowing from coil into cabinet or ADS.	Adjust fan speed, clean coil fins, ensure OEM-supplied deflectors are in place, or replace coil as necessary to eliminate water carryover.
Shall inspect condensate drain piping (and traps) for proper operation.	Clean, insulate, repair, or replace as necessary.
Shall inspect drain pan and accessible drain line for biological growth.	Clean as needed to remove bio growth and ensure proper operation, add algae tablets or strips as necessary. Ensure algae tablets and cleaning agent are compatible with the fin and tube material.
Evaporator Coil Section	
Shall inspect coil fins.	Ensure fins are straight and open. Clean and straighten as required.
Shall inspect for condensate blowing from coil into cabinet or ADS.	Adjust fan speed, clean coil fins, or replace coil as necessary to eliminate water carryover.
Shall measure and record TD across indoor coil.	Evaluate this measurement with airflow, refrigerant charge, and operating conditions.
Refrigeration	
Shall inspect accessible refrigerant connecting lines, joints, and coils for oil leaks.	Test all oil-stained joints for leaks; clean or repair as necessary.
If indoor airflow is within OEM specifications but TD is not, shall measure and record system refrigeration charge, in COOLING mode.	Evaluate metering device for proper installation and operation, then add or recover refrigerant as necessary.
Shall test reversing valve operation.	Record findings; repair or replace as necessary.
If indoor airflow is within OEM specifications but TD is not, shall measure and record system refrigeration charge, in HEATING mode.	Evaluate metering device for proper installation and operation, then add or remove refrigerant as necessary.
Source Loop	
Shall test pressure of the loop without the unit operating.	Add solution or water to meet industry standards.
Should test closed-loop solution for antifreeze concentration.	Add appropriate antifreeze if needed.
Hydronic Loop	
Shall inspect water pump.	Clean or clear as needed to reduce cavitation and ensure proper operation.
Shall measure and record TD of water entering to water leaving coil/ heat exchanger. BTUH = 500*GPM*TD	Add or remove refrigerant as necessary.
Shall measure and record pressure drop (PD) of the water loop across the refrigerant water heat exchanger.	Adjust water pump or control valve as necessary.
Shall inspect screen on reducing valve, pressure reducing valve, and "Y" strainer if available.	Clean or replace as necessary.
Shall test bladder expansion tank for proper air cushion or proper air cushion on expansion tank if applicable.	Adjust to provide proper air cushion on expansion tank as per manufacturer's specifications.

*All corrective actions should be performed in accordance with the applicable OEM's instructions. Corrective actions that involve health and safety should follow the applicable building codes.

Table 12–9 Additional Tasks for Hot Water Recovery Checklist

Inspection Task	*Recommended Corrective Actions
Shall measure and record amperage to domestic hot water (DHW) heat recovery pump.	If outside OEM rating or specification, inspect for cause and repair as necessary.
Shall measure and record TD of water entering and leaving DHW at the heat recovery pump.	Check for improper plumbing or insulation of DHW lines if the water temperature exceeds OEM specifications or local codes.
Shall measure resistance of 120°F water temperature limit switch.	Replace if shorted or out of OEM specifications.

*All corrective actions should be performed in accordance with the applicable OEM's instructions. Corrective actions that involve health and safety should follow the applicable building codes.

QUALITY INSTALLATION STANDARDS

ACCA has developed Quality Installation (QI) guidelines that are recognized by ANSI. The following list is excerpted from those standards and is provided as an insight to the standard. It is recommended that the standard be downloaded from the ACCA website. For that reason, only the topic headers are featured below. It should be noted that if each of the following headers and subheaders are followed, a quality system installation will result. Eliminating or neglecting any part of these may cause the system installation to be flawed.

- 3.0 Equipment Aspects
  - 3.1 Building Heat Gain/Loss Load Calculations
  - 3.2 Proper Equipment Capacity Selection
  - 3.3 Matched Systems
- 4.0 Equipment Installation Aspects
  - 4.1 Airflow across Indoor Heat Exchangers
  - 4.2 Refrigerant Charge
  - 4.3 Electrical Requirements
  - 4.4 On-Rate for Fuel-Fired Equipment
  - 4.5 Combustion Venting System
  - 4.6 System Controls
- 5.0 Duct Distribution Aspects
  - 5.1 Duct Leakage
  - 5.2 Airflow Balance
- 6.0 System Documentation and Owner Education Aspects
  - 6.1 Proper System Documentation to the Owner
  - 6.2 Owner/Operator Education

SUMMARY

In this chapter we have discussed QM and QI standards as they apply to heat pumps. We observed that many general checks for standard systems also apply to heat pumps. Where additional checks for heat pumps are required, they are included in additional tables. We also noted that to have a good maintenance program, there needs to be a good installation. Without a good installation, additional work would be required to fix or bring the system up to standard before the quality maintenance program could continue.

REVIEW QUESTIONS

1. Describe the need for standards.
2. Explain how standards relate to heat pump maintenance.
3. Discuss how the general inspection tasks relate to heat pump outdoor units.
4. Describe the difference between Quality Maintenance (QM) and Quality Installation (QI) standards.
5. Describe what might happen if system controls were neglected in the QI standards.

CHAPTER 13

Troubleshooting ASHPs

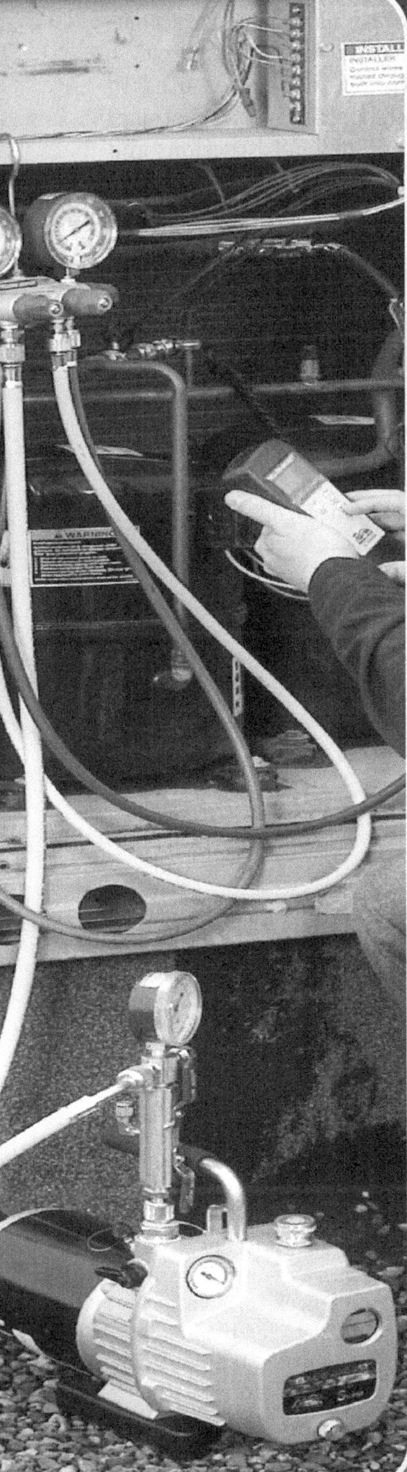

LEARNING OBJECTIVES

The student will:

- Describe the principles of troubleshooting
- Describe how symptoms lead to possible causes
- Relate what is to be done with each cause
- Describe the "ruling-out" process
- Explain how verification of a cause is performed
- Describe the difference between mechanical and electrical troubleshooting

INTRODUCTION

Troubleshooting is a skill that is developed over time and an analytical process of systematic problem solving. Additionally, there is no standard troubleshooting process for electromechanical devices. There are, however, industry-accepted practices and procedures for determining if electrical or mechanical devices are functioning properly. Sometimes, the manufacturer requires the device to be checked for operation or nonoperation in a certain way.

Troubleshooting skill is also directly related to familiarity with the unit, components, applications, and electrical/mechanical sequence of operation. A person that is skilled in troubleshooting one type of system may be able to troubleshoot another system, but would take significantly more time to accomplish the task as compared to a person who was very familiar with the system.

This chapter will explore a few scenarios where troubleshooting is used to determine the possible cause of symptoms that have been listed from the customer complaint, discussion with the customer, and observation of what functions and what does not function relative to system operation.

Field Problem

The ASHP was 5 years old and having scheduled maintenance performed. It had been giving great service and the customer was very happy with the system. It was a split system with a fossil fuel back-up. The customer noted that the back-up system operated very little and the fuel bills were well within the original fuel estimates each year.

Symptoms:
None

Possible Causes:
None

While the system check was being done, the technician conducted a voltage-drop test across the contactor contacts. The voltmeter showed 3 volts less on the load side than the applied line voltage. Turning off the system, the technician removed the contactor for a visual verification of the problem. Noting severely pitted contacts, the technician discovered discolored contacts, something that can cause voltage drop. One contact was so deformed that the technician wondered how it conducted electricity at all. Speaking with the customer, the technician suggested that the contactor be replaced and the customer concurred. The technician installed a new contactor and finished the system checks, giving the ASHP a clean bill of health.

Tech Tip

When inspecting contacts, keep in mind that they may be discolored just due to age or operating cycles, but other causes should be considered that may cause contact overheating. Improper control voltage is one of the possible causes—too low and the contactor may chatter, causing excessive arcing; too high and the contactor coil may overheat and swell, causing the contactor mechanism to bind in a partially open or shut position (insects stuck in the works of the contactor have also been known to cause this); and worn contacts in the control circuit may have excessive voltage drop, causing the applied control voltage to the contactor to be low (i.e., relay contacts, thermostat contacts, pressure control contacts, etc.) Also check the control wiring for pinching, crimping, loose wire nuts or connectors, or anything else that may cause an intermittent connection (check for damaged insulation, especially between the building and the outdoor unit, as the plastic insulation may be deteriorated by the

Tech Tip (Continued)

weather exposing the conductors; check for chewed-up insulation, as rodents like to chew on this). Other things to consider include hard-starting situations; remember that all motors draw locked-rotor amps the instant they start, but if they do not come up to speed quickly as designed, this may also cause contact arc to be excessive. Check other starting components, wiring, and connections (i.e., starting relays, capacitors, etc.) as well as the motor itself (i.e., check for binding or worn bearings, endplay on fans, etc.)

PRINCIPLES OF TROUBLESHOOTING

There are some basic aspects of troubleshooting that are sometimes referred to as principles. These aspects involve gathering enough information to make judgments about the system. An experienced technician will take this information and list the symptoms of the problem from the customer complaint and discussion, and, with an understanding of the electrical/mechanical sequence of the system, list the possible causes. Generally, the principles of troubleshooting can be listed as steps:

1. Obtaining information about the problem by listening to the customer
2. Obtaining information about the system operation
3. Determining the symptoms
4. Developing a list of possible causes for the symptoms
5. Ruling out possible causes
6. Identifying the problem
7. Verifying the problem
8. Correcting the problem
9. Verifying that the problem is corrected and system operation is normal (take a full set of readings)

Obtaining Troubleshooting Information from the Customer

Most of the important information comes from the customer. The technician is required to develop his or her customer relations technique as it applies to obtaining information from the customer. Let the customer describe the condition and the operation of the system as you listen. Clarify what was said by repeating what is heard and letting the customer verify. Listen to what the customer wants or needs. Empathize by feeling what the customer is saying.

For example, the customer may tell the technician the following in a service call related to the heating mode: "The system is always running. Sometimes the emergency heat light comes on. It never seems to stay warm."

OBTAINING INFORMATION ABOUT THE SYSTEM OPERATION

A complete set of system performance measurements should be made. This is a record of where the technician started or how the system was operating when the technician arrived. The technician should take temperatures, airflow measurements, and electrical measurements. With these measurements, the technician obtains critical information that points out symptoms exhibited by the system. These symptoms can be compared to customer information, which will verify or eliminate information provided by the customer. The customer is not expected to be a knowledgeable expert, even with his or her own system. The technician must attempt to confirm or reject information provided by the customer that will help in diagnosing the system.

For our example in the heating mode, the technician takes system measurements and finds that the airflow is low, the indoor airflow is low, supply-air temperatures are higher than expected, indoor coil temperatures are higher, and outdoor conditions are above the balance point, which means that the heat pump should adequately carry the load.

Determine the Symptoms

Every complaint has a set of symptoms. Most of these come from the customer's explanation of the problem and the technician's understanding of system operation and the electrical/mechanical sequence. It is important for the technician to record these symptoms because they form the basis for the customer complaint, and each must be resolved in order to satisfy the customer. Determine what works and what doesn't: Is the thermostat calling for heating or cooling? Is the indoor blower operating? Is the outdoor fan operating? What works and what doesn't? Symptoms point to possible causes, and causes can be checked. With an understanding of the electrical/mechanical sequence, we can rule out possible causes.

An example of a symptom might be that indoor temperature is dropping to the auxiliary thermostat setting and the auxiliary electric heat strips are operating. Also, low air discharge at the diffuser is another symptom. Low indoor temperature and low airflow may point to several causes, and each cause must be checked in order to come to some troubleshooting solution.

Developing a List of Possible Causes

Each symptom has a possible cause. Some symptoms may have several causes. Our example system has the symptoms of low indoor temperature and low air discharge, indoor coil temperature high, and discharge temperature of the indoor coil high.

The possible causes are:

- Duct blockage
- Indoor coil blockage
- Dirty filters
- Leaky or disconnected ductwork
- Defective fan motor
- Wrong fan motor speed tap selected
- Loose duct insulation
- Dirty indoor coil
- Improper duct sizing or installation

All of the possible causes for a particular symptom should be noted. A list of possible causes may be generated by mental notation when done by an experienced technician. As a person learns troubleshooting, it is suggested that a written list be generated and checked off as the troubleshooting procedure continues.

Ruling Out Possible Causes

As each cause is checked, the technician must be able to verify or rule out the possible cause as the reason for the system condition. Ruling out is an important step in the process and eliminates potential causes until the cause of the problem can be determined. The cause leads to identifying the problem and eventually a repair.

In our example, the technician would need to start with one of the possible causes, investigate it, and either determine that it was a cause or that it was not—ruling out the possible cause. The investigation should start with the easiest check. In this case, the filter would be examined. If it was determined that the filter was plugged, a new filter would be inserted and the technician would check (verify) that the system was operating properly by taking a new set of measurements. If the filter was not sufficiently dirty, the coil would be inspected and, in turn, the fan motor, ducting system, and so forth.

Identifying the Problem

When a possible cause cannot be ruled out, it becomes a troubleshooting problem (or one of the problems) that needs to be fixed or repaired. The technician usually will identify one problem that relates to a possible cause; however, there may be more than one system problem. As each problem is discovered, it is noted and marked for service or repair.

As mentioned in our example, if the filter was determined to be the cause, the technician would replace the filter and take a new set of readings. If not, the technician would continue the investigation (troubleshooting process) until the problem was found. If the filter was good, the technician would continue down the list of possible causes, ruling out the filter and checking the next cause on the list. Let us say that the filter was plugged, but not with just dirt. In this case, the technician found that the filter was plugged with duct insulation. With this information, the technician might not stop with the filter as being the problem. Instead, the investigation would turn to the condition of the duct liner and a complete inspection of the interior of the duct system would be done.

Verifying the Problem

After the problem is identified, the technician must verify that the problem exists by conducting more than one test. If the problem can be verified by a second test, the problem has been properly identified. If the problem cannot be verified with a second test, the diagnosis must be suspect. This means that the initial identification of the problem was incorrect and that there is another problem that is causing the symptom.

Let us say that in our example, the technician determines that the fan motor is the problem. The motor is noisy, indicating that it may have worn bearings. Verifying that the bearings are worn requires that the technician check for bearing play (excessive movement of the motor shaft in the bearing). Both the motor noise and excessive bearing play are the first and second test or verification of the problem. When the technician replaces the motor and conducts another airflow test, it should show that the airflow is now at specified volumes.

Verification does not stop with the problem being identified. In many cases, the troubleshooting process begins again in an attempt to discover any other problems the system may have. A good troubleshooter generally follows these nine steps, but develops his or her own method. This means that some technicians may repeat the process from the beginning, while others will go back a step or two to continue with the troubleshooting of a particular system.

Relating a set of symptoms to a set of possible causes helps to connect what a system is doing to what the problem might be. It is important to remember that there may be a number of possible causes for each general symptom, and it is the responsibility of the technician to rule out or identify and verify each possible cause.

At the end of the verification process, a full set of readings is recorded to document system performance. These readings can be compared with the initial set of readings to show system changes that result from repairs. These records are shared with the customer and placed in the customer file for future reference.

TROUBLESHOOT PROBLEM 1

The service order indicated that the air-source heat pump (ASHP) was not cooling properly. The technician greeted the customer, introduced himself, and let the customer describe the problem. The customer explained that only one room in the house was experiencing cooling problems; the other rooms were fine.

Symptoms: Poor cooling as reported by the service order and verified by the customer, but for only one room; the outdoor unit was running; indoor blower operating in "auto" with thermostat calling for cooling.

Possible Causes: Low airflow; mechanical refrigeration issues; ductwork problems.

Checking each room, the technician found that it was warm in the kitchen, compared to the rest of the rooms in the house. The technician also noticed that the thermostat was located in a hallway just off the living room. The living room was between the kitchen and the thermostat. Making a record of these readings, the technician continued to record measurements of the discharge- and return-air temperature, indoor coil temperature, outdoor temperature, and outside coil temperature until there was a complete record of initial readings.

During the process of taking a full set of readings, the technician made note that the insulation on the exposed refrigerant lines was missing and that the system air filter needed to be replaced. After recording these things, the technician returned to the room that was warm to determine if airflow to the room was the problem. In the kitchen there were two floor diffusers. One diffuser was mostly open and the other was closed. The technician determined that this could be the cause of the warm room problem. Checking the other rooms, the diffusers were all completely open. Opening both kitchen diffusers created an immediate drop in temperature. He replaced the system air filter, insulated the suction line that was missing insulation, and performed a system check to verify proper system operation.

TROUBLESHOOT PROBLEM 2

The heat pump was not cooling. The customer explained that the system stopped the day before and she called the service company. It had been working fine during the hot days and quit on a cooler day.

Symptoms: No cooling; thermostat set correctly; indoor blower operating; power was on.

Possible Causes: Compressor problems; control circuit problems; safety lock-out; mechanical/refrigeration problems.

The technician checked the thermostat and found that the blower would operate when switched to manual—what does it do in "auto"? This ruled out the power for both line and low voltage for the indoor blower and low voltage to the thermostat. Going to the outdoor unit, the technician removed the service panel and verified line voltage to the contactor Next, the technician checked the load side of the contactor and found no line voltage. Then the technician verified that the contactor was also receiving a normal control voltage signal. Shutting off the power and removing the contactor, the technician checked the operator (coil) for continuity and found none. Satisfied that he had identified the cause and verified that contactor coil was open, the technician discussed the repair option with the customer and replaced the contactor. After replacing the contactor and checking for proper contactor operation, the technician took another full set of readings to verify proper system operation and to log system conditions.

TROUBLESHOOT PROBLEM 3

The air-source heat pump with electric back-up heat was not doing a good job of heating, as reported by the customer. It had slowly been getting worse, so a service call was made. The supplemental heat had not come on, but the customer was sure that it was not keeping up.

Symptoms: Room temperature was 4°F below setpoint; supplemental heaters were not on; outside air temperature was low and below thermal balance point (first stage of electric supplemental heat should be on); air-source heat pump outdoor unit and indoor unit are operating.

Possible Causes: Inefficient compressor; compressor short-cycling; long defrost; inoperable back-up strip heat; insufficient capacity; ductwork problems; defrost problem; slow-running outdoor fan; blocked outdoor coil; worn bearings in outdoor fan.

The technician used an accurate thermometer to measure the outdoor air temperature close to the outdoor thermostat. The outdoor temperature was measured at 20°F. The balance point, as noted on the outdoor thermostat by the installing technician, was 26°F. The technician verified that the outdoor thermostat was set at 26°F. The back-up heat should have been turning on. Checking the outdoor thermostat, it was set for 26°F. Moving the temperature dial to a higher temperature eventually turned on the back-up heat. Resetting the outdoor thermostat to 26°F and waiting 15 minutes, the technician noted that the supplemental heat had turned off and not come back on. The outdoor thermostat was defective or out of calibration and needed to be replaced. After the system was operating at a steady state, a system check was performed to verify proper operation.

TROUBLESHOOT PROBLEM 4

No heat was the complaint for a commercial ASHP system. Neither the supplemental heat nor the heat pump was working.

Symptoms: No heat; breakers were on; building power was on; indoor blower not operating either in auto or on positions.

Possible Causes: Bad breakers; blown fuses.

Going to the service panel next to the indoor coil, the service technician verified line power to the control board. Low voltage was not present. Checking the transformer, there was a reading of 24 volts. Tracing the wiring, a fuse was blown. Knowing that a fuse only blows for a reason, the technician checked for shorts, but did not find a short. Installing another fuse allowed the system to operate. Shutting off power again, the technician started a systematic visual inspection to uncover the problems. As the low-voltage wires wrapped around the control board, they ran against a metal portion of the system cabinet. There, one wire had rubbed through the insulation to the wire, and ground out on the cabinet. By repairing the wire, the technician placed the system back into service.

All possible causes should be checked. The technician should not stop at the first possible cause that he or she finds or discovers. All possible causes must be investigated. Usually, there is only one cause for a symptom, but sometimes there can be more. It is the technician's responsibility to check all possible causes.

TROUBLESHOOT PROBLEM 5

The ASHP was intermittently cooling. The owner would find the system locked out and reset the system at the thermostat. The system would operate well for a varied length of time and then lock out again.

Safety Tip

When dealing with line-voltage power before the disconnect, extreme caution needs to be taken to ensure that power is off. Both lock-out/tag-out procedures and arc flash precautions should be followed.

Symptoms: Intermittent lock-out.

Possible Causes: Faulty wiring or connections; contactor problems; limit controls causing short-cycling; insufficient airflow; possible system charge problem.

The technician arrived during a time when the system was operating. All system conditions seemed to be within specifications. Checking across the contactor and then from the line connection to the contactor showed a voltage drop. The voltage drop was not across the contactor, but from the disconnect to the contactor. Shutting off power at the main panel, the technician investigated the service disconnect. When the inner panel wires were moved, they fell off the terminals. The terminals were arced and contact surfaces burned. The disconnect and all associated wiring that was burned or showed signs of overheating were replaced and the intermittent problem was solved.

TROUBLESHOOT PROBLEM 6

The ASHP was operating on emergency heat. The owner indicated that the emergency heat light had been on for the last day and the outside coil was iced over, and that the outdoor fan and indoor fan were running.

Symptoms: Emergency heat light is on; outside coil iced; compressor and fans operating.

Possible Causes: Drainage problems; defrost problems; system charge; airflow problems.

The technician opened the service panel of the outdoor unit and found that the compressor and outdoor fan were operating. Inside, the system was calling for defrost, but the reversing valve was not operating. A small screwdriver confirmed that the solenoid was energized (was magnetic). Power was going to the valve, which was correct for defrost, and defrost was not occurring. Operating the reversing valve solenoid three times made the solenoid move and the system went from heating to defrost mode (cooling position for the reversing valve)—the valve was intermittently sticking. Knowing it would happen again, the technician explained the problem to the customer and discussed replacement considerations.

TROUBLESHOOT PROBLEM 7

The ASHP was operating and the outdoor coil was mostly covered with frost. The customer complained that the supplemental heat was staying on too long during defrost and defrost was occurring too often.

Symptoms: Frequent defrost cycles.

Possible Causes: Defrost sensors; defrost control board; outside coil drainage; system charge; airflow problems

The technician checked the system and then placed the system into defrost manually and monitored the operation; defrost occurred and operated properly. Outside the coil cleared at the top, but the bottom ice did not. The technician noticed that there was nowhere for the water to drain easily. Using a shovel, he created a small hole in the ground to drain the water away from the coil. The outside temperatures had been below freezing, but that was unusual and a warmer period was coming. The technician alerted the customer to the problem and discussed several ways to eliminate the problem. The technician would return with another or taller pad to set the outdoor unit up higher. The drainage hole would work until the weather broke and the new pad could be installed.

MECHANICAL

ASHPs are very similar to split-system air conditioning. Many troubleshooting steps are similar in both of these systems. If the technician is very familiar with air conditioning, ASHPs should only take a little more time to learn.

Mechanically, an ASHP has several additional components compared to an air conditioning system including:

1. Reversing valve
2. Bidirectional filter drier
3. Two metering devices or one device that can operate both ways
4. Defrost cycle
5. Back-up or emergency heating source (electric, propane, natural gas, or oil)
6. Accumulator (sometimes found on air conditioning systems)
7. Requirement for outdoor drainage

The rest of the mechanical system is the same:

1. Compressor
2. Indoor heat exchanger
3. Outdoor heat exchanger
4. Interconnecting refrigerant piping
5. Outside cabinet or enclosure
6. Outdoor fan
7. Indoor blower
8. Indoor filter

The following are mechanical symptoms that could occur and related possible causes for common ASHP problems. Start with the symptoms and read for possible causes. Each of the possible causes needs to be ruled out or confirmed (see troubleshooting chart, Table 13-1).

The mechanical system includes the refrigeration system. The chart in Table 13-2 can be used to determine the cause for refrigeration problems using the symptoms or conditions of the system.

TROUBLESHOOT PROBLEM 8

The heat pump had been cooling until just recently, when the amount of cooling decreased and running time increased. The customer further indicated that high-pitched squealing noises were also coming from the outside unit.

Symptoms: Insufficient cooling; unusual noises at the outside unit; indoor fan operating.

Possible Causes: Condenser fan motor; refrigerant charge low; reversing valve leaking; insufficient airflow.

The technician opened the service panel of the outdoor unit and switched the system on and off (waiting the correct period of time to equalize pressures). When the system was turned on, the condenser fan made squealing noises and slowly started. Shutting down the power, the technician moved the condenser fan by hand and noticed that it did not move easily. Checking further, the fan bearings had gotten hot and were seizing. After discussing the need for a fan motor change-out with the customer, the motor was replaced and the system was placed back into service.

Table 13–1 Troubleshooting

Symptoms	Possible Electrical Causes	Possible Mechanical Causes
Compressor not running (contactor pulled in; compressor is cold to the touch)	Compressor burned out Failed open internal overload Broken wires between contactor and compressor	Mechanical problems that cause electrical to fail
Compressor hums but will not start Compressor operates intermittently	Run capacitor Start capacitor Start relay Hard-start kit Loose terminals Pitted contactor contacts Loose terminals or bad contact Single-phasing of a three-phase unit	Compressor mechanically seized or bearing surface is bad
Compressor running all of the time (frosty or icy)	Contacts are welded closed and cannot open Bad thermostat (shorted control wiring or improper setting)	Contactor mechanically stuck closed Thermostat stuck closed Airflow restriction Defrost control malfunctioning Restricted suction line Low system charge
Compressor running all of the time (hot to the touch)	Shorted control wiring	Low refrigerant charge, partial restriction, or malfunctioning metering device causing high superheat High load condition Airflow restriction at (condenser) coil
Compressor cycles on overload	Low voltage (high amperage) Bad condenser fan motor Compressor windings are defective Compressor overload is defective Compressor internal problems	Compressor internal problems Low charge Liquid-line restriction Dirty coils Low suction pressure Dirty outdoor coil (cooling) Low air volume Filter plugged (heating)
Compressor is noisy	Start relay does not open	Bad compressor mount Liquid slugging (low superheat) Refrigerant charge is too high Defrost system malfunctioning Compressor internal problems Refrigerant lines touching or rattling
Insufficient system capacity	N/A	Reversing valve leaking Incorrect refrigerant charge Insufficient airflow (indoor or outdoor coil) Unusual load conditions Structure addition or change Ductwork leaking (see Figure 13–1) Outdoor balance point thermostat set wrong Inefficient compressor
Defrost system malfunction	Defective defrost sensor or control Control wiring disconnected Reversing valve solenoid is bad	Sticky or stuck reversing valve

Table 13–1 Troubleshooting (*Continued*)

Symptoms	Possible Electrical Causes	Possible Mechanical Causes
Outdoor fan motor does not run	Run capacitor Wiring or fan motor windings defective or open Failed defrost control or defrost relay	Bearings are seized or worn Fan blade is impinged and cannot turn Something has fallen into the motor or fan and is preventing the shaft from turning
Low indoor air volume	Wrong blower motor speed tap Wrong blower motor (previous replacement)	Improperly designed or installed ductwork Duct blockage (air delivery problems) Dirty/plugged coil Blower motor problems Speed selection Filter plugged Duct connections loose Duct leaks
Indoor blower does not run	Run capacitor is bad Wiring or fan motor windings defective or open Blower relay not pulling in	Bearings are seized or worn Fan belt slipping or broken Motor drive coupling is broken Blower shaft is scored or broken Blower wheel is impinged and will not turn
Indoor coil is iced (indoor temperature is too high)	Blower motor is burned out Indoor coil freeze protection thermostat failed or improperly located (if equipped)	Refrigerant is low (cooling mode) Low air flow Refrigerant restriction
Outdoor coil is iced	Fan motor operating while in defrost	Defrost system malfunction Improper drainage Low charge Defrost cycles are too short Blocked air flow (heating mode) Malfunctioning reversing valve
Indoor humidity too high	Economizer control board	Incorrect equipment design or equipment selection Refrigerant charge is incorrect Expansion valve out of adjustment Short cycling Capacity is too great Blower speed too high Duct leakage Inefficient compressor Building pressurization problem (too negative) Structural drainage problems
Emergency/supplemental heat is on	Refrigerant system lock out or malfunction Defrost system is stuck on	Refrigerant system lock out or malfunction Outdoor balance point thermostat set wrong Defrost system is stuck on Outdoor temperatures too low

Table 13–2 Refrigeration Troubleshooting Table

Symptoms						
Amp Draw	**Suction Pressure**	**Head Pressure**	**Super- Heat 1 = Fix Bore 2 = TXV**	**Sub cooling**	**Indoor Air Temp. Diff.**	**Possible Cause**
Low	Low	Low	High 1&2	Norm	Low	Suction line restriction between service port and coil
Low	Normal/ High	Low	High1&2	Norm	Low	Suction line restriction between service port at compressor and metering valve
Normal/ Low	Low	Low	High 1&2	High	Low	Plugged orifice. Metering valve starving. Liquid-line restriction
Normal/ Low	High	Low	Low 1&2	Norm/Low	Low	Evaporator orifice oversized or bypassing; or metering valve overfeeding—normal charge
High	High	High	High 1&2	Normal	Low	Hot vapor discharge. Line restriction
Low	High	Low	High 1&2	Low	Low	Inefficient compressor
Low	Low	Low	Low 1 Normal 2	Low	High	Insufficient evaporator airflow
Normal/ Low	Normal/ High	Depends on Load	Low 1 Normal 2	High	Low	Refrigerant overcharge
Low	Low	Low	High 1&2	Low	Low	Insufficient charge
High	High	High	High 1 Normal 2	Low	Low	High evaporator load—latent and sensible
High	High	High	Low 1 Normal 2	Low	Low	High-temperature ambient air entering condenser. Dirty condenser. Low condenser airflow

Note: This chart is a general chart of conditions that will depend on the metering device and system application. The chart should not be used for all systems and all applications.

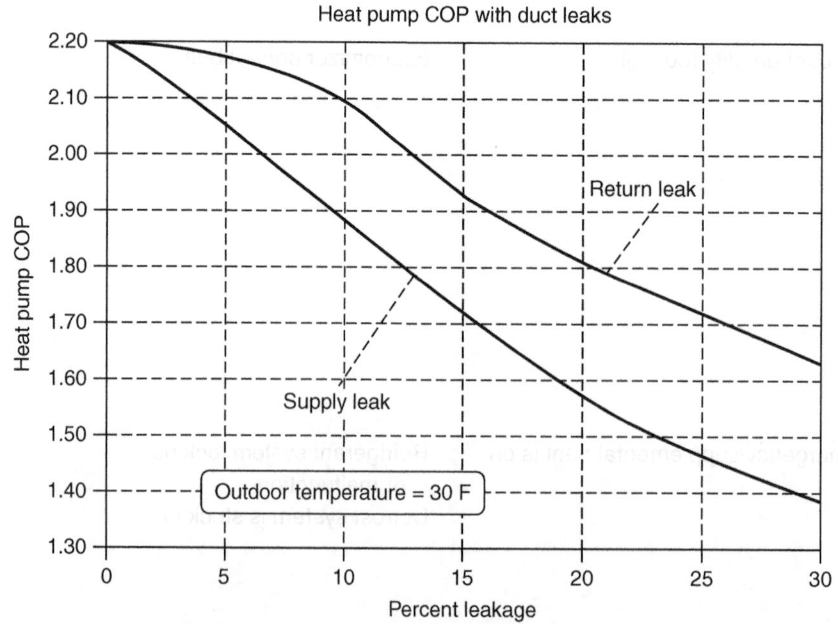

Figure 13–1
This graph shows the effect of duct leakage on the coefficient of performance (COP) of the heat pump. Efficiency decreases as duct leakage increases. (Courtesy of ACCA)

ELECTRICAL

Electrically, an ASHP is very similar to a split-system air conditioning system with a few additional components: a defrost system, reversing valve, balance point thermostat, thermostat capable of operating back-up heat, and back-up heating control. The defrost system is primarily electrical and incorporates a defrost timer or demand sensor system, defrost sensors on the outdoor coil, and a means of controlling the reversing valve while turning on back-up heat. The defrost system is operated intermittently or as needed during the heating mode. If it is not working, the outdoor coil ices and the output drops. If it works when not necessary, the owner will experience higher electrical bills because the back-up heat comes on too often.

After conducting a general check using the senses of touch, smell, hearing, and vision, the technician may have made some mechanical observations. When those observations require electrical diagnosis, test meters must be used. Electrical problems account for 70% or more of the problems associated with ASHPs. Notice that some symptoms in the electrical table are the same symptoms as found in the mechanical table (see troubleshooting chart, Table 13–1).

TROUBLESHOOT PROBLEM 9

The heat pump had been cooling until it quit. The customer reported that the outdoor unit buzzes and then stops, but there is no cooling.

Symptoms: No cooling; compressor hums (buzzes) but doesn't start; indoor blower running.

Possible Causes: Compressor internal mechanical problems; bad overload; motor start devices are bad; low line voltage

The technician checked line voltage and found it to be correct (see Table 13–1). Turning the power off and checking the compressor winding, the technician found continuity, proving the windings were not open or shorted. The compressor was hot. Checking the run capacitor, he found an open capacitor. Discussing the repair with the customer, the technician also suggested replacing the overload. With the run capacitor and the overload replaced, the system was back in operation. In all cases the compressor should not be condemned unless the compressor windings are checked when the compressor is cool to the touch.

TROUBLESHOOT PROBLEM 10

The heat pump was short-cycling and not cooling. There was no air coming from the floor diffusers, according to the owner.

Symptoms: Blower not running; compressor running but short-cycling.

Possible Causes: Blower motor problems; control system problems.

The technician attempted to operate the blower at the thermostat, but the blower would not operate on the manual setting. Going to the indoor section, the technician assumed there was normal control voltage, because the compressor was operating. Operating the fan by closing R to G at the system terminal board did not bring the blower on. Checking the power terminals for the blower, the technician did not find power going from the blower relay to the blower. Control voltage was going to the relay. Pulling the relay from its socket, the technician confirmed that the coil was intact. The relay contacts were not closing. After discussing the repair with the customer and replacing the relay, the technician operated the blower and again checked for proper voltage and amp draw of the blower motor (blower motor problems may have been the cause of the failed relay contacts).

SHOFFNER MECHANICAL SERVICES

TROUBLE SHOOTING CHECK LIST

REQUIRED MEASUREMENTS

Hi-Pressure (H) _______ Saturation temperature (1) _______

Lo-Pressure (L) _______ Saturation temperature (2) _______

Liquid line temperature (3) _______ (at condenser outlet)

Suction line temperature (4) _______ (at evaporater outlet)

Condenser ambient air temperature (5) _______

Evaporater air in temperature (6) _______

Evaporater air out temperature (7) _______

Compressor running amps (8) _______ Rated amps (9) _______

CALCULATIONS

Subcooling	(1) _______	–	(3) _______	=	(SC) _______
Superheat	(4) _______	–	(2) _______	=	(SH) _______
Delta "T"	(6) _______	–	(7) _______	=	(DT) _______

NORMAL OPERATING CONDITIONS

Subcooling	5 - 20 degrees
Superheat	8 - 12 degrees
Delta "T"	15 - 20 Degrees

Liquid line temperature (3) should never be lower than
ambient air temperature (5) (unless the reading is taken after an external subcooler)

CONDITIONS	PROBLEMS
If (SC) > 30	Overcharged or restricted
If (SC) < 0	Undercharged
If (SH) > 12	Restricted or TXV out of adjustment
If (SH) < 8	Flooding or TXV out of adjustment
If (DT) > 20	Restricted air flow
If (DT) < 15	Too much airflow, undercharged, freon restriction, high humidity level or low incoming air temperature.

If (H) is low and (L) is high and amperage is below rated then the compressor may have bad valves, pumpdown compressor to verify. It should pull down below 5" of vacuum.

If liquid temp. (3) < air temp. (5) then the condenser coil is probably restricted. (refrigerant side restriction)

Figure 13–2
Example of a checklist used in the field to record and troubleshoot heat pump conditions. (Courtesy of Shoffner Mechanical Services)

SUMMARY

In this chapter we have discussed troubleshooting for ASHPs. We have noted that there are a few differences between air-source and split-system air conditioning systems. Chapter 23 and this chapter are related in terms of information. This chapter begins the explanation of troubleshooting for ASHPs, and Chapter 23 expands on the information, providing troubleshooting information on ground-source heat pumps (GSHPs).

The chapter began with the principles of troubleshooting:

1. Obtaining information about the problem by listening to the customer
2. Obtaining information about the system operation
3. Determining the symptoms
4. Developing a list of possible causes for the symptoms
5. Ruling out possible causes
6. Identifying the problem
7. Verifying the problem
8. Correcting the problem
9. Verifying that the problem is corrected and system operation is normal (take a full set of readings)

Symptoms and possible causes drive the troubleshooting process. The development of a list of symptoms from information obtained mostly from the customer and determining the possible causes is the basis for good troubleshooting practice (see Figure 13–2 for an example of a system check sheet).

The chapter concluded with a table for troubleshooting ASHPs. Troubleshooting was divided into mechanical and electrical. Common mechanical and electrical symptoms and possible causes were provided (Figure 13–2).

REVIEW QUESTIONS

1. Describe the principles of troubleshooting.
2. Describe how symptoms lead to possible causes.
3. Relate what is to be done with each cause.
4. Describe the "ruling-out" process.
5. Explain how verification of a cause is performed.

CHAPTER 14

System Sizing/Design

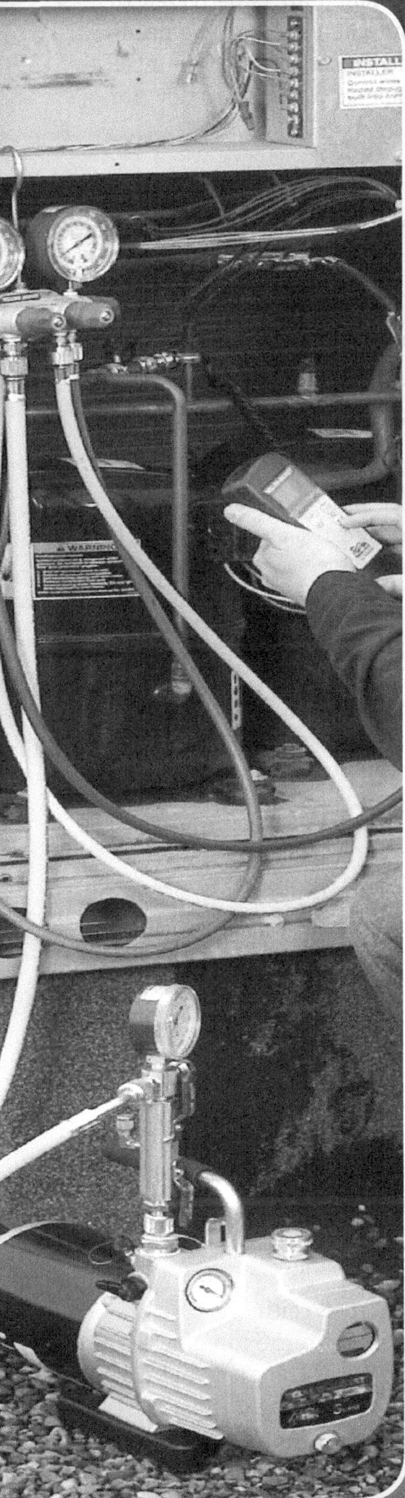

LEARNING OBJECTIVES

The student will:

- Describe the importance of conducting a thorough heat gain and loss calculation
- Relate how air-source heat pumps (ASHPs) are selected after the heat gain is known
- Describe the function of a thermal balance point graph
- Explain why moisture removal from the conditioned air is of particular concern for ASHPs
- Explain how heat is pulled from cold air
- Describe how to determine and adjust cubic feet per minute (CFM) for an individual room

INTRODUCTION

In this chapter we will review the calculations needed to determine the structure heat gain and loss. Most air-source heat pump (ASHP) systems are designed using heat gain (cooling load). Heating capacity of a selected piece of equipment is plotted on a graph along with structure heat loss to establish a thermal balance point. The thermal balance point also helps in selecting the amount of supplemental heat that will be needed to maintain occupant comfort levels.

Duct sizing requires accurate heat gain calculations and proper equipment selection. The amount of air necessary for the entire structure will be distributed to individual rooms. How that air is handled and how much air is delivered to the room will change as the system operates throughout the year. Cooling needs may not be the same as heating requirements, and the homeowner or the technician must be able to know how to set the duct dampers to adjust airflow for the season.

Field Problem

During a maintenance check, the customer mentioned that the heating bill was higher last month than the bill of a year ago, but he didn't seem that concerned. As the technician checked and cleaned the system, the higher electrical bill stuck in his mind—something could be going wrong with the system. Some problems that could cause this might be:

1. Dirty filters
2. Blower motor going bad
3. Blocked heat exchanges
4. Increased defrost events
5. Incorrect settings of the duct dampers
6. Improper refrigerant charge
7. Supplemental heat thermostat not set to proper balance point

It was February and the system seemed like it was functioning well during this midseason call. By the end of the maintenance check, the technician did not find anything unusual that would cause higher electrical bills. He decided to check something that he had forgotten to list: loose electrical connections. On the truck he had an infrared imager that would easily identify hot spots. He checked the inside unit connections and operated the emergency heat. Everything looked good; no hot connections. Going outside, he checked the outdoor connections and they looked good too.

While he stood there, the imager picked up a white flash—heat. He looked again and as he moved the imager across the wall of the house, one section of wall lit up! This was obviously an area where heat was leaking out of the house at a large rate. He touched the wall but could feel no difference; the wall was as cool as other parts of the structure.

He told the homeowner what he had found in his investigation of higher electrical bills and the homeowner smiled. The technician was asked if he wanted to take a look at the customer's new hobby—tropical plants. Inside the room was a bank of lights, and a space heater with a pan of water on top. The room temperature was 85°F and the humidity was very high. Next time, thought the technician, I will ask if there have been any changes in the structure or use of the rooms, like I should have.

HEAT GAIN/HEAT LOSS CALCULATIONS REVIEW

Heat gain occurs during the cooling season when outside ambient conditions are high (Figure 14-1). Heat loss occurs in the heating season when indoor temperatures are higher that outdoor temperatures. Heat always moves from high areas of concentration to low (hot to cold).

During the cooling season, heat enters a space being cooled by:

1. Conduction—through walls, doors, windows, and roof/ceiling.
2. Convection—heated air displaces cooler air and rises to the top of a room. When a door is opened, the cool air drops out of the room and is replaced by warm air from outside. This process also occurs, to some degree, inside insulated walls, the attic, and the basement of a structure.

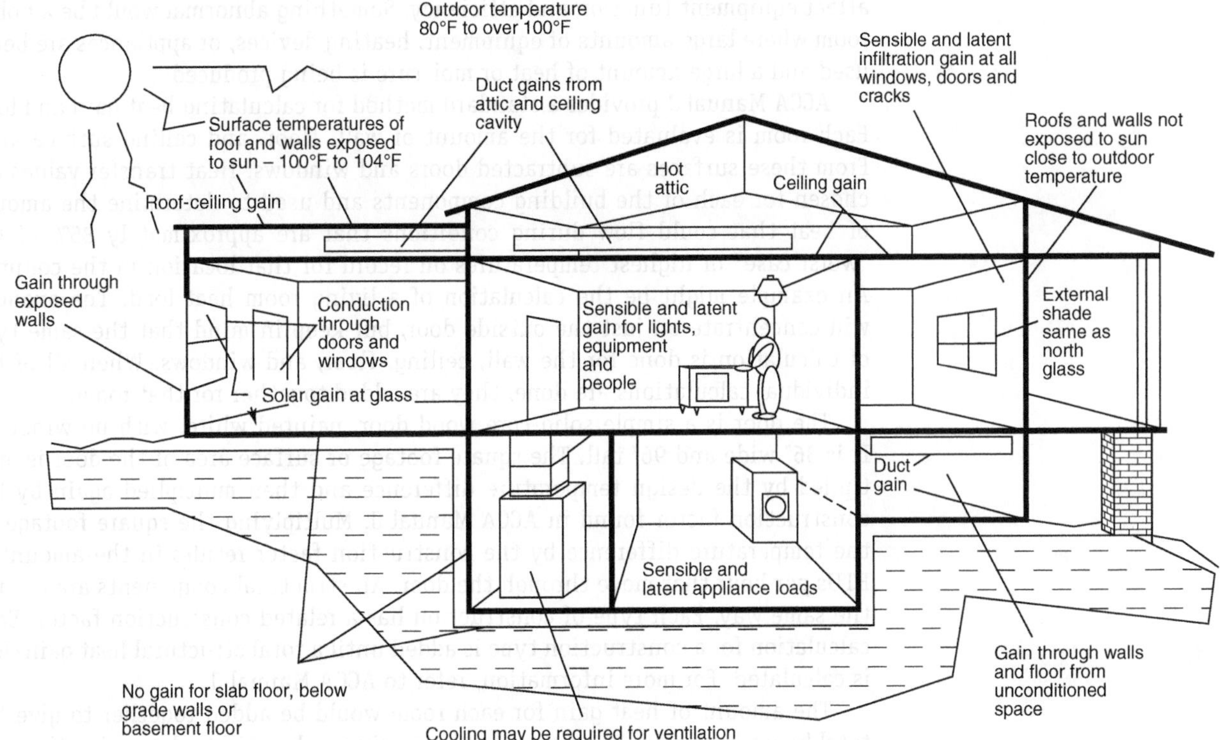

Figure 14–1
This figure shows many of the sources of heat gain that contribute to the cooling load. (Courtesy of ACCA)

3. Radiation—heat is received directly from the sun through windows to heat interior carpeting and furnishings. They, in turn, conduct heat to the indoor air. Radiant energy also travels, to some degree, through walls, doors, and attic areas. Radiant energy travels in a straight line from an object or surface that has a higher heat content to a lower-heat-content surface or substance.

Each of these is not exclusive to the cooling season. Conduction and convection heat loss occur in the heating season, as well. However, for proper summer dehumidification most ASHPs are sized and selected based on heat gain, so it is important to emphasize the importance of heat gain.

In addition to the structural heat gain in calculating cooling loads, there are internal sources of heat that need to be factored into the total amount of heat to be removed. The heat pump system will be working to move the combined indoor heat to the outside. Sources of internal heat are:

1. People.
2. Cooking.
3. Bathing.
4. Appliances.
5. Other motors, equipment, and devices that use energy.

If all of the sources of heat were identified and counted, there would be a long list of very small values. Everything that uses energy gives off heat. All living things give off heat—people, pets, and plants. Anything that converts energy gives off heat—cell phones, chargers, computers, and calculators. For all of these smaller values, there is some tolerance provided in the calculation process that can account for the normal amount of activity that occurs in a living space. However, the technician should be looking for abnormal situations that could affect equipment function and efficiency. Something abnormal would be a hobby room where large amounts of equipment, heating devices, or appliances are being used and a large amount of heat or moisture is being produced.

ACCA Manual J provides a standard method for calculating heat gain and loss. Each room is evaluated for the amount of wall, floor, and ceiling surface area. From these surfaces are subtracted doors and windows. Heat transfer values are chosen for each of the building components and used to determine the amount of heat that could flow during conditions that are approximately 85% of the "worst case" or highest temperatures on record for that location in the country. An example might be the calculation of a living room heat load. The example will concentrate on just the outside door, but keep in mind that the same type of calculation is done for the wall, ceiling, floor, and windows. When all of the individual calculations are done, they are added together for that room.

The door is a simple solid-core wood door, painted white, with no windows. It is 36″ wide and 96″ tall. The square footage or surface area of the door is multiplied by the design temperature difference and then multiplied again by the construction factor found in ACCA Manual J. Multiplying the square footage by the temperature difference by the construction factor results in the amount of BTUs per hour that move through the door. All structural components are treated the same way. Each type of construction has a related construction factor. Each calculation for a construction type is added until a total structural heat gain/loss is calculated. For more information, refer to ACCA Manual J.

The amount of heat gain for each room would be added together to give the total house heat gain. The amount of heat gain can be used to determine the cooling load for the entire house. ACCA Manual D provides a method for duct design. Every room uses a percentage of the total load or the total amount of cooling that the equipment can provide. Cubic feet per minute (CFM) for each room is determined by dividing the amount of CFM delivered by the equipment to each room, matching the room needs with the equipment.

OVERVIEW OF THE SYSTEM

Most ASHPs are installed as split systems. The indoor unit contains a coil, an air handler, a filter rack, and room to put electric resistance strip heaters. Some systems come with everything the installer needs pre-mounted, in a single housing. Ductwork is attached to the indoor unit that supplies air to each room and returns air from each room or a central location. Other systems might include an ASHP installed in conjunction with a fossil-fuel furnace.

The ductwork can be made of metal or duct board, an insulated duct material. If the ductwork is metal, the metal is lined or wrapped with insulation, per local energy code. Attached to the ductwork in the room are the second most visible part of the HVAC system—diffusers and grilles (the first most visible part is the thermostat). All of the joints between the indoor unit and the supply diffusers and return grills have been sealed so that controlled and conditioned air reaches each intended room.

The designer begins with how the system should be installed and determines the location of the indoor unit and how the ducts will be installed. A sketch is done of the intended location of the indoor unit and ductwork, then a final drawing is made for the installer to use.

Equipment Selection

Equipment size, for most ASHP applications, is based on load calculations. This seems to be true for both moderate and colder climates. After the equipment has been selected, a balance point diagram can be used to evaluate potential performance.

Table 14–1 Tabulated Performance Data for a Generic Heat Pump (Courtesy of ACCA)

Model 25 1,200 CFM		
Outdoor Temperature °F	**Heating BTUH**	**Input (Watts)**
60	42,200	3,735
50	37,100	3,485
40	29,500	3,230
30	23,800	3,080
20	21,200	2,725
10	19,700	2,490
0	15,800	2,255
−10	11,900	2,025

1) BTUH @ 70°F air into coil. . . . (+2% @ 60; −2% @ 80).
2) Heating capacity derated for defrost cycle (70 % RH).
3) Input power for compressor, add 400 Watts for blower.

The manufacturer's performance data is critical to equipment selection, as explained in ACCA Manual S (Table 14–1 provides an example). It is important to obtain detailed equipment performance data. This data is usually presented in a table format. Design load is compared with performance data to determine which system will work best for the application.

THERMAL BALANCE POINTS

After the system is selected based on cooling load, some determination needs to be made about the operation of the heating mode. A thermal balance point diagram is drawn (Figures 14–2 and 14–3).

A thermal balance point is a point on a graph where the heating-load line crosses the equipment-capacity line. Where they cross is the point where the

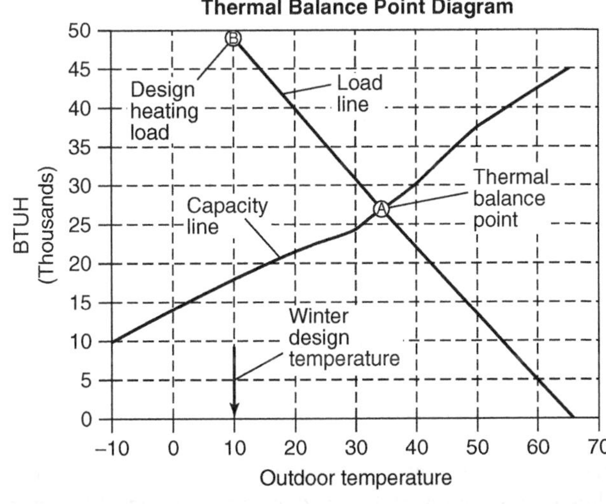

Figure 14–2
This thermal balance point diagram shows the design heating-load line and the capacity line of the heat pump. Where the two lines intersect is labeled the thermal balance point. (Courtesy of ACCA)

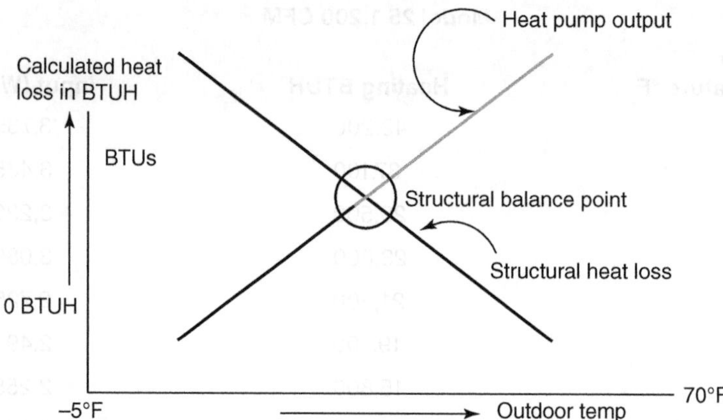

Figure 14–3
This thermal balance point diagram shows the design heating-load line and the capacity line of the heat pump. Where the two lines intersect is labeled the thermal balance point. (Courtesy of Delmar/Cengage Learning)

heat pump cannot deliver all of the heat necessary for the structure. The optimum thermal balance point would be a very low temperature point. The example chart in Figure 14–2 shows a temperature of approximately 35°F. Some equipment manufacturers have systems that are capable of moving the economic thermal balance point below the freezing point.

As the outdoor air temperature drops, the capacity of the ASHP in the heating mode drops while the structural heating requirement increases. If these two lines are plotted on a graph, that point where they intersect is known as the building or structural thermal balance point, indicating the outdoor air temperature at which the output from the heat pump matches exactly the needs of the building. Above that temperature the heat pump can adequately heat the structure, and below that temperature would require additional or supplemental heat. Supplemental heat can be provided with electric resistance heaters and, with some limitations, a fossil-fuel furnace.

The output from the heat pump in the heating mode is compared to the same amount of heat output from electric resistance heat utilizing the same amount of electrical energy. This comparative term is called the coefficient of performance (COP). If a heat pump has a COP above 1, it is providing more heat for the energy used than electric resistance heat. If the COP drops below 1, the heat output from the heat pump is less than what could be obtained with electric resistance heat utilizing the same amount of electrical energy. When comparing the cost of operation between electric resistance heat with the output from the heat pump, that point at which the COP equals 1 would be the economic thermal balance point.

In many parts of the country other fuel sources such as natural gas will be less expensive in producing heat than electric resistance. For example, if electric costs 3.5 times as much as natural gas, then the temperature that achieves a COP of 3.5 would be the economic thermal balance point for that option.

In general, an ASHP will not produce enough heat to provide all of the heat required for the application. Supplemental heat will most often be required, and what you install for supplemental heat will determine how it is operated. For example, electric resistance heating elements are often used with heat pumps, and are installed downstream of the indoor coil so that they can additionally warm the air coming off the indoor heat pump coil. If a fossil fuel is used, it is generally in the form of a furnace that provides the air movement blower as well. Because the heat pump indoor coil would be mounted downstream of the furnace heat exchanger, it would be subject to very high temperatures coming off the furnace. This would cause extremely high head pressures on what is acting as the condenser coil, so if a heat pump is used in conjunction with a fossil-fuel furnace, it most often locked out whenever the furnace is operating. Carrying that thought one step further, you would use the fossil-fuel furnace and lock out the heat pump whenever the heat pump cannot produce enough heat (below the structural thermal balance point) and whenever the fossil fuel would operate less expensively (economic thermal balance point).

Tech Tip

As the outside air drops to a very low temperature, the heat pump capacity reduces to a point where supplemental heat must handle the entire load. This also is the case if the compressor fails. Until that time, this source of heat is often referred to as supplemental heat or second-stage heat. Second-stage heat is also activated when the ASHP goes into defrost mode.

After the heating equipment has been sized and selected, the thermal balance point graph is used to select and size electric resistance heat—supplemental and emergency heat. These are ACCA-recommended steps 4 and 5. The sixth or last step is to select the resistance heating system. This selection will be based on the BTUH. Depending on the system and controls, it is possible to have outdoor thermostats set at each of these temperatures to signal the system to bring on more electric heat to supplement heat being produced by the heat pump. Using the graph in Figure 14-4, two 5-Kw heat strips could be installed and would cover the amount of heat needed. Each strip could be staged to come on at a different balance point.

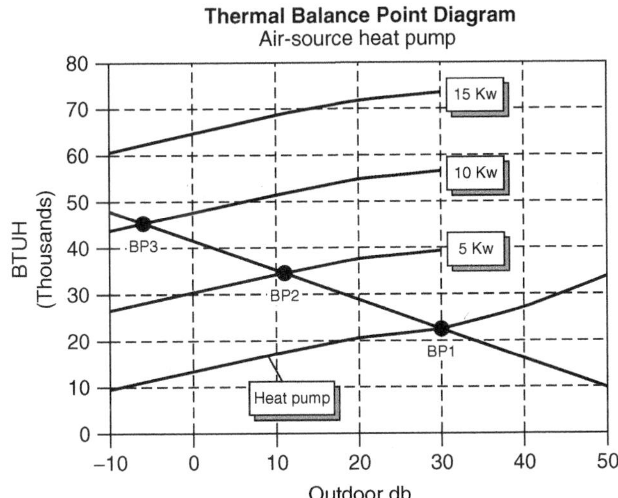

Figure 14-4
This diagram represents the system thermal balance point at BP1 and the operation of three strips of electric resistance heat. At BP2, 5-Kw cross the design load line, and 10 Kw cross at BP3. (Courtesy of ACCA)

EFFECTS OF SYSTEM SIZING

Depending on location, improper equipment sizing can cause:

1. Higher energy and installation costs.
2. Undesirable indoor air conditions, such as high humidity.
3. In the case of undersizing, systems may run continuously and not meet the load requirements, requiring additional supplemental heat.

EVALUATING AIR DISTRIBUTION

After the delivery equipment is selected and before the delivery system is sized, the next step is to evaluate each room for diffuser placement and airflow needs. The challenge is to select a diffuser that will deliver the right amount of air, with the right static pressure drop, and have a pattern/throw that will get room air to move without creating drafts in the occupied zone.

Diffuser placement is needed to complete the ductwork design. Some forethought and insight is needed to locate a place in the room that would not be objectionable for the diffuser and allow the diffuser to do the work of moving air (Figure 14-5). For instance, placement of the diffuser under a window might

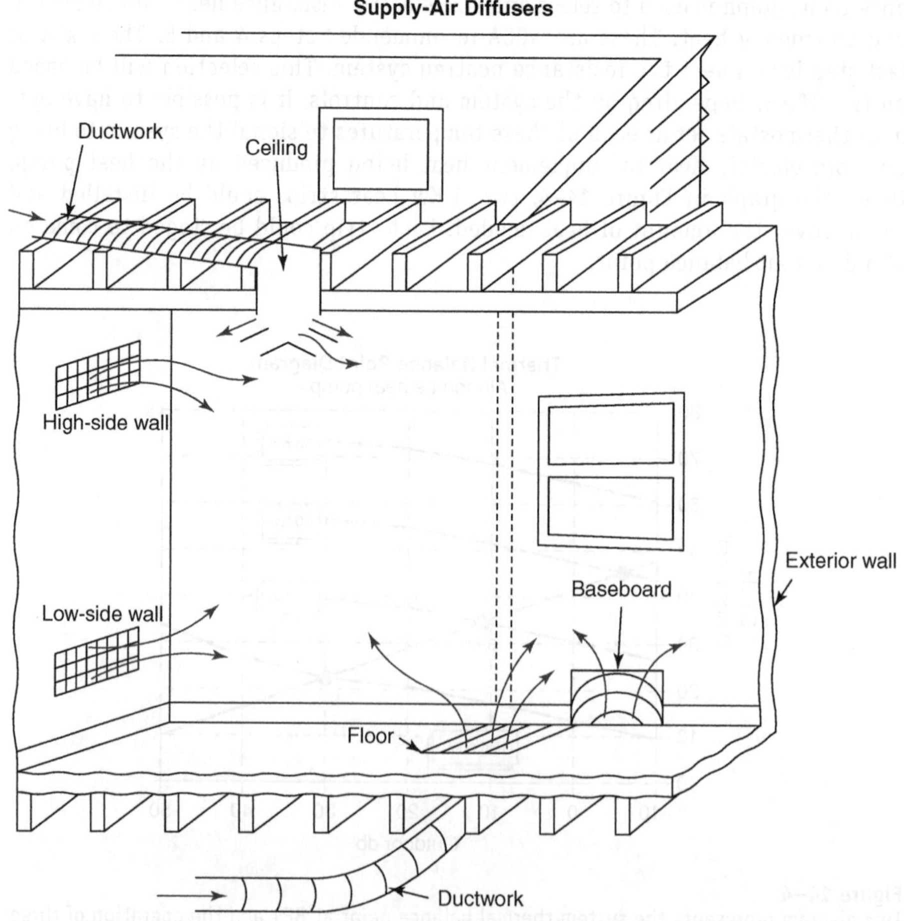

Figure 14–5
There are many types and styles of diffusers as well as placement options. The right diffuser for the room is always a design concern. (Courtesy of Delmar/Cengage Learning)

Table 14-2 Air Diffuser Selection Considerations

	Diffuser Performance		Interference with Airflow			Cost
Diffuser Type	Heating	Cooling	Room Decor	Room Furnishings	Drapes	Installation
Ceiling	Good, except for slab houses in northern climates	Good	Decorative styles are available	N/A	N/A	High cost due to attic duct insulation requirements
High-Side Wall	Fair, except for slab houses in northern climates	Good	Visible and few choices of style and color	N/A	N/A	Low if ductwork is placed in conditioned spaces
Low-Side Wall	Excellent when used with perimeter systems	Excellent when air pattern is rising	Visible and few choices of style and color	Interference is likely	Drapes may cover	Medium because wall plates need to be cut for installation
Baseboard	Excellent when used with perimeter systems	Excellent when used with perimeter systems	May be hidden if baseboard is the same color and height	Minimal interference	Drapes may cover	Low if ductwork is attached from below
Floor	Excellent	Excellent	May be hidden if color is the same as the floor covering	Minimal interference	Drapes may cover, unless diffuser is placed away from the wall	Low if ductwork is attached from below

look like a great place, until it is realized that the picture window may have full-length drapes that would prevent the diffuser from working. Another location should be chosen that won't be covered by furniture.

The amount of air being delivered to the room will determine the number of runs and diffusers that will be needed. ACCA Manual D will guide the designer in determining how many diffusers are needed (Table 14-2).

Room Cubic Feet per Minute (CFM)

Room airflow is the result of careful calculation and prudent system design. ACCA Manuals J, D, and S provide a wealth of information that helps the designer with pre-planning for a heat pump installation. The amount of air delivered to a room depends on the following:

1. Heat loss and heat gain—calculated by using Manual J.
2. Equipment selection—guided by using Manual S.
3. Equipment and duct delivery size—calculated by using Manual D.

The amount of air delivered by the blower to the whole house is subdivided in the duct system and directed to a room. The amount of air delivered is a product of the heat gain calculation and equipment capacity. When the duct calculation is complete, the amount of air for each diffuser will be stipulated. If this information is available for the technician, it would be easy for the technician to troubleshoot airflow problems. For example, if the technician tests for room air supply by measuring the face velocity of the diffuser, the amount of air delivered to the room could be determined. If the average air velocity is 600 fpm and the diffuser manufacturer's Ak number is .19, the volume would be 114 CFM (600 × .19 = 114). If there is only one diffuser, the amount of air entering the room from the system is 114 CFM. If the condition of the air coming from the diffuser is 60°F @ 50% relative humidity (RH) and the return air leaving the room is 70°F @ 50% RH, the technician knows that mostly sensible cooling is being accomplished. The temperature difference is 10°F and the electronic psychrometric chart shows a total heat difference of 11.7 BTU/CFM. Multiplying the CFM, 114, times the BTU difference per CFM, 11.7, lets us know that 1333.8 BTUH of total heat gain is being removed.

If this room were 72°F (with the same conditions as above) and needed to be 70°F, the technician would adjust the diffuser to allow more air into the room. The trick is to know how much more. If the outdoor temperature is 92°F, there would be a 20°F difference. Divide 20 into the heat gain (heat being removed, or 1,333.8) and then multiply by the new temperature difference of 22°F (92 − 70 = 22). The new heat gain is 1,467.2 BTUH and the difference between the old and the new is 133.4 BTUH. (1467.2 − 1333.8 = 133.4). Divide 11.7 BTU/CFM into 133.4 BTUH and the answer is 11.4 CFM more (133.4/11.7 = 11.4). The damper is opened and a new velocity reading is taken. The new volume should be 125.4 CFM (114 + 11.4 = 125.4) and the new velocity should be close to 660 fpm (125.4/.19 Ak = 660).

Tech Tip

If only one diffuser is adjusted, all of the other diffusers may be affected, or, depending on the ductwork, one other diffuser will be affected. When adjusting for flow, start with all of the dampers open and adjust the flow so that each damper is delivering design CFM.

SUMMARY

In this chapter we discussed load estimation and how heat pump systems are selected and sized for an application. ACCA Manual J is used to determine the heat gain and loss for a structure. Because most ASHPs are sized for cooling, the measurement of heat gain becomes the focal point for the selection of the right-sized heat pump. ACCA Manual S assists with the selection of heat pump equipment. It guides the selection by providing a sizing sequence.

Thermal balance point graphs are developed to help determine the relation between equipment selection and the equipment's ability to perform in the heating mode. Supplemental electric resistance heat is determined by using the thermal balance point graph and the amount of emergency heat (if in addition to supplemental heat).

An effect of system sizing is the ability to remove moisture. Low fan speeds help to reduce moisture by allowing the coil temperature to drop. When that is done, the equipment efficiency also drops. To combat this problem, manufacturers are offering heat pumps that incorporate capacity control. In this way, the heat pump can operate at reduced capacity to remove moisture and high capacity, when needed, to reduce the temperature of the occupied space.

After the heat gain and loss are calculated, the equipment is selected, and the air delivery system is designed, air to individual rooms is adjusted to meet cooling and heating needs. Face velocity readings of the diffuser and use of the diffuser manufacturer's data can easily determine the amount of CFM delivered to the room. Because the amount of air delivered to a room changes with the season, heating and cooling CFM to an individual room may be different. This difference will require that the duct dampers be set in one position for cooling and in another for heating.

REVIEW QUESTIONS

1. Describe the importance of conducting a thorough heat gain and loss calculation.
2. Relate how air-source heat pumps (ASHPs) are selected after the heat gain is known.
3. Describe the function of a thermal balance point graph.
4. Explain why multi-speed or variable speed blowers are beneficial in the removal of moisture for ASHPs.
5. Explain how heat is pulled from cold air.
6. Describe how to determine and adjust cubic feet per minute (CFM) for an individual room.

CHAPTER 15

Typical Heat Pump Systems

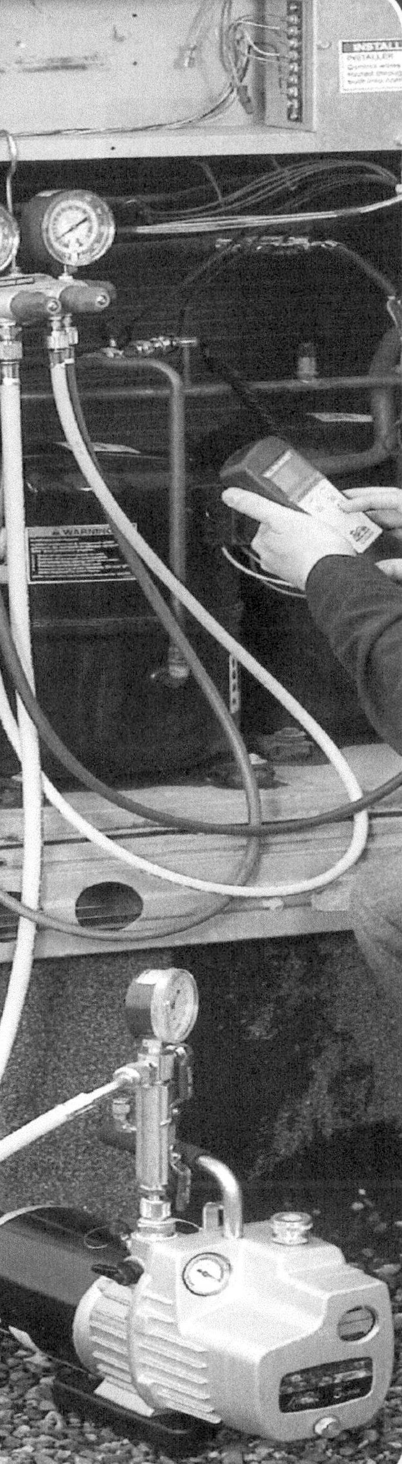

LEARNING OBJECTIVES

The student will:

- Describe how a split-system heat pump is properly installed. At least two different configurations should be compared
- Describe where mini-split (ductless) systems can be installed and their purpose
- Describe a packaged system and the two different places it may be installed
- Describe a dual-fuel installation and why a fossil fuel might be used
- Explain what "geothermal" means
- List and describe two methods to extract heat from the ground or water using a geothermal heat pump

INTRODUCTION

Heat pumps come in many different types of configurations. That is because heat can be pumped from various sources. There needs to be three things: (1) a source of heat, (2) a heating medium to transfer the heat, and (3) and a place to put the heat. Heat that is absorbed from various sources can be delivered where needed or disposed of (as in air conditioning systems).

In this chapter, the major types of heat pump configurations and associated heat sources will be discussed. Along with air-source heat pumps, there are geo-source and water-source heat pumps. Even within these classifications, there are several configurations that make the heat pump a versatile and easily adapted heating and cooling system.

Field Problem

A homeowner with a geothermal unit is complaining that the heat pump is running constantly in the heating mode. He is also indicating that his electrical bills are higher this year than last. A third complaint is that a noise like marbles rolling around is coming from the unit. The technician gathers all of the information he needs and moves to the basement location of the heat pump.

Symptoms: Noises; high electrical bills; constantly operating.

Possible Causes: Refrigerant charge; mechanical problems; airflow problems; pump/water flow problems.

He recognizes that the sound of marbles is air trapped in the circulator. The electric strip heaters are also operating. The water loop is a multilayer, horizontal. If air is in the circulator, it is possible that one or more of the loops is air-locked and the circulation has stopped. The technician measures the water temperature difference and finds 8°F. The loop should have a temperature difference that is nearer to 5°F. (Note: The technician would use the formula Temperature difference × Gallons per minute × Constant (485 [constant for 35% glycol/H20 solution]) = BTUH.) This is an indicator that possibly one or more of the loops is not circulating. If air has entered the system, there must be a leak. Air leaking into the system occurs on the suction side of the pump, so the most logical check is the circulator coupling flange. The technician finds that the flange bolts are loose and the coupling has leaked. Power flushing removes the air, and after reconnecting the circulating system, the technician tests the repair. No noise is coming from the circulator, and after an appropriate amount of time, the temperature difference settles to nearly 5°F. The electric strip heaters are off because they do not need to supplement. There is enough heat being produced by the heat pump.

Tech Tip

Power flushing can be of two types: flushing to remove contamination and flushing to remove air in the loop. Flushing to remove air pockets involves using a container of water and glycol (antifreeze solution) to push glycol solution through the loop and having the other end of the loop dump back into the container. As the loop dumps glycol solution back into the container, air bubbles indicate that air is being removed from the loop. This process does not waste the glycol solution. Residual glycol solution is kept in the container and used in the next flushing process.

SPLIT SYSTEMS

A split-system heat pump consists of an outdoor unit containing the compressor, outdoor coil, outdoor fan, the reversing valve, and the controls that operate the unit (Figures 15-1 through 15-3). The indoor section is generally an air-handling

Figure 15–1
The outdoor unit contains the outdoor coil, compressor, reversing valve, and other components for an ASHP. (Courtesy of Carrier Corporation)

1. Electronic system diagnostics monitor
2. Direct-drive fan
3. Outdoor coil
4. Insulated compressor compartment
5. High-efficiency two-stage scroll compressor

Figure 15–2
Cutaway view of the ASHP revealing the inner parts of the outdoor unit. (Courtesy of RSES)

unit that contains the indoor coil, the system blower, and electric heat strips. The electric heat strips are used for heating whenever the heat pump cannot provide adequate heat, either because it is too cold outside for the heat pump to produce adequate heat, or in the defrost mode to compensate for the "cooling" effect of operating the heat pump in the cooling mode for a short defrost period.

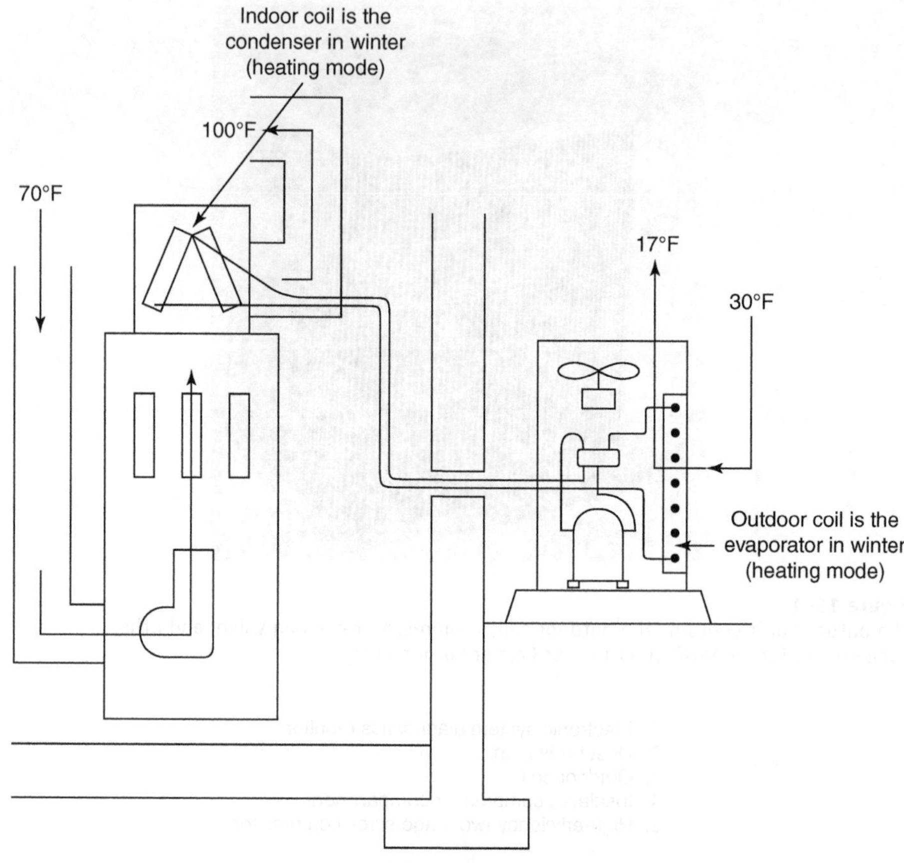

Figure 15–3
The installation of a split heat pump system is generally the same as that of a split air conditioning system. In the heating mode, outside air is the source of heat for the structure. (Courtesy of Delmar/Cengage Learning)

Air handlers most often have the electric strip heaters downstream of the indoor coil, so you can use both the heat pump and the electric heat simultaneously. If the electric heating coils were upstream of the indoor coil, the heat from the electric heaters would go across what is the condensing coil for the heat pump in the heating mode, and that would send the head pressures way up.

We use the terms "supplemental" and "emergency" heat when talking about heat pumps. Generally, they are the same heating elements but are controlled differently. As "supplemental" heat, the electric strip heaters are staged to make up the difference between what the heat pump can produce and the needs of the space. In "emergency" heat, the heat pump is bypassed and the electric heat responds to the initial call for heat. In many cases the electric heat operates from the second stage of the thermostat in "supplemental" mode, and the control point of the electric heat is shifted to stage one of the thermostat in the "emergency" mode.

The indoor unit (indoor coil) could be mounted in similar fashion as a split air conditioning system: in the basement, crawlspace, garage, alcove, or attic. Air volumes will run slightly higher than those of an air conditioning system. Heat pumps will require between 350 and 450 cubic feet per minute (CFM) per ton of capacity. Installation requires adequate service space, lighting, and electrical service outlets. Because the indoor unit also functions as an air conditioning system, the installation may require overflow pans to protect against condensate damage.

MINI-SPLIT (DUCTLESS) SYSTEMS

A mini-split (ductless) system is generally considered a room-sized heat pump (Figure 15–4). These systems can be used for small apartments and individual rooms to condition the space. The indoor coil and outdoor unit are remote-connected

Figure 15–4
Mini-split heat pump systems are ductless units that are designed to heat/cool small areas. This unit shows a remote control. (Courtesy of RSES)

with piping and electrical runs. These systems are designed so that the indoor section can be mounted in several locations: floor, wall, or ceiling. They may come complete with supplemental heat. More than one indoor unit may also be connected to a single outdoor unit in some models. Some systems also come with swinging vanes to direct airflow, wireless remote controls, and in various colors.

PACKAGED SYSTEMS

A packaged system is a system contained in one enclosure or cabinet (Figure 15–5). The cabinet houses all of the components and controls that create a functioning heat pump system. The packaged unit will contain the heat pump refrigerant

Figure 15–5
A packaged heat pump system looks similar to a packaged air conditioning system. (Courtesy of Delmar/Cengage Learning)

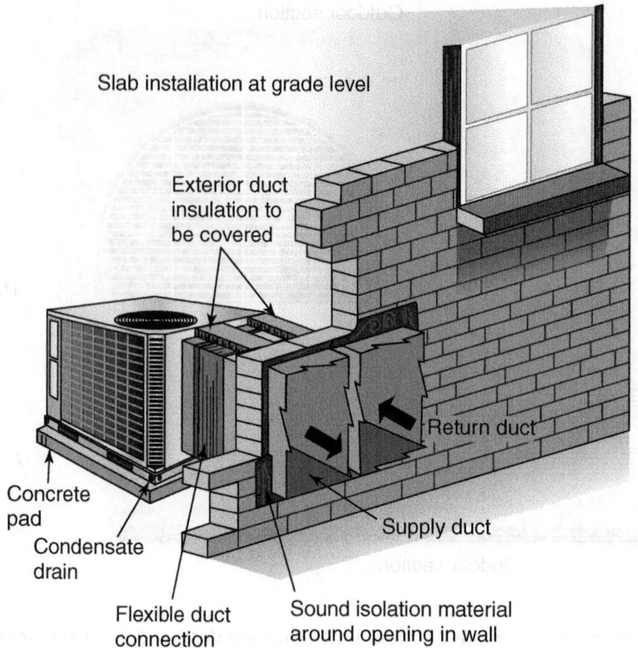

Figure 15–6
The installation of a packaged heat pump is the same as that of a packaged air conditioning system. The unit can be installed on-grade or on the roof as an RTU (rooftop unit). (Courtesy of Climate Control Technologies, Inc.)

system, electric heat strips; and connections for external power and thermostatic control. Packaged systems come in several configurations to allow the installer options for rooftop unit (RTU), slab, and through-the-wall installations (Figure 15–6).

DUAL-FUEL SYSTEMS

In many areas of the country, the cost of operating electric strip heaters is more costly than fossil fuels such as natural gas, propane, and fuel oil. In areas where you have lower outside design conditions, and hence a greater need for heating and less opportunity to produce this heat with an air-to-air heat pump, heat pumps are often installed with a fossil-fuel furnace to make up the difference (Figure 15–7). A typical furnace contains the blower and the heating section, and the indoor coil of the heat pump would be mounted at the discharge of the furnace. Heated air coming off the furnace going across the indoor (condenser) coil would create high head pressures. Up until recently, you could not operate the furnace and heat pump simultaneously, so you would have to lock out the heat pump and use strictly the fossil-fuel furnace once the outdoor air temperature got down to a point (thermal balance point) where the heat pump could not produce enough heat to satisfy the space. With the fairly recent advent of modulating gas furnaces, some manufacturers can control the temperature of the air leaving the furnace to a point that they can use the heat pump and fossil-fuel furnace simultaneously. In any case, whenever you integrate an air-to-air heat pump with a fossil-fuel furnace, you must follow the manufacturer's instructions and use an interface connection typically called a "fossil-fuel kit." Conduct an operating cost analysis to determine which is most economical (economic balance point). Fuel and utility rates vary from region to region.

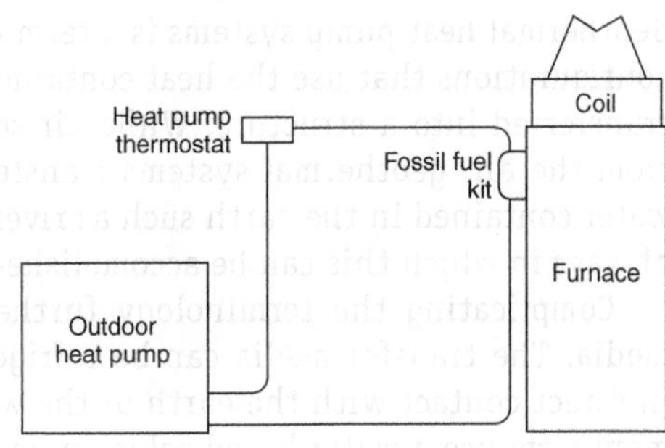
Heat pump thermostat
Coil
Fossil fuel kit
Furnace
Outdoor heat pump

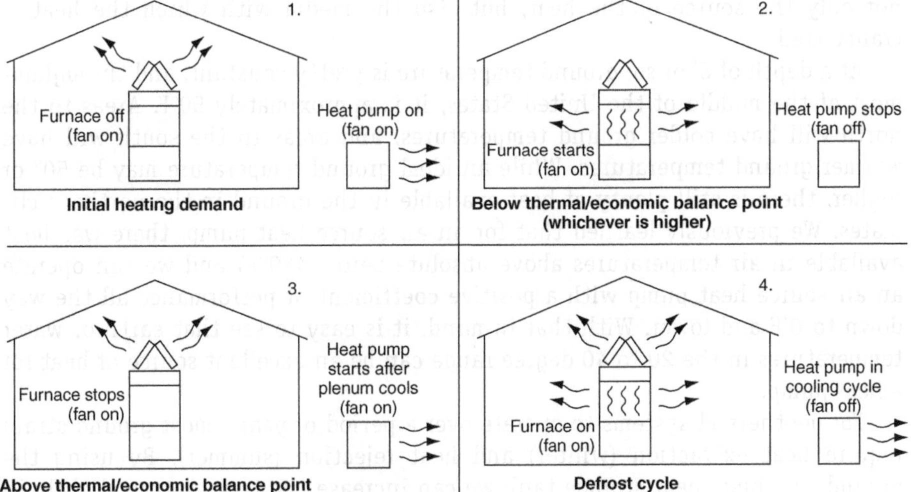
1.
Furnace off (fan on)
Heat pump on (fan on)
Initial heating demand
2.
Heat pump stops (fan off)
Furnace starts (fan on)
Below thermal/economic balance point (whichever is higher)
3.
Furnace stops (fan on)
Heat pump starts after plenum cools (fan on)
Above thermal/economic balance point
4.
Heat pump in cooling cycle (fan off)
Furnace on (fan on)
Defrost cycle
Note: The furnace fan stays on whenever heating is called for.
The heat pump fan cycles depending on the mode of operation.

Figure 15–7
Dual-fuel system operation. The furnace blower turns on whenever there is a heating need. The heat pump outdoor fan operates depending on the mode of operation. (Courtesy of RSES)

GEOTHERMAL SYSTEMS

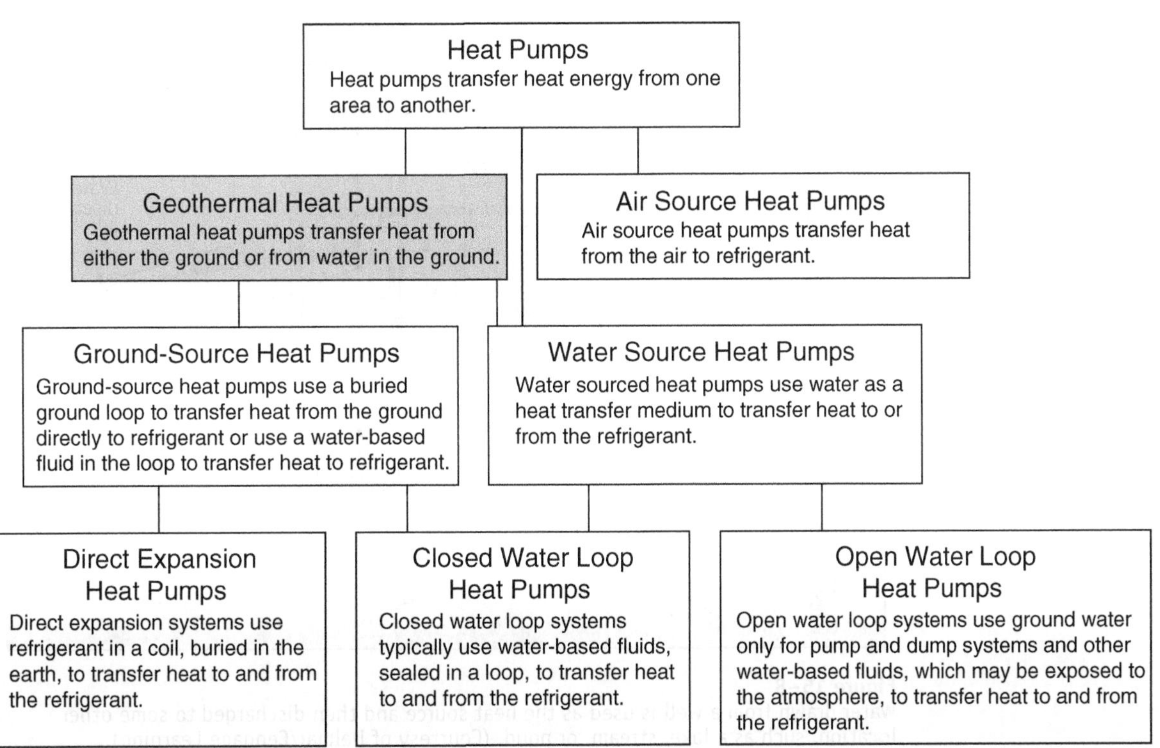
Heat Pumps
Heat pumps transfer heat energy from one area to another.
Geothermal Heat Pumps
Geothermal heat pumps transfer heat from either the ground or from water in the ground.
Air Source Heat Pumps
Air source heat pumps transfer heat from the air to refrigerant.
Ground-Source Heat Pumps
Ground-source heat pumps use a buried ground loop to transfer heat from the ground directly to refrigerant or use a water-based fluid in the loop to transfer heat to refrigerant.
Water Source Heat Pumps
Water sourced heat pumps use water as a heat transfer medium to transfer heat to or from the refrigerant.
Direct Expansion Heat Pumps
Direct expansion systems use refrigerant in a coil, buried in the earth, to transfer heat to and from the refrigerant.
Closed Water Loop Heat Pumps
Closed water loop systems typically use water-based fluids, sealed in a loop, to transfer heat to and from the refrigerant.
Open Water Loop Heat Pumps
Open water loop systems use ground water only for pump and dump systems and other water-based fluids, which may be exposed to the atmosphere, to transfer heat to and from the refrigerant.

Geothermal heat pump systems is a term used to describe a variety of heat pump configurations that use the heat contained in the earth as a source of heat to be transferred into a structure. While air source heat pumps (ASHP) transfer heat from the air, geothermal systems transfer heat from the ground itself, or from water contained in the earth such as rivers, lakes, and wells. There are a number of ways in which this can be accomplished. See Figures 15-8 and 15-9.

Complicating the terminology further, we also consider the heat transfer media. The transfer media can be refrigerant where the heat exchanger comes in direct contact with the earth or the water in (or on) the earth, or more commonly, we use a water-based solution heat exchanger in contact with the earth or the water in (or on) the earth. From a terminology point of view, we consider not only the source of the heat, but also the media with which the heat is transferred.

At a depth of 6′ or so, ground temperature is pretty constant, and throughout most of the middle of the United States, it is approximately 50°F. Areas to the north will have colder ground temperatures, and areas to the south will have warmer ground temperatures. While an ideal ground temperature may be 50° or higher, there is still plenty of heat available in the ground in the northern climates. We previously learned that for an air source heat pump, there was heat available in air temperatures above absolute zero (-460°F) and we can operate an air source heat pump with a positive coefficient of performance all the way down to 0°F and lower. With that in mind, it is easy to see that earth or water temperatures in the 20 to 40 degree range can be an excellent source of heat for a heat pump.

For geothermal systems to operate over a period of years, most ground strata require heat extraction (winter) and heat rejection (summer). By using the ground as a huge heat storage tank we can increase the temperature slightly in the summer and lower the temperature in the winter and not allow permafrost or extremely warm ground conditions to prevail. With this in mind, loops should not be installed close to septic tanks as they could reduce the temperature and stop the breakdown of waste, or close to building foundations that could be damaged by frozen expanded ground.

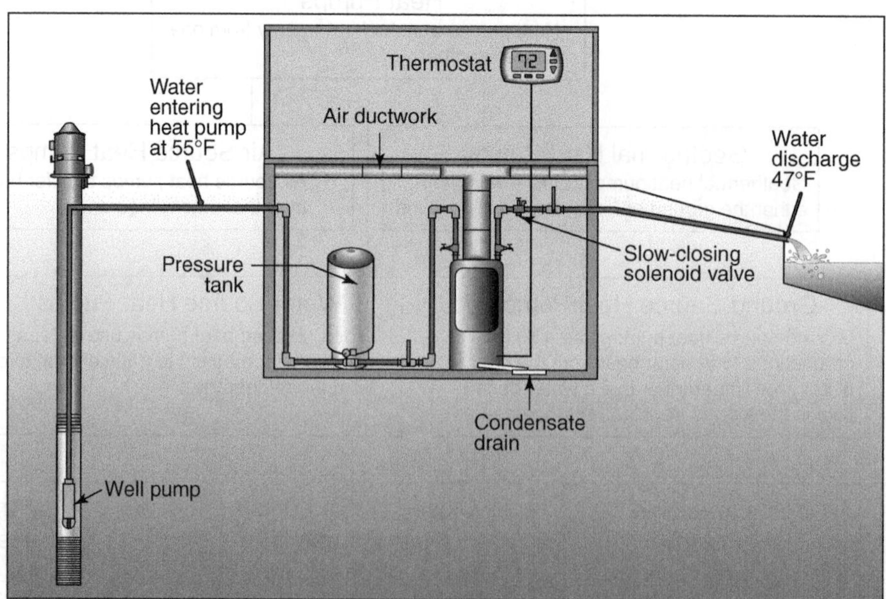

Figure 15–8
Water drawn from a well is used as the heat source and then discharged to some other location, such as a lake, stream, or pond. (Courtesy of Delmar/Cengage Learning)

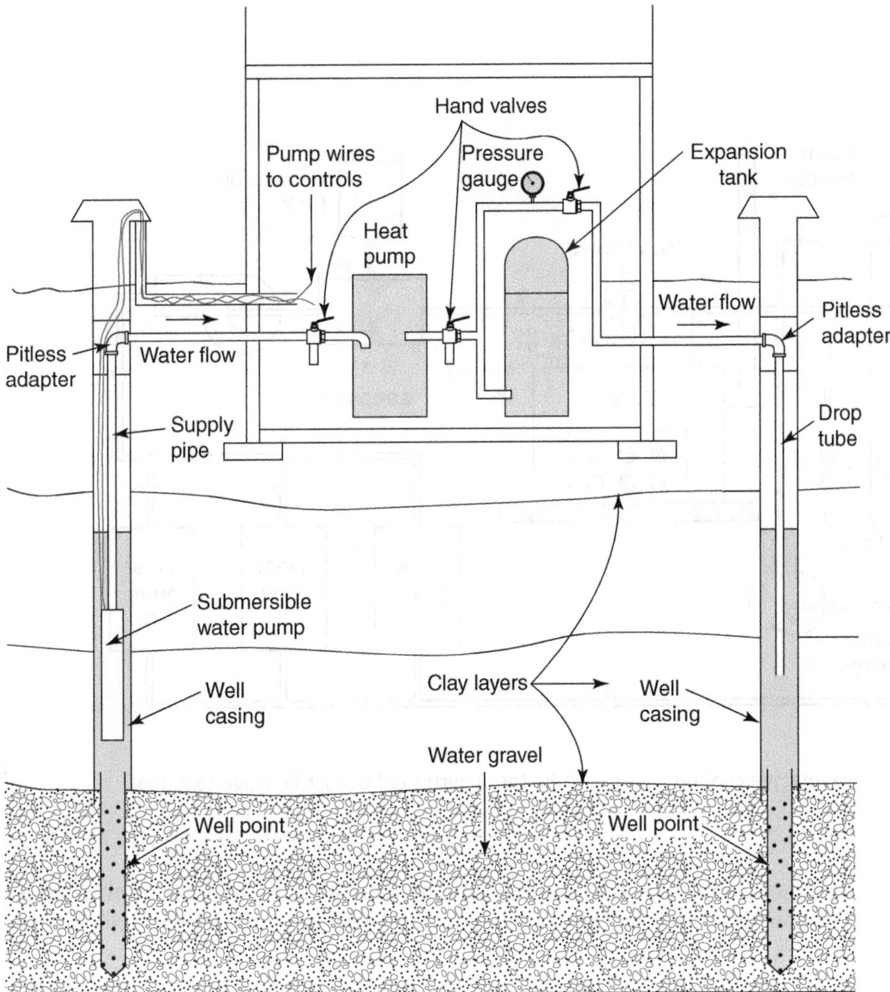

Figure 15–9
In some instances the water is injected into a return well that is driven to the same depth as the supply well (as per local code requirements). (Courtesy of Delmar/Cengage Learning)

Ground Source Heat Pumps

Ground source heat pumps (GSHP) use a buried ground loop to transfer heat from the ground directly to a refrigerant or use a water-based fluid in the loop to transfer heat to the refrigerant. Those systems that use refrigerant as the media are referred to as "direct expansion" systems, and those that use a water based solution as the media are referred to as "water loop" systems. Water loop systems can be either "open" or "closed", meaning that the loop itself is either open or closed to the atmosphere. We consider this a ground source heat pump because the source of the heat is the ground itself. The fact that we may be using a water-based solution to transfer the heat does not alter the fact that it is still a ground source heat pump.

Closed Water Loop Heat Pumps

Ground source heat pumps using closed loop water based solution as the media are by far the most common systems today. These systems are very often confused with water source heat pumps, but remember, the original source of the heat is the ground, not ground water, and therefore is a ground source heat pump. See Figure 15-10.

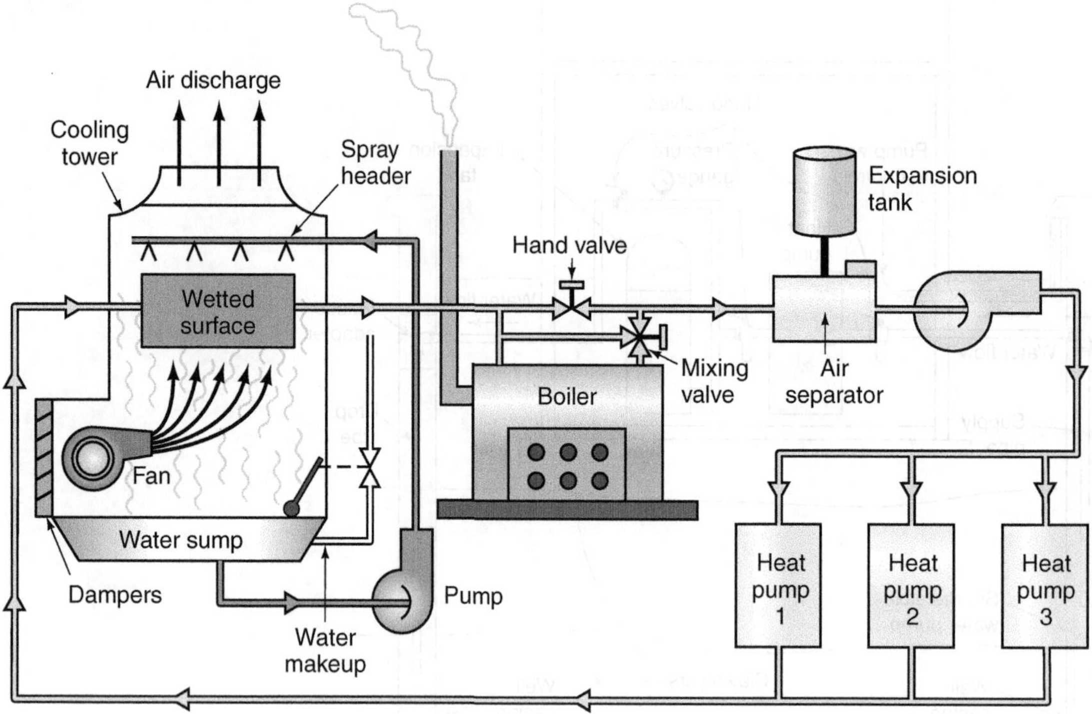

Figure 15–10
Closed-loop heat pump system coupled to an evaporative cooling tower with boiler. (Courtesy of Delmar/Cengage Learning)

An "open" loop system is far less common than a closed loop for ground source heat pumps. In this case, an "open" cooling tower may be used in the cooling mode to transfer heat out of the system.

Direct Expansion Heat Pumps

Prior to the development of polybutylene tubing for direct burial of water loops, ground source heat pumps buried refrigerant tubing directly in the ground. These refrigerant tubes were connected together in a header (Figures 15-12 and 15-13) and then brought to the heat pump in common piping. Another way of burying the tubing was with overlapping loops (see Figure 15-11) commonly referred to as a "slinky" because of its resemblance to the toy.

These systems do not use water in a secondary loop; they bury the refrigerant tubing directly in the ground. These refrigerant tubes are connected together in a header (see Figures 15-12 and 15-13) that contains a unique TXV-like expansion device allowing the heat exchanger to act as an evaporator in the winter and a condenser in the summer. These systems eliminate the heat transfer of a closed or open loop to a water-to-refrigerant coaxial heat exchanger and thus can operate at higher efficiencies with a small footprint installation.

Water Source Heat Pumps

As the name implies, water source heat pumps use rivers, lakes, and wells as a source of heat. The relatively "warm" water is pumped across a water-to-refrigerant heat exchanger where the heat contained in the water is absorbed into the refrigerant (evaporator). That heat-containing refrigerant is then compressed and sent to the indoor refrigerant-to-air heat exchanger (condenser coil) where the air from the structure is heated and delivered to the space.

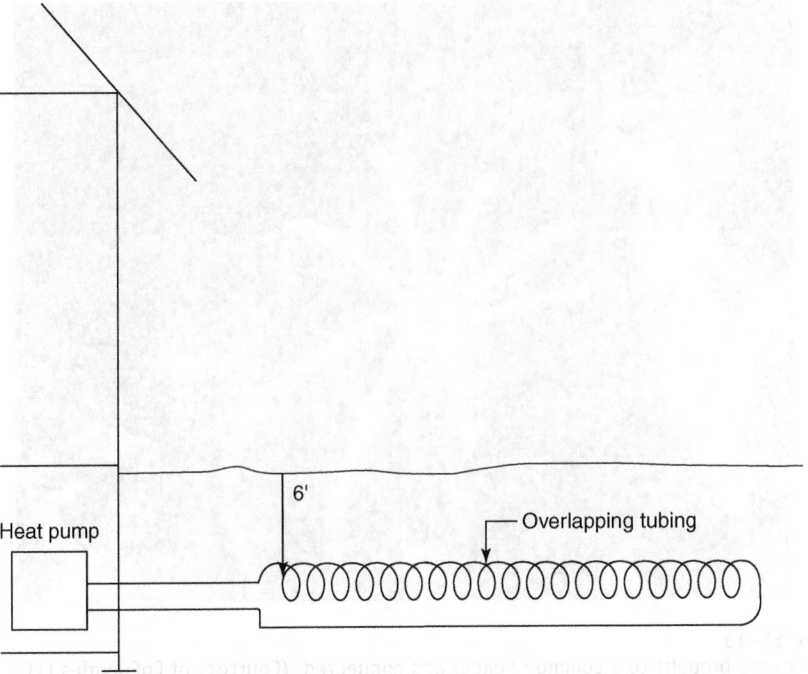

Figure 15–11
A bundled loop for a closed-loop or geo-loop system. This installation is sometimes called a "slinky." (Courtesy of Delmar/Cengage Learning)

Figure 15–12
A horizontal DX loop being dug. (Courtesy of CoEnergies LLC. Traverse City, Michigan)

A water source heat pump, meaning that the source of heat is ground water, is an open loop system where water is pumped from the source and returned to the environment. These are often referred to as "pump and dump" systems.

Commercial Systems

Commercial buildings often have an internal heat gain (lights, people, computers, machinery etc) that exceeds their heat losses (walls, window, roof, etc) and therefore require cooling even in the winter months. At the same time, there

Figure 15–13
DX loops are brought to a common header and connected. (Courtesy of CoEnergies LLC. Traverse City, Michigan)

are likely to be some areas of the building that need heat. There is a type of heat pump system, often referred to as a water source heat pump system, that is able to transfer heat from a warm zone that needs cooling to the zones that need heat.

This system has a closed loop water system pumped throughout the building and individual heat pumps are connected along the way. The water in the loop acts as a source of heat for those heat pumps that are supplying heat, and a place to reject heat from those heat pumps in the cooling mode. The water loop temperature is maintained at approximately 50° F, so it is both an excellent source of heat for the heat pump as well as providing for very efficient operation in the cooling mode.

If the loop temperature gets too cold because the need for heat exceeds the need for cooling, a hot water boiler is added to the system to bring the loop temperature back up. If the loop temperature gets too warm because the need for cooling exceeds the need for heating, a cooling tower is employed to cool the loop water temperature back down. A system like this affords the customer with the utmost in operational efficiency as well as a great deal of flexibility in building layout due to the use of many smaller heat pumps rather than one central system.

SUMMARY

Various types of heat pump systems have been discussed in this chapter. All of these different configurations use nearly the same components. The major distinctions are the way they transfer heat. Some transfer heat from the air, others from water directly, and still others transfer heat from the ground indirectly through a closed-loop system or directly from the ground using a DX loop. All of these systems have indoor coils, outdoor coils (or water, or ground coils), compressors, reversing valves, metering devices, and associated controls. They also have different operating characteristics, but that information will be presented in other chapters.

REVIEW QUESTIONS

1. Describe how a split-system heat pump is properly installed. At least two different configurations should be compared.
2. Describe where ductless (mini-split) systems can be installed and their purpose.
3. Describe a packaged system and the two different places where it may be installed.
4. Describe a dual-fuel installation and why a fossil fuel might be used.
5. Explain what "geothermal" means.
6. List and describe two methods to transfer heat from the ground or water using a geothermal heat pump.

CHAPTER 16

Geothermal Heat Pumps

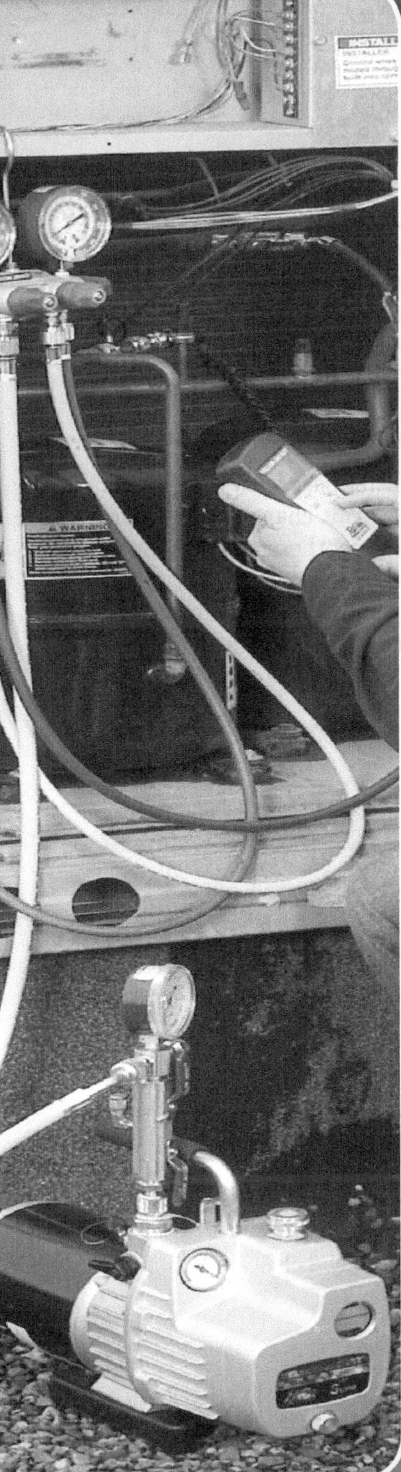

LEARNING OBJECTIVES

The student will:

- Describe the difference between geo-power and geo-source systems
- Describe how a geothermal heat pump transfers heat from the ground to the occupied space
- Relate what ground sources can be used for geothermal heat pumps
- Describe the difference between ground-source heat pumps (GSHPs) and water-source heat pumps (WSHPs)
- Explain the difference between open and closed loops

INTRODUCTION

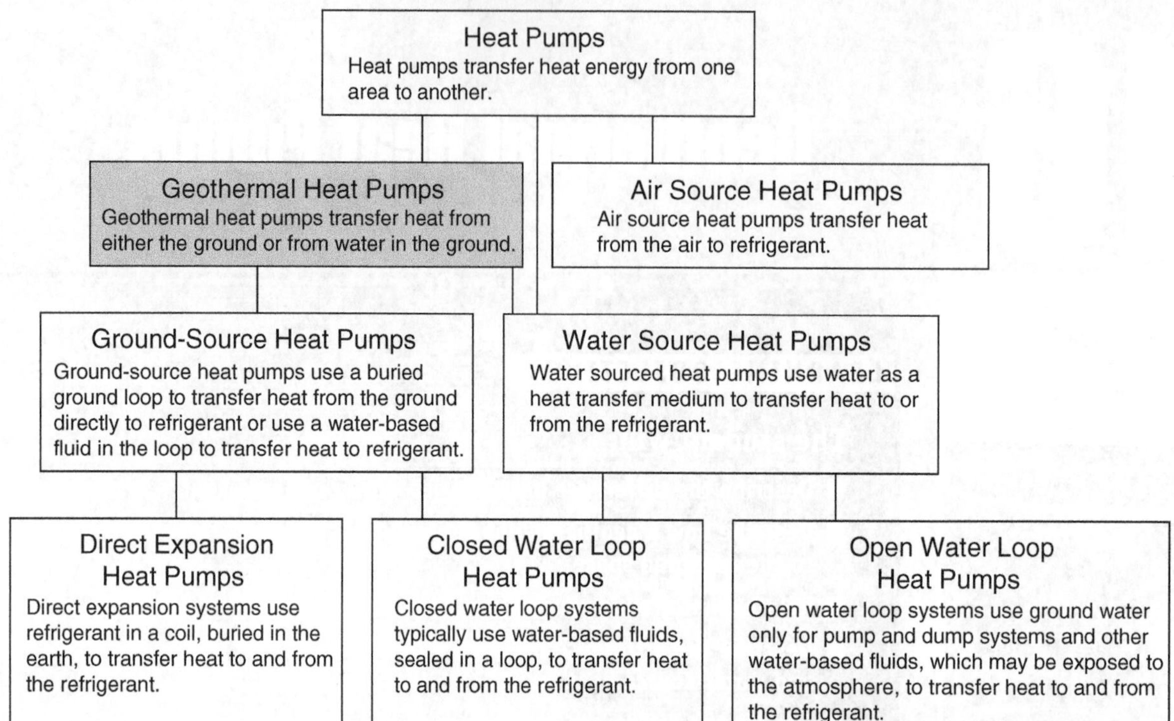
Heat Pumps
Heat pumps transfer heat energy from one area to another.
Geothermal Heat Pumps
Geothermal heat pumps transfer heat from either the ground or from water in the ground.
Air Source Heat Pumps
Air source heat pumps transfer heat from the air to refrigerant.
Ground-Source Heat Pumps
Ground-source heat pumps use a buried ground loop to transfer heat from the ground directly to refrigerant or use a water-based fluid in the loop to transfer heat to refrigerant.
Water Source Heat Pumps
Water sourced heat pumps use water as a heat transfer medium to transfer heat to or from the refrigerant.
Direct Expansion Heat Pumps
Direct expansion systems use refrigerant in a coil, buried in the earth, to transfer heat to and from the refrigerant.
Closed Water Loop Heat Pumps
Closed water loop systems typically use water-based fluids, sealed in a loop, to transfer heat to and from the refrigerant.
Open Water Loop Heat Pumps
Open water loop systems use ground water only for pump and dump systems and other water-based fluids, which may be exposed to the atmosphere, to transfer heat to and from the refrigerant.

The geothermal heat pump is unique in its ability to transfer heat from the ground. However, it is sometimes confused with geothermal power. This chapter will explore and define these two sources of earth energy. After defining geothermal, this chapter will concentrate on the various types of geothermal heat pumps and their applications.

Additional terms associated with this type of heat pump are: open and closed loop, single-use, desuperheater, primary and secondary loops, and heat transfer fluid. All of these will be explored in detail.

Field Problem

The homeowner was complaining of a "banging" noise coming from the geothermal heat pump. She wasn't sure that it happened when the unit started or when it stopped. She did know that it started a day ago and was keeping her up at night.

Symptoms:
Noise

Possible Causes:

1. Compressor mounts
2. Unit housing rubbing or touching the compressor or other component
3. The blower or pump motor mounts
4. The duct system
5. The pump packs
6. Air in system

Starting and stopping the system, with the required delays, allowed the technician to listen for the source. It came from the pump packs. This could mean:

1. The pump pack mount is loose.
2. The pump mount is loose.
3. Air in the loop is creating the noise.

Investigating further, all of the mounts were tight, but when a hand was placed on the piping, there was a noticeable thump that could be felt—water hammer. This is where the water that is flowing in the system wants to continue to flow, but is suddenly stopped. The stopped water creates a noise, like a hammer, when it comes to a stop. The valve that is designed to act slowly, in this case, was acting too quickly. The "slow-acting" solenoid was closing too quickly. This valve uses an internal fluid to slow the action of the valve and the fluid had leaked from the valve. When the valve was replaced, the noise was gone!

GEOTHERMAL

Geothermal is a term that literally means "heat from the ground." Scientists still do not know exactly what the center of the Earth is made of, but we have ample evidence that it is hot—magma spews from volcanoes and hot springs shoot geysers skyward with violent force.

There are two types of geothermal energy available from the Earth: geo-power and geo-source. The first, geo-power, is geothermal, which refers to the heat and volcanic type of activity that is connected to the center of the Earth. Most often this form of energy can be harnessed as steam and is referred to as geothermal power. Geothermal power plants tap the heat from the center of the Earth by sending water from the surface to interact with the hot rock strata deep below the Earth's surface. The general process is to pump water down to bedrock (granite) or rock structures that are very hot. The water pumped down 3 miles or more is forced under pressure. As it arrives at the bedrock level, it flashes to steam and is transported through fissures in the rock to a steam reception pipe. The steam comes to the surface, where it is used to drive a steam turbine. After driving the turbine to produce electricity, the steam is condensed and sent back down under pressure as water to begin the process again (Figure 16–1).

The second form of geothermal energy, geo-source, is lower-level geothermal heat. Because the rock strata is thicker and bedrock is further away, this form of geothermal energy is fed by two sources of energy: the core of the Earth and the sun. The sun warms the ground during the summer and the ground releases the sun's energy during the winter. Warming and cooling of the geo-strata serves to act as a thermal flywheel, delaying the seasons. We can see this as warm weather and cool weather seasons do not exactly coincide with the solar calendar. For example, March 20th is supposed to mark the first day of spring, but for many in the country, there is still winter-like weather. September 20th is supposed to be the first day of fall, but oftentimes the weather has been delayed and it is more like summer. This shift is due to the thermal flywheel that is absorbing and releasing heat energy as the solar cycle changes. We notice this shift as we move through the year, marking this shift between the solar calendar and the Earth calendar as a shift of about 45 days. In other words, it takes 45 days for the sun to warm the ground to begin spring and about 45 days to lose the heat to begin fall.

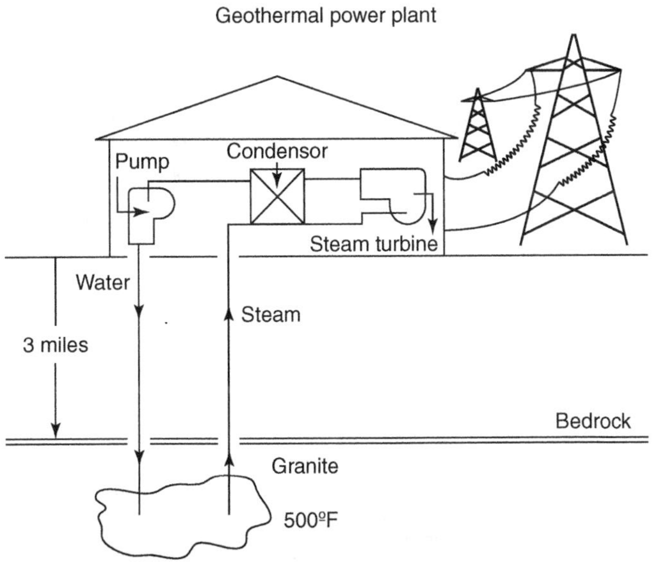

Figure 16–1
Geothermal heat from the center of the Earth turns water to steam to drive a steam turbine. (Courtesy of Delmar/Cengage Learning)

It is the second form of heat that geothermal heat pumps exploit. Ground temperatures hover at about 50°F throughout the country at a depth of approximately 6 feet. The exact temperature depends on the part of the country, the terrain, and the soil strata. A refrigerating machine is used to transfer the heat from the geo-strata into a structure. First outlined in a paper by Lord Kelvin in 1852, the refrigeration system was further developed by Peter Ritter von Rittinger in 1855. Robert C. Webber built the first direct-exchange (DX) system that transferred heat from the ground in the 1940s. The first successful geothermal system was installed in the Commonwealth Building in Portland, Oregon, in 1946 and has since received national historic status. Open-loop systems were popular prior to the development of polybutylene pipe in 1979, which made closed-loop systems possible. A geothermal system generally operates the same as an air-source heat pump (ASHP)—except that all of the components could be in the same unit, and the air coil is replaced with a water/refrigerant heat exchanger coil and a water loop or supply. The water supply provides the medium for heat transfer. A geothermal unit does not require a defrost system.

GEOTHERMAL HEAT PUMP

The geothermal heat pump differs from an ASHP in several ways. First, it does not have a defrost cycle. There is only a heating and a cooling mode. Defrost is not necessary because the geothermal temperature is above freezing and there is no concern about moisture accumulating or the need to get rid of moisture on the outdoor coil. Another difference is a primary and secondary loop. The

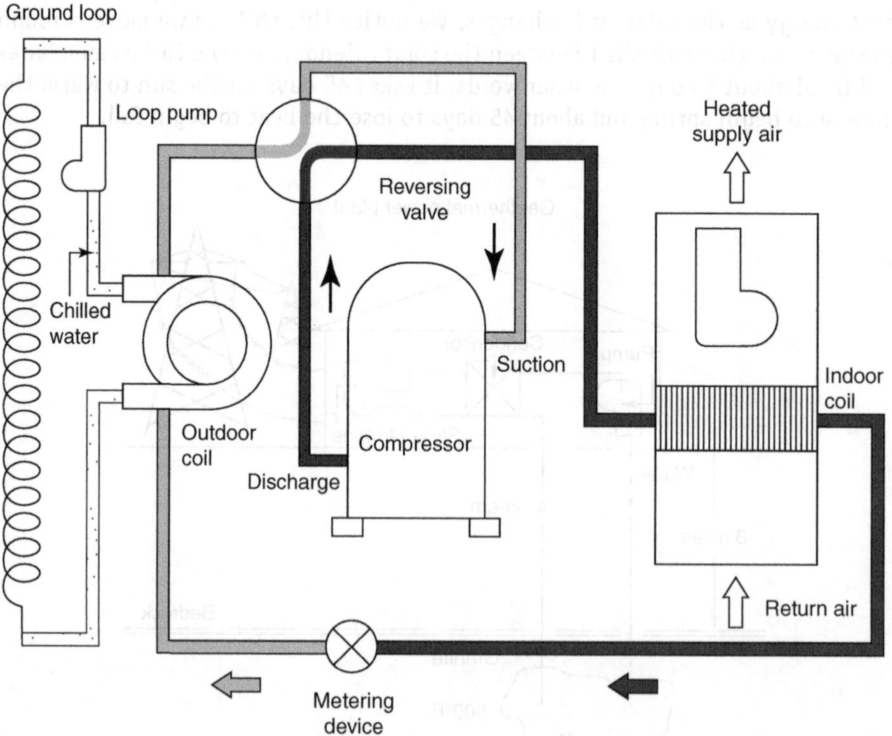

Figure 16–2
Geothermal system in the heating mode. (Courtesy of Delmar/Cengage Learning)

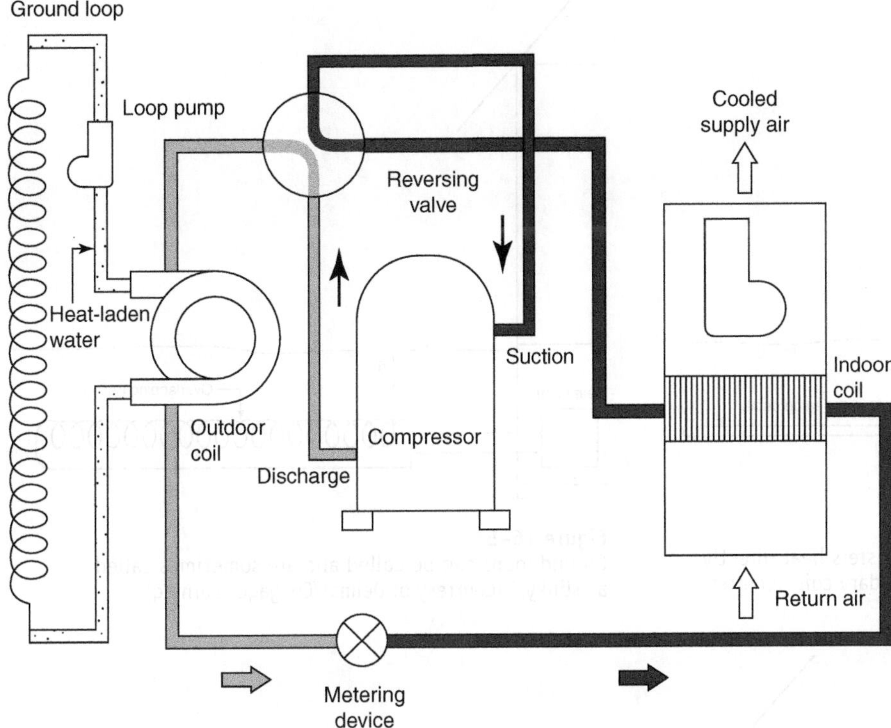

Figure 16–3
Geothermal system in the cooling mode. (Courtesy of Delmar/Cengage Learning)

primary loop is the refrigerant exchange loop. This loop transfers heat with water in a water-to-refrigerant heat exchanger. This heat exchanger takes the place of the air-to-refrigerant coil of an air-source system. It is still referred to as the outdoor coil, but there is also an outside loop. This is the secondary loop and contains water or a water-glycol solution. The water in the secondary loop can be closed or trapped. Closed-loop systems use the same water to exchange heat with the ground and the refrigerant. An open-loop system brings new water to the refrigerant heat exchanger and discharges the used water. It is called "open" because the supply or discharge water (or both) is open to the atmosphere and atmospheric pressure. The diagrams in Figures 16–2 and 16–3 show a geothermal system in the heating and cooling modes.

Geothermal versus Water-Source Heat Pumps

There are two subcategories of geothermal heat pumps. Both of these subcategories transfer heat from the ground. One transfers it from the soil; the other transfers it from water in the ground. The two types are referred to as the ground-source (or geo-source) heat pump and the water-source heat pump (see Figures 16–4 through 16–8).

Both the GSHP and the WSHP use the same refrigeration components and heat exchangers. In most cases, the manufacturer has designed the system to be installed as either a GSHP or a WSHP. The difference to remember is that the GSHP is a closed-loop system and the WSHP is an open-loop system. Another way to think about the difference is that a GSHP uses the same water, because it is a closed system. The WSHP is supplied with, in most cases, new water that is open to the atmosphere.

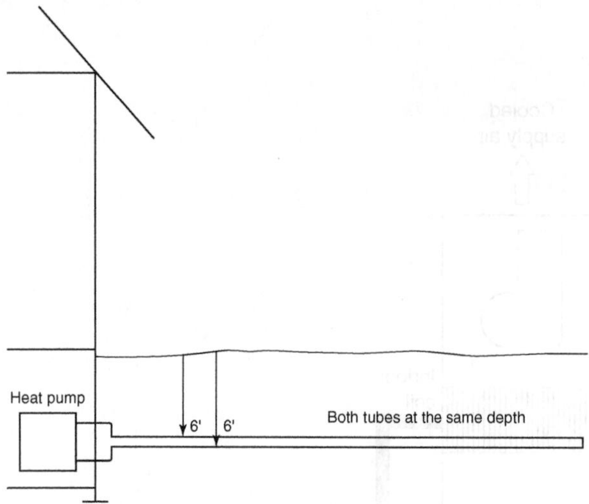

Figure 16–4
A ground-source heat pump (GSHP) transfers heat directly from the soil using a closed-loop secondary coil. (Courtesy of Delmar/Cengage Learning)

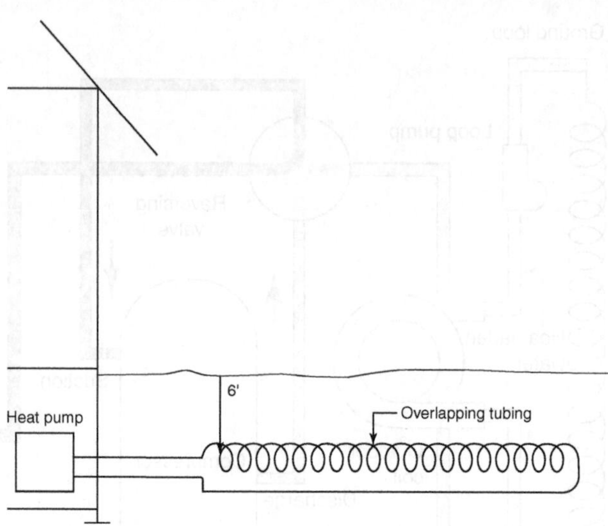

Figure 16–5
Ground loops can be coiled and are sometimes called a "slinky." (Courtesy of Delmar/Cengage Learning)

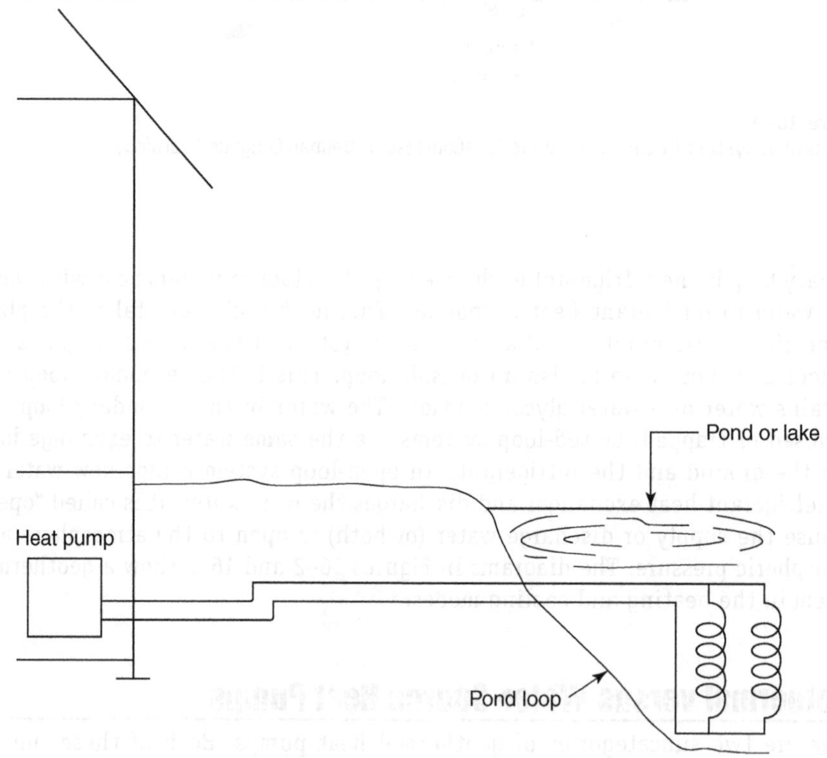

Figure 16–6
Ground systems also include pond loops. The closed-loop system is submerged in a lake or pond. (Courtesy of Delmar/Cengage Learning)

GEOTHERMAL—DIRECT EXPANSION

There is one other type of geothermal heat pump system that does not use water in a closed or open secondary loop. This type of geothermal system buries the refrigerant heat exchanger directly in the ground. This system is called a direct-expansion (DX) system because refrigerant is allowed to change state directly in the ground, absorbing and rejecting heat directly with the earth (Figures 16–9 and 16–10).

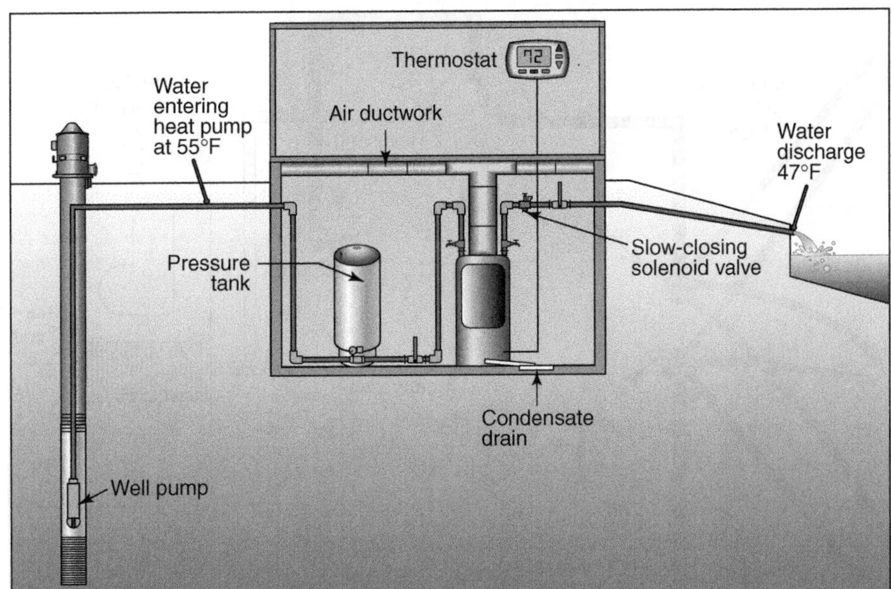

Figure 16–7
A water-source heat pump (WSHP), open-loop system draws water from a well and discharges it to a lake, pond, or ditch. (Courtesy of Delmar/Cengage Learning)

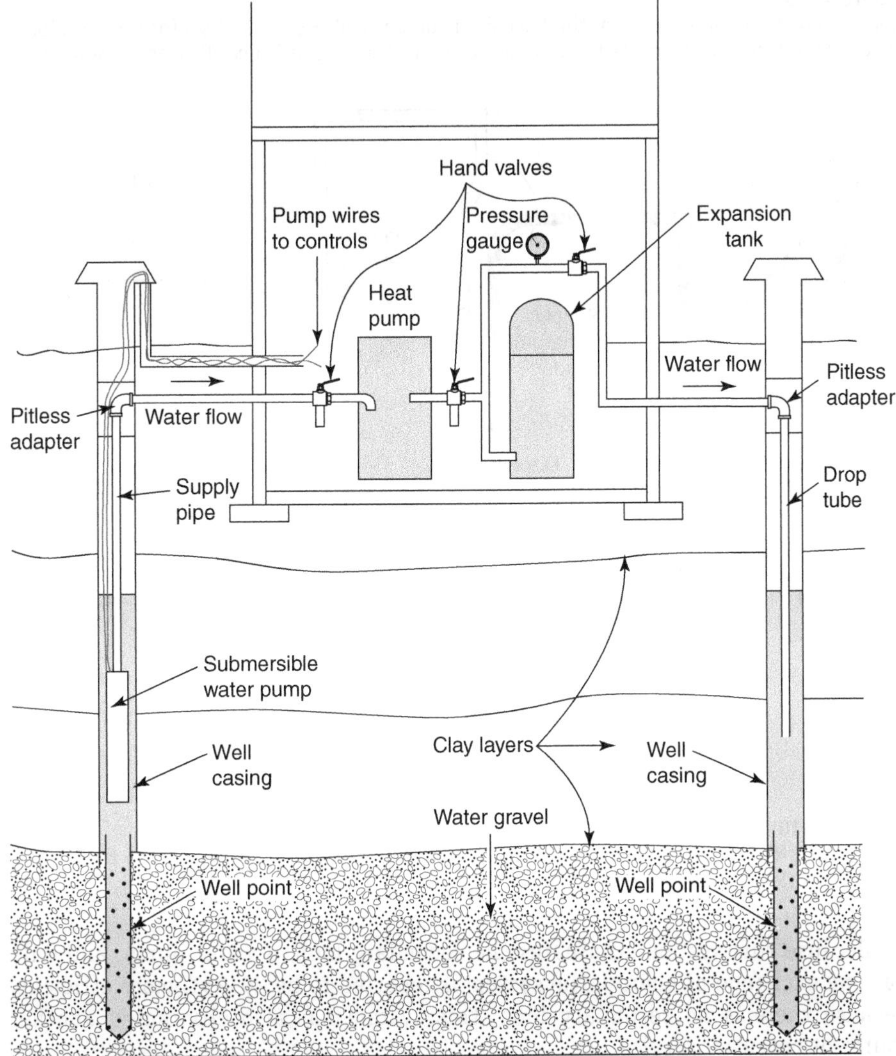

Figure 16–8
Another type of WSHP, open-loop system in which water is drawn from one well and discharged to another well. This system is called an injection system. Follow local code requirements. (Courtesy of Delmar/Cengage Learning)

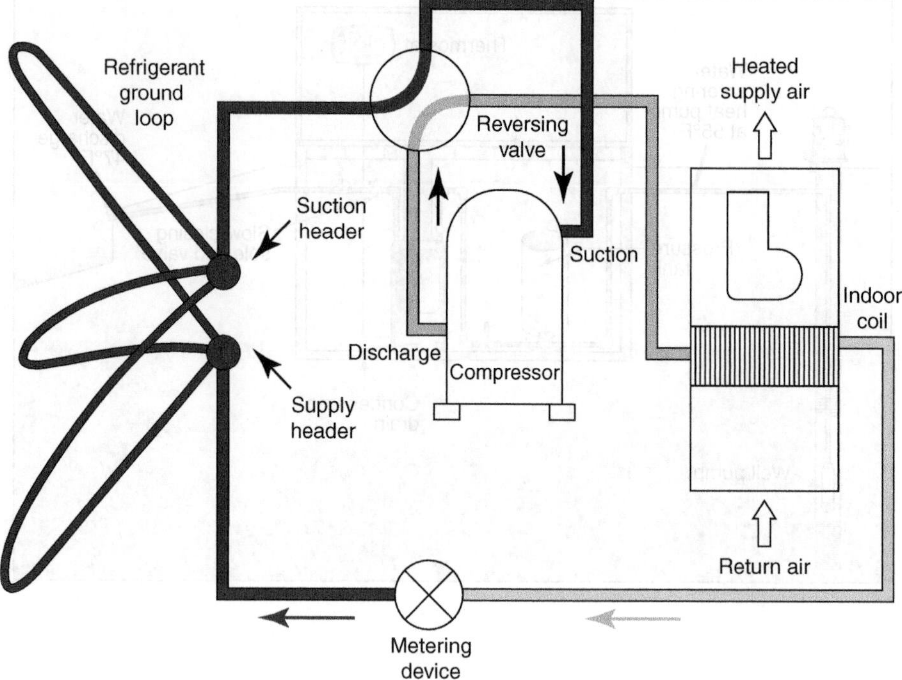

Figure 16–9
Heat is transferred directly from the buried refrigerant coil. Headers allow for separate loops of refrigerant lines to be buried in a radial fashion. (Courtesy of Delmar/Cengage Learning)

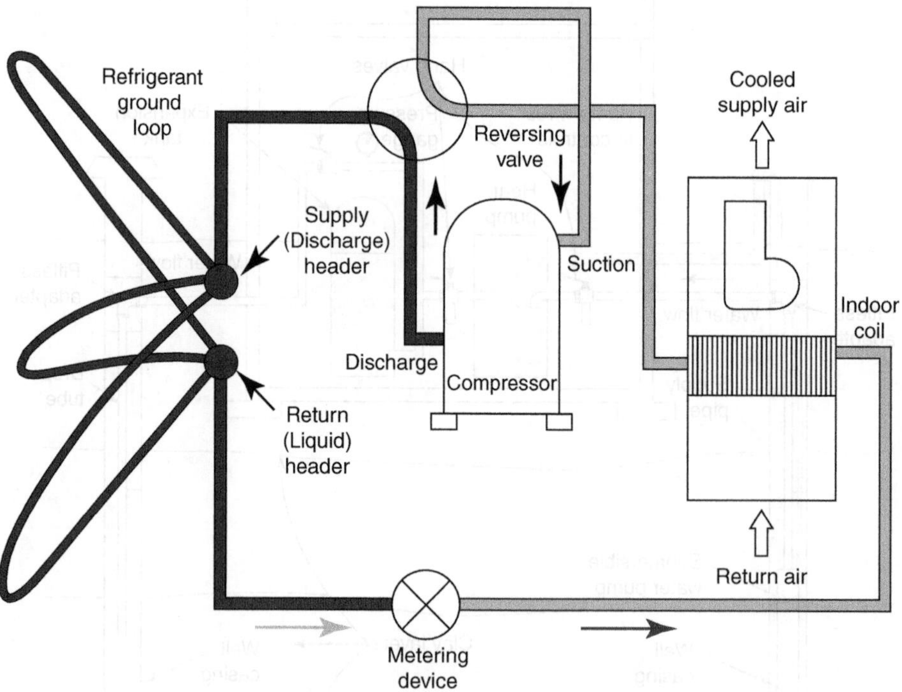

Figure 16–10
Heat is transferred directly into the ground as vaporous refrigerant condenses to a liquid. (Courtesy of Delmar/Cengage Learning)

Because this system eliminates the secondary loop, a water pump, and only has one exchange of energy, the system is reported to be 20–25% more efficient. Because there is no secondary loop, this system has also eliminated another source of potential service.

Composition of a Geothermal Heat Transfer Medium

The heat transfer medium or solution for a geothermal system is generally a water-based fluid. Water and antifreeze solutions are mixed to keep water from freezing under certain circumstances. Anti-corrosives are also added to some systems to keep circulating components from rusting or corroding.

Single-Use Systems

Single use refers to how the heat transfer medium is used. In this system, single use means that the transfer medium is used only once. Water is the transfer medium and is pulled from a water well. Ground water is a good source for geothermal heat pumps and, where the water is plentiful, the water is pulled from the well, heat is transferred by the heat pump, and the water is released to a surface water location. Ponds, lakes, rivers, and streams are used to "dump" water after the heat is transferred. Because the water goes through the heat pump and only heat is rejected, it can be released without worry that it has become contaminated. For the reasons stated, the single-use system is commonly known as a "pump-and-dump" WSHP.

Another version of the single-use system is one that tends to replenish or maintain the aquifer (water-bearing strata) by returning the water to the same level as the supply pump. This system is called an injection WSHP. In this version, water is pulled from the water well, heat is removed by the heat pump, and the water is returned to the same aquifer by using a well that is drilled to the same depth as the supply well (as per local codes dictate). The theory is that water returned to the same aquifer will eventually return to the supply well, but because of the immensity of the aquifer and the path water flows in the aquifer, the injected water may never be used again. See Figures 16-8 and 16-9.

Open Loop

WSHPs are generally open-loop systems (Figure 16-11). "Open loop" means that the water being used is at atmospheric pressure. Water pulled from a well is under the influence of the atmosphere, because there are vent holes in the well that allow pressure to equalize to atmospheric pressure. If water is either taken or released to atmospheric pressure, the system is considered to be an open-loop design.

Closed Loop

System with water contained either under pressure or no pressure; within a vessel, pipe, or tubes; and with no opening to the atmosphere are considered closed-loop designs. Closed loops contain water-based transfer solutions for heat pumps. Water and an antifreeze solution are mixed and circulated through the closed system, being used to repeatedly transfer heat without losses of fluid. Only if the closed-loop system leaks will the fluid leave the system.

BOILERS, EVAPORATIVE COOLING TOWERS, AND CHILLERS

Geothermal heat pumps are used in residential, commercial, and industrial HVAC applications. The water-to-refrigerant coil can transfer heat with cooling tower water that acts as a water source for the heat pump (Figure 16-12). This system is a water-source, closed-loop system. Notice that the water is taken and returned

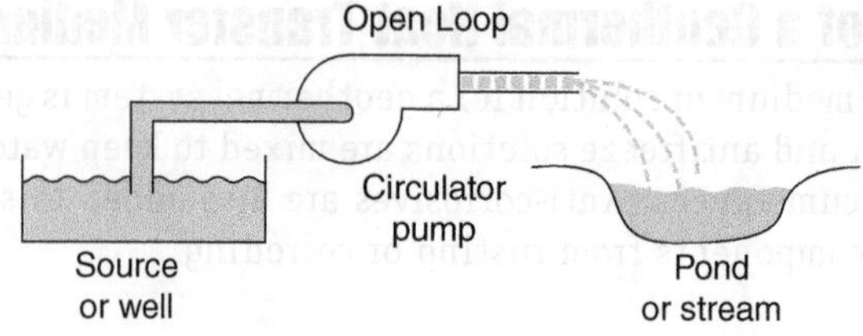

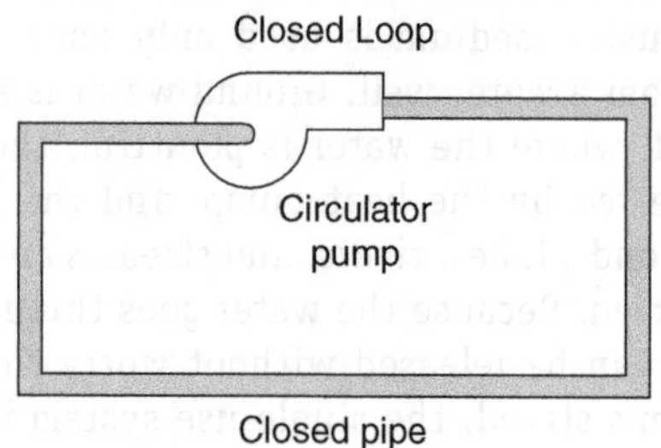

Figure 16–11
Open loops are connected to atmospheric sources or discharges. Closed loops contain fluid that can be under pressure or no pressure, but in either case, the fluid does not connect with atmospheric pressure. (Courtesy of Delmar/Cengage Learning)

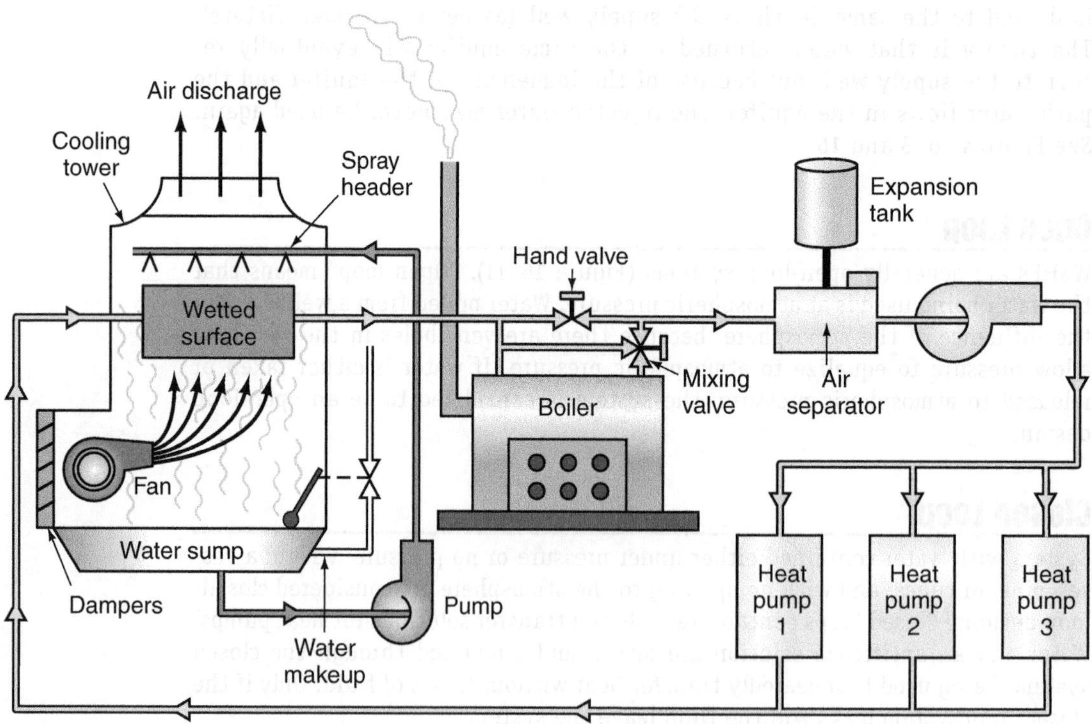

Figure 16–12
An closed-loop WSHP with boiler and cooling tower to maintain the loop temperature. (Courtesy of Delmar/Cengage Learning)

to the cooling tower and is used continually; it becomes the exception to a single-use system where the water does not return. A boiler can be used for additional heat in the winter to add heat to the loop. Water in the closed loop is returned, and additives and antifreeze solution can be added to this system, which allows the system to work through the winter months.

SUMMARY

In this chapter we have introduced the term "geothermal." Geothermal power production (geo-power) creates power directly from high-temperature sources of heat in the Earth. Geothermal heat pumps exploit lower temperatures of heat in the ground strata (geo-source). This chapter was devoted to discussing geo-source heat pumps.

Geothermal heat pumps are systems that transfer heat from geo-sources, primarily in the ground. Most geothermal systems require a primary and secondary loop to accomplish the heat transfer from the ground. The first loop (primary) is the refrigerant-to-water loop. The second loop is a ground-to-water loop. Combined, they form the heat transfer bridge from ground to refrigerant.

Geo-source systems transfer heat from the ground or ground water. WSHPs use ground water as a heat source. GSHPs transfer heat from the ground, which may contain water. The WSHP is generally considered an open-loop system in which water is either being pulled from or released to a well, pond, or lake at atmospheric pressure. GSHPs are generally considered closed-loop systems, because water-based heat transfer solution is sent out through piping that is buried in the ground and is then returned, with no loss of fluid.

Geothermal heat pump systems are suited to transfer heat from any water-based heat transfer fluid. They are successfully connected to commercial cooling towers that act as sources and discharge points for the water-based solution.

REVIEW QUESTIONS

1. Describe the difference between geo-power and geo-source systems.
2. Describe how a geothermal heat pump transfers heat from the ground to the occupied space.
3. Relate what ground sources can be used for geothermal heat pumps.
4. Describe the difference between ground-source heat pumps (GSHPs) and water-source heat pumps (WSHPs).
5. Explain the difference between open and closed loops.

CHAPTER 17

Ground-Source Heat Pumps

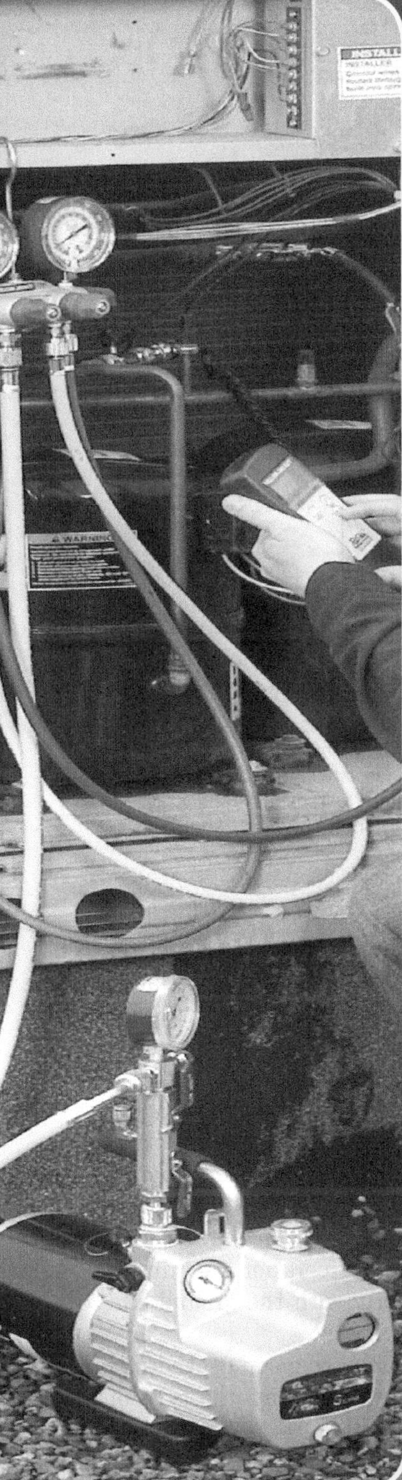

LEARNING OBJECTIVES

The student will:

- Describe the difference between an open- and closed-loop system
- Describe the function and use of geothermal plastic pipe material
- Relate why the heat transfer solution must be protected from freezing in some installations
- Describe why "slinky" systems use less ground area
- Explain why vertical loops might be used rather than horizontal loops
- Describe what would influence the final installed cost of a geothermal system

INTRODUCTION

A ground-source heat pump (GSHP) is a geothermal heat pump. The designation "ground source" has more to do with where the source of heat is located. In this chapter, the emphasis will be on how heat is transferred from the ground, through all heat transfer processes, and is moved into the structure.

Please note that all explanations of GSHPs in this chapter are explained in terms of the "heating mode." The heating mode is where heat is transferred from the ground to the structure. Because GSHPs are also cooling systems, in the "cooling mode" the system uses a reversing valve to switch cycles to transfer heat from the structure to the ground.

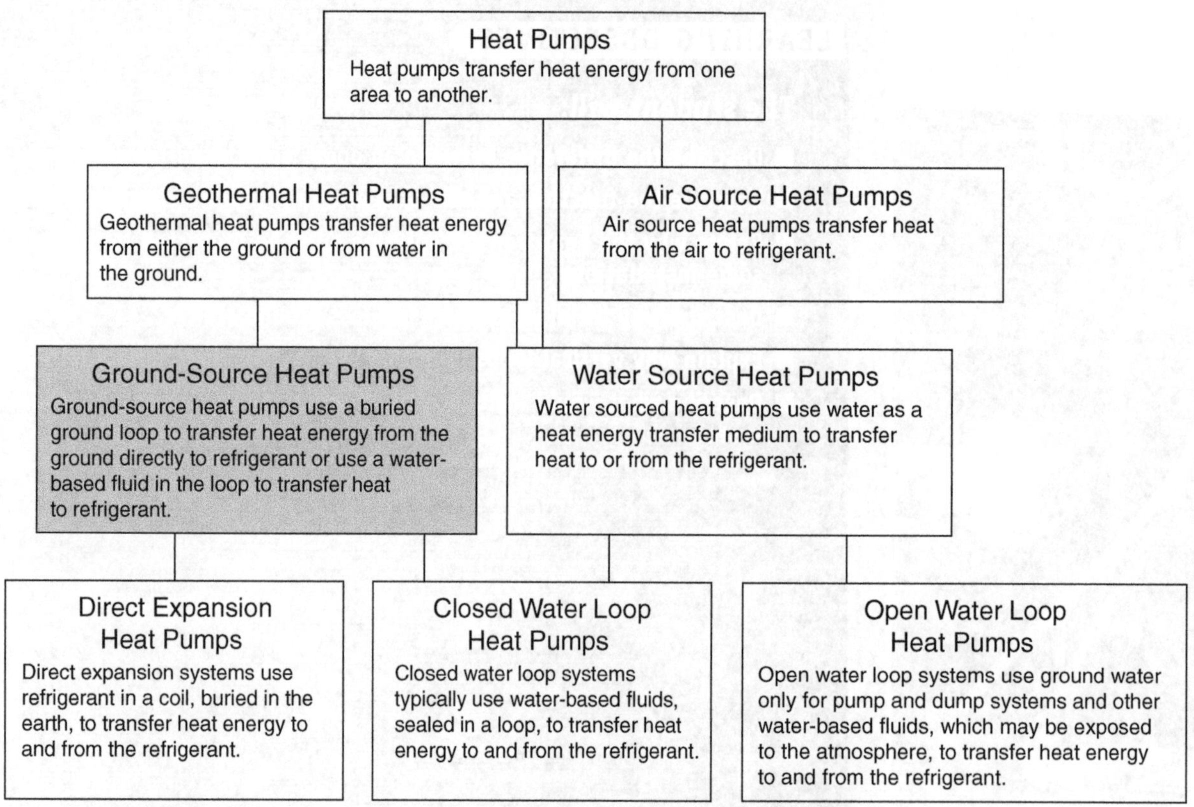
Heat Pumps
Heat pumps transfer heat energy from one area to another.
Geothermal Heat Pumps
Geothermal heat pumps transfer heat energy from either the ground or from water in the ground.
Air Source Heat Pumps
Air source heat pumps transfer heat from the air to refrigerant.
Ground-Source Heat Pumps
Ground-source heat pumps use a buried ground loop to transfer heat energy from the ground directly to refrigerant or use a water-based fluid in the loop to transfer heat to refrigerant.
Water Source Heat Pumps
Water sourced heat pumps use water as a heat energy transfer medium to transfer heat to or from the refrigerant.
Direct Expansion Heat Pumps
Direct expansion systems use refrigerant in a coil, buried in the earth, to transfer heat energy to and from the refrigerant.
Closed Water Loop Heat Pumps
Closed water loop systems typically use water-based fluids, sealed in a loop, to transfer heat energy to and from the refrigerant.
Open Water Loop Heat Pumps
Open water loop systems use ground water only for pump and dump systems and other water-based fluids, which may be exposed to the atmosphere, to transfer heat energy to and from the refrigerant.

A GSHP transfers heat from the ground either directly or indirectly from water or air. One of the efficiency advantages of both the ground-source and water-source heat pumps over an air-source heat pump (ASHP) is that they do not require a defrost cycle. Instead, they have freeze sensors that keep the water, used as a heat transfer solution, from freezing.

Pumps work to deliver water or a water-based solution from the outside loop, through the coil (located in the heat pump, inside the structure), and back to the outside loop. Two types of loops are used, an open loop and a closed loop. In the heating mode, the heat pump cools the water and sends it back to the ground to pick up more heat. Direct expansion heat pumps are different in that they do not use a loop of water. Instead, the outside coil of the refrigeration machine is buried in the ground. This system makes contact directly with the ground and removes heat. Figures 17-1 and 17-2 show a closed-loop WSHP.

The chapter will close with factors that affect the cost of common geothermal systems. Maintenance is also a concern, and the chapter ends with the type of maintenance that is related to each system type.

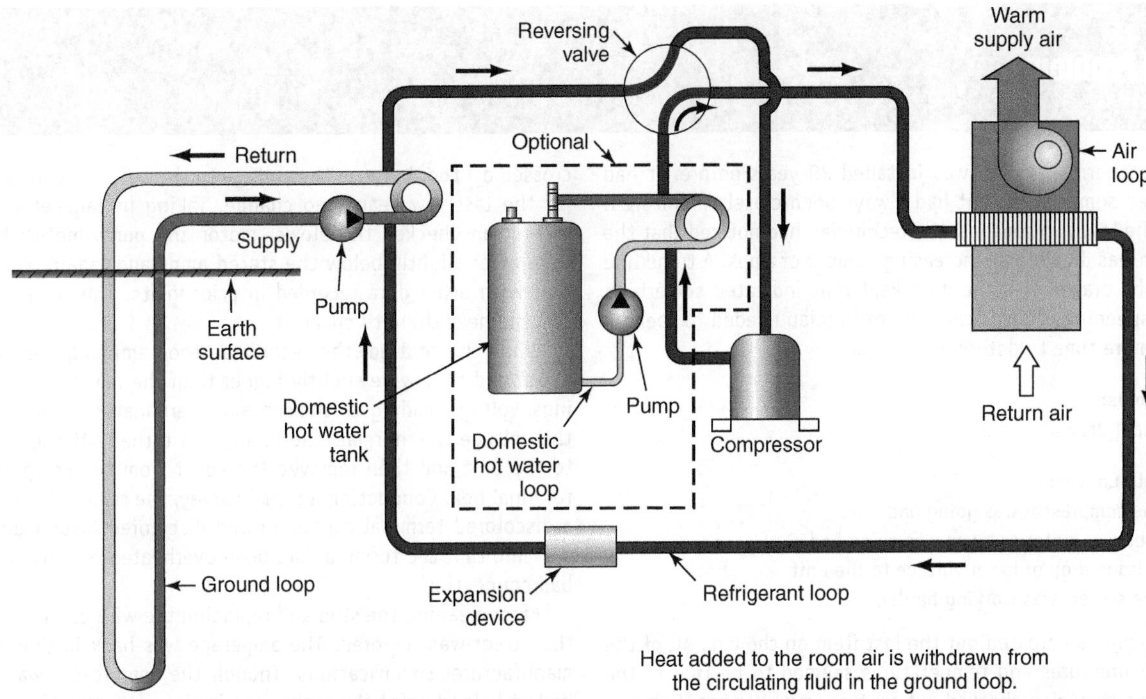

Figure 17–1
The closed-loop WSHP is shown in the heating mode. (Courtesy of Delmar/Cengage Learning)

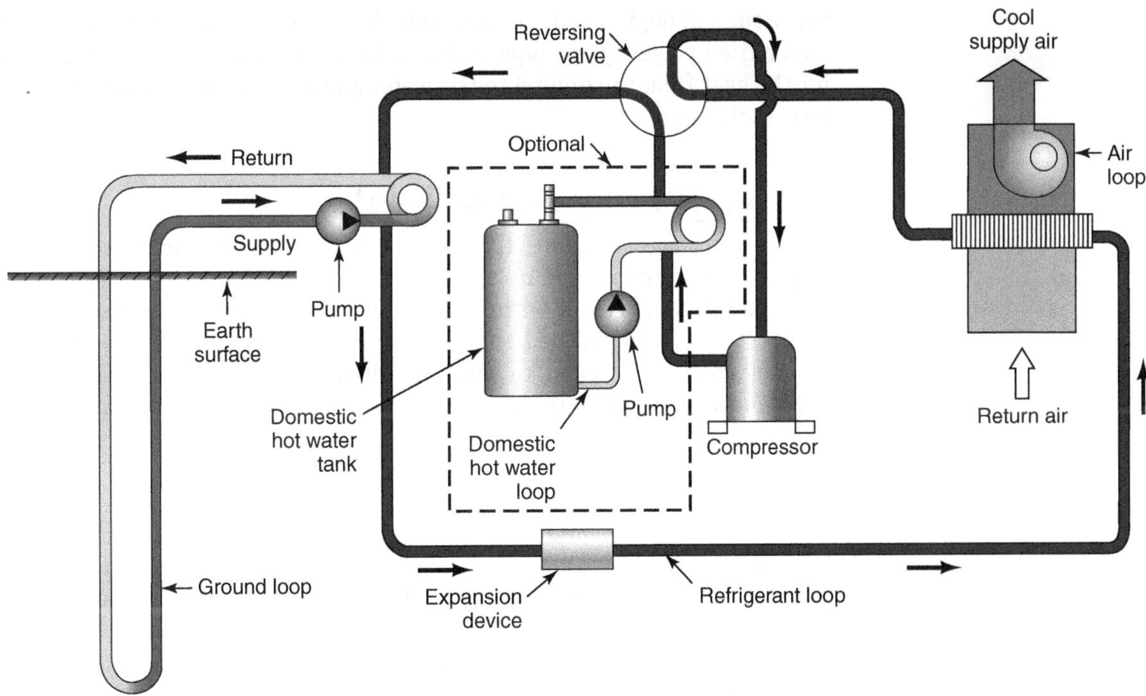

Figure 17–2
The closed-loop WSHP is shown in the cooling mode. (Courtesy of Delmar/Cengage Learning)

Field Problem

The geothermal system was installed 20 years before. It had received some repair, but had always been closely monitored. Over the last two seasons, the technician had noticed that the system was drawing an increasing number of amps. A trend line could be drawn with the data kept that indicated something was happening. On this call, the technician needed to spend a little more time to determine the cause.

Symptoms:
High amp draw.

Possible Causes:
1. The compressor was going bad.
2. A blower motor or pump was going bad.
3. Voltage drop or lower voltage to the unit.
4. The system was working harder.

The technician crossed out the last item on the list. All of the system pressures and temperatures seemed to be fine in the past and they also checked out for this visit. Voltage drop was crossed off too, because the voltage to the unit was consistent for the last few years—no change. Taking his ammeter, the technician checked the blower motor and pump motor; both were at or slightly below the stated amps and, again, consistent with other data recorded in prior visits. The compressor was the next thing to check.

While it operated, the technician took amperage readings and found they were slightly higher than the nameplate readings. Voltage readings at the contactor terminals indicated that the voltage was normal. The technician turned off the power to the unit and then removed the cover from the compressor terminal box. Conducting a visual survey, the technician found a discolored terminal connector and discolored insulation indicating that the terminal had been overheated, resulting in a bad connection.

After cleaning the stud and replacing the wire connections, the power was restored. The amperage was back in line with manufacturer specifications. Though the compressor was not bad, it helped point the technician in the right direction.

CLOSED-LOOP SYSTEMS

One of the most popular geothermal installations is the closed-loop or ground-source heat pump. It is called "ground source" because it transfers the heat from the ground at depths of 6′ or more, transferring that heat into the living space. It uses buried polyethylene pipe that is filled with a water-glycol solution to transfer the heat from the ground to the water and bring it to the water-to-refrigerant heat exchanger.

High-density polyethylene (HDPE)

High-density polyethylene (HDPE) plastic pipe is recommended for use as a ground loop. Polyethylene is used in and for a variety of products, such as crates, racks, and spacers, and mixed with wood it becomes structural material for outside decks. It is typically labeled as #2 plastic and can be recycled. The "high-density" label means that it is able to withstand higher pressures and temperatures. This plastic is made from petroleum and can be used at temperatures of 230°F (110°C).

Heat Transfer Solution

The solution used in ground loops is usually a combination of water and propylene glycol. Although 100% water would be the best transfer solution, water will freeze at 32°F (0°C). Considering that the ground temperature is around 50°F and the temperature difference from inlet to outlet could be as much as 10°F, the coil temperature would, theoretically, be 40°F. Also, keep in mind that typically the refrigerant temperature is 10°F colder then the fluid temperature. However, the ground temperature will be lower in the vicinity of the ground coil and could get as low as 30°F. This would mean that the heat transfer fluid could be operating below freezing.

To counteract the problem of freezing, an antifreeze solution is added to the water. Adding anything to the water will decrease the ability of water to transfer heat. By itself, water has a specific heat of 1. One pound of water will rise 1°F when 1 BTU is added. When propylene glycol is added to water, the specific heat of the solution drops as does the freezing point (see Table 17-1).

There are three common water-based antifreeze solutions on the market that are easily obtained and are used in the HVACR industry: ethylene glycol, propylene glycol, and methanol alcohol. Their properties are as follows:

- Ethylene glycol is a substance that is toxic and requires special handling.
- Propylene glycol is less toxic and safer to handle.
- Methanol alcohol, or wood alcohol, is more toxic than propylene glycol.

Tech Tip

The technician needs to write down/record the fluid in the loop and percentages of solution. This information is recorded and left with the machine. The percentage of solution can be measured with either a refractometer or hydrometer. When the fluid type is unknown, a sample of the fluid can be taken to a water-testing service to determine the type and percentage.

Ethylene glycol is an antifreeze solution that is used in boilers and other HVAC applications. It is generally used in above-ground systems where leaks can easily be seen and repaired. Use below ground is not recommended, because it could leak directly into the ground and contaminate the water table.

Propylene glycol is a good match for geothermal systems. It is easy to purchase and easy to use. It doesn't react with the propylene pipe and is relatively less toxic than other antifreeze solutions if spilled or leaked into the environment. The downside is that when mixed with water, it lowers the heat transfer ability of water by as much as 15% compared to pure water at 40°F. It also changes the viscosity (thickness) of water, requiring pumps to work harder to push the solution. Loop sizing tables take both of these factors into consideration by lengthening the loop, but pumping power needs to be increased by 30–40% to compensate for the longer length and the more viscous fluid.

Table 17–1 **Propylene Glycol Table**

Percent of Propylene Glycol in Water	Freezing Point		Burst Protection	
	Fahrenheit	Celsius	Fahrenheit	Celsius
0%	32	0	32	0
10%	26	−3	22	−6
20%	20	−7	10	−12
30%	10	−12	−20	−29
36%	0	−18	−60	−51
40%	−5	−20	−60	−51
43%	−10	−23	−60	−51
48%	−20	−29	−60	−51
52%	−30	−34	−60	−51
55%	−40	−40	−60	−51
58%	−50	−46	−60	−51
60%	−60	−51	−60	−51

Tech Tip

Freeze point is the temperature at which the mixture will start to freeze (slushy mixture), and at this point it will become difficult to pump, causing pumps to draw excessive amps. Although the system may not be operable below the freeze point, the system is still protected from bursting piping because even though the mixture will become slushy below the freeze point, it will not freeze solid until colder temperatures. Information and percentage charts for burst protection as well as corrosion inhibitors used are available from the manufacturer or supplier of the glycol used.

Methanol (wood) alcohol is another antifreeze solution. It comes from wood and is found in auto windshield washer fluid to keep it from freezing and to help clean the glass. It can be mixed with water and only affects the heat transfer coefficient by less than 5% at 40°F water temperature. Like propylene glycol, methanol does not react with the propylene pipe. But when mixed with water, it is not as viscous and requires an additional 15–25% more pumping power than pure water. The cost of methanol alcohol is lower than that of propylene glycol, making this antifreeze solution a viable choice. The downside is that it is toxic.

Tech Tip

Ethylene glycol is not a recommended antifreeze solution. It has properties that do not make it an environmentally safe substance. If this substance is discovered in a system, it is recommended that the solution be changed to another, recommended transfer fluid. Other fluids may be used, but may require manufacturer recommendations and in some cases specialized equipment.

Horizontal Loop

One type of loop for a GSHP is the horizontal loop. Horizontal runs of HDPE are done in trenches that are 6 or more feet deep. The runs of pipe can be laid at an excavation site and then covered with soil. Large areas that are to be parking lots can be excavated to a depth of 6 feet, the piping loop laid, and then covered, tamped, and paved. If done in an urban backyard, straight runs of piping could take up all of the yard.

Two or more pipes can be laid in the trench, and there are design suggestions as to the distances apart. These horizontal runs can also be called "single-layer" loops (Figure 17-3), because they are all at the same depth. Another design places one pipe above the other. This is called a two-layer loop (Figure 17-4). The first loop is laid and then the trench is backfilled to a certain level before the pipe is laid and returned to the origination point (Figure 17-5).

The approximate length of pipe per ton is 300 to 500 feet. The actual length will depend on ground conditions and the experience of contractors in an area. The length of the run will have an impact on pumping energy and the amount of heat that can be absorbed. Designers attempt to minimize the length of the run and match it to the heat pump capacity (Figures 17-6 and 17-7).

Earth coil type:	horizontal-single pipe
Flow pipe:	series
Typical pipe size:	1 1/4 to 2 inches
Nominal length:	350 to 500 feet/ton
Burial depth:	3.5 to 5 feet
Maximum heat pump size:	5 tone

Figure 17–3
A single-layer horizontal loop. (Courtesy of Delmar/Cengage Learning)

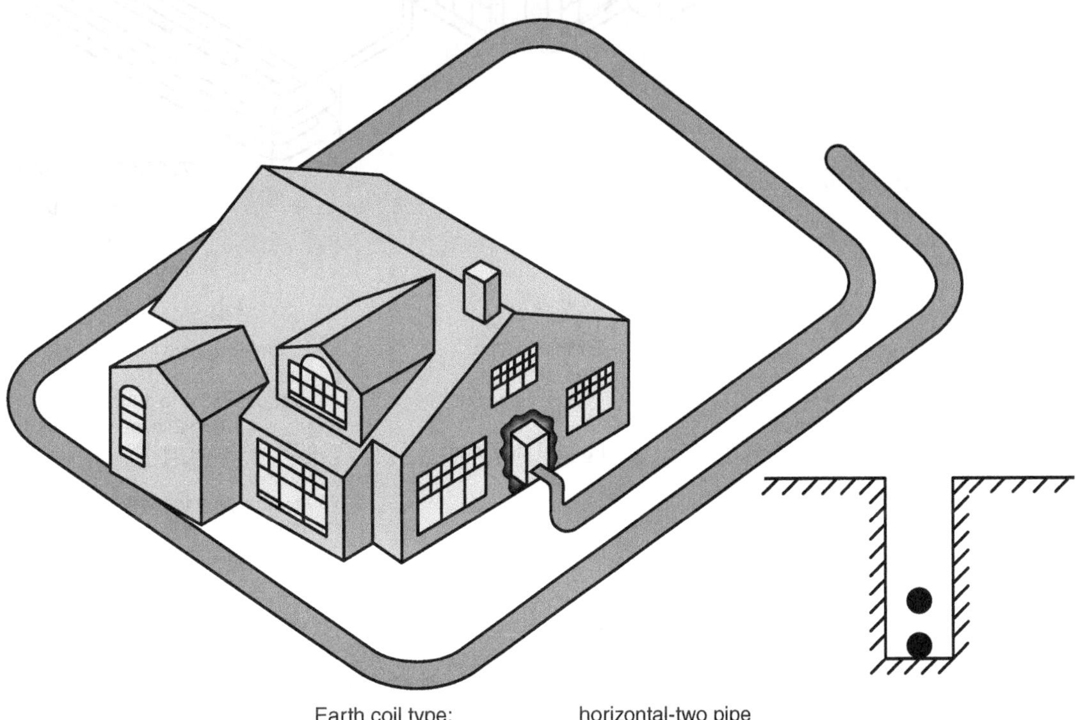

Earth coil type:	horizontal-two pipe
Flow pipe:	series
Typical pipe size:	1 1/4 to 2 inches
Practical length:	210 to 300 feet of trench/ton 420 to 600 feet of pipe/ton
Burial depth:	4 feet and 6 feet 3 feet and 5 feet

Figure 17–4
A two-layer horizontal loop. (Courtesy of Delmar/Cengage Learning)

Figure 17–5
Horizontal trenches being dug for a horizontal loop. (Courtesy of CoEnergies LLC, Traverse City, Michigan)

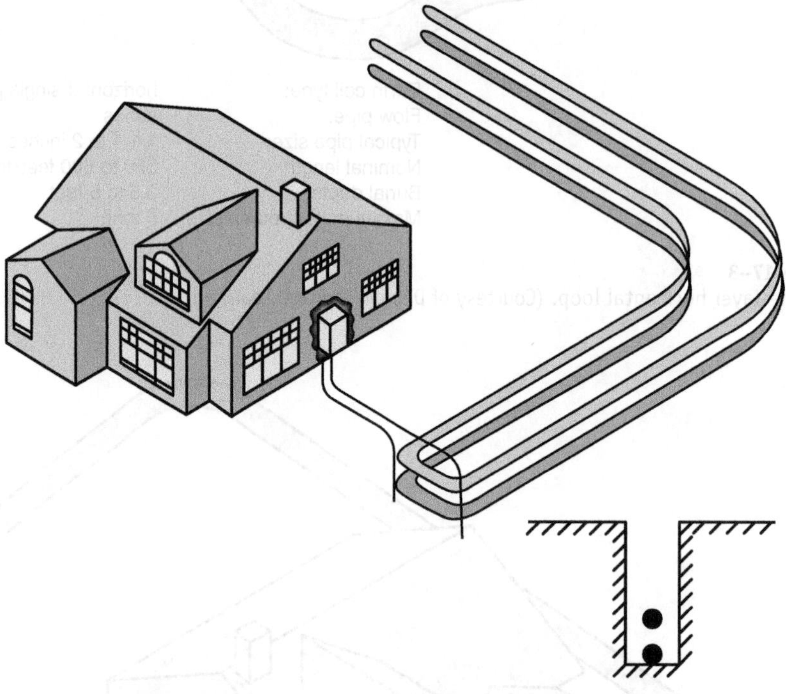

Earth coil type: horizontal-four pipe
Flow pipe: parallel
Typical pipe size: parallel paths 3/4 to 1 inches
headers 1 1/2 to 2 inches
Parallel pipe length: 500 ft. max. pipe length (3/4 inch)
750ft. max. pipe length (1 inch)
Burial depth: 7 feet, 12 inch spacing

Figure 17–6
When multiple loops are required, as for a 4-ton system, four individual loops can be laid with a header that ties them all together. In this drawing the ground loops are laid in a multilayer fashion. (Courtesy of Delmar/Cengage Learning)

Slinky

A slinky, as it is more commonly known, is also known as a coiled-loop, overlaid loop, and a bundled horizontal loop system (Figure 17–8). The design of the loop allows it to be placed in trenches that are a lot shorter than horizontal straight-pipe systems. Each circle of pipe is approximately 5′ to 6′ diameter and is laid over itself before continuing to the next circle. It is called a slinky because it looks like a toy Slinky™ that has been pushed flat, sideways. The length of the run is generally

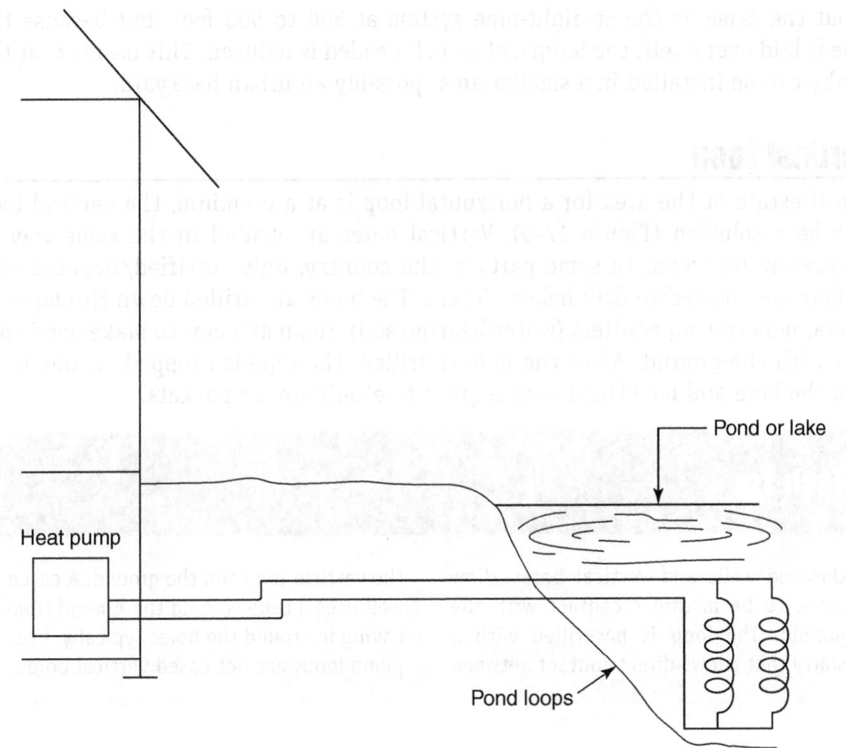

Figure 17–7
A pond loop or coil pack is also considered a horizontal-loop configuration. This loop is formed and tied as a bundle of pipe and then sunk, as a unit, to the bottom of a lake or pond. (Courtesy of Delmar/Cengage Learning)

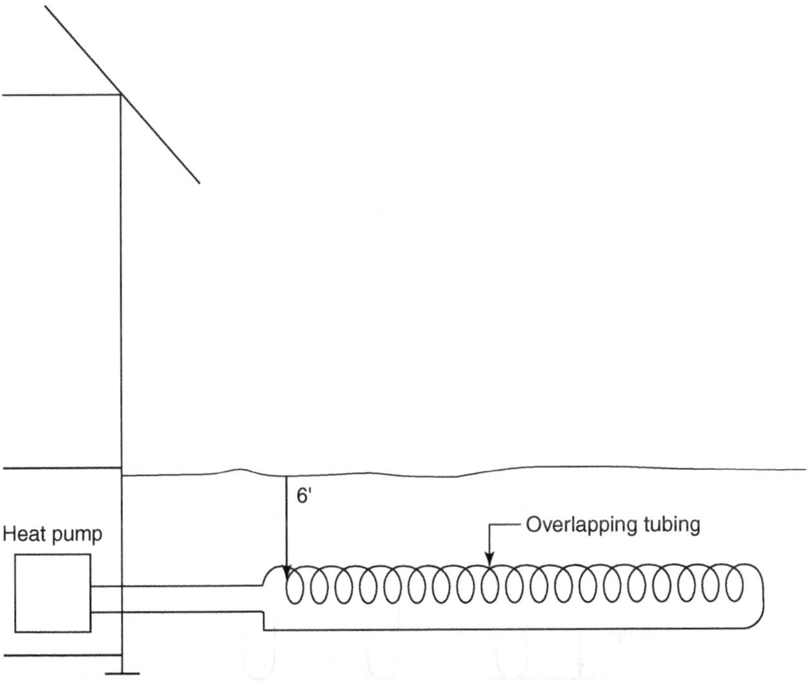

Figure 17–8
A slinky is laid flat in the trench. An optical illusion, the drawing can make it look like a coil, but the coils are laid flat in much the same way a Slinky™ toy lays when it is pressed flat on its side. (Courtesy of Delmar/Cengage Learning)

about the same as the straight-pipe system at 300 to 500 feet, but because the pipe is laid over itself, the length of trench needed is reduced. This means that the slinky can be installed in a smaller area, possibly an urban backyard.

Vertical Loop

If real estate or the area for a horizontal loop is at a premium, the vertical loop may be a solution (Figure 17–9). Vertical bores are drilled in the same way as uncased water wells. In some parts of the country, only certified/licensed well drillers are allowed to drill below 15 feet. The bores are drilled down through the strata, penetrating aquifers (water-bearing soil), in an attempt to make good contact with the ground. After the hole is drilled, the pipe is dropped, as one loop, into the bore and backfilled with a grout to eliminate air pockets.

Tech Tip

Uncased wells and vertical bores allow loops to be in direct contact with the ground. The loop is backfilled with a slurry that allows direct contact between the vertical loop and the ground. A cased well uses a tube to hold the ground from caving in around the bore. Typically, heat pump loops are not cased vertical bores.

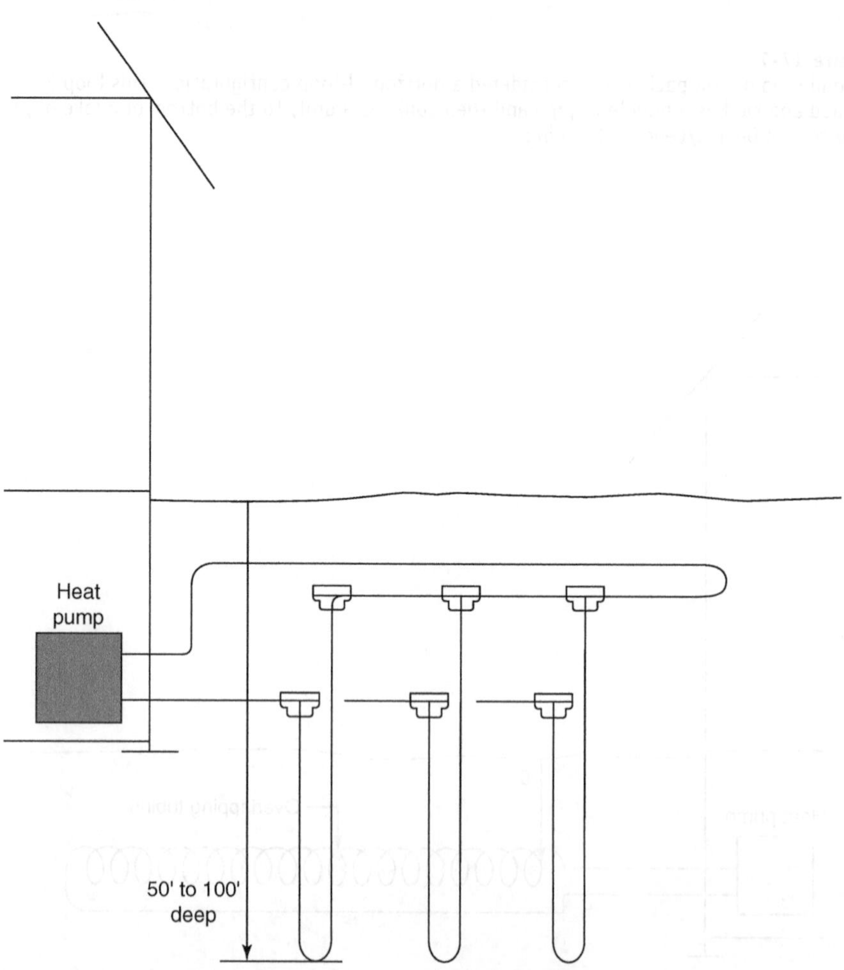

Figure 17–9
Vertical loops are dropped down a bore shaft that is typically 100 feet deep. This diagram shows five vertical loops connected to a common header with five bores. (Courtesy of Delmar/Cengage Learning)

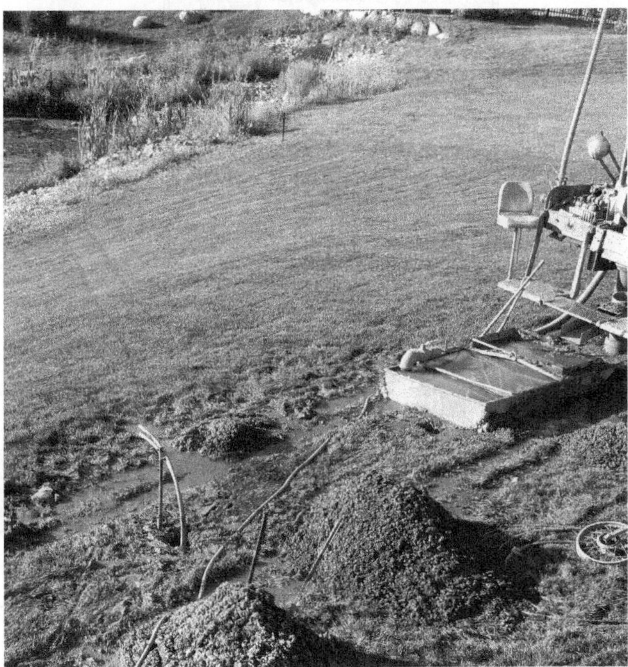

Figure 17–10
After the vertical bore is complete, the polyethylene tubing is sent down the hole and the ends remain visible. (Courtesy of Jamie Simpson)

Figure 17–11
A trench is dug alongside the vertical bores and the vertical pipes are connected to return and supply header pipes. (Courtesy of Jamie Simpson)

Vertical loops are typically drilled to 100′ depth (Figure 17–10). Bore holes have been drilled from 100′ to 300′ deep, and loops are 100′ to 300′ per ton/loop. The cost of drilling the holes and the size of the equipment help to decide how the vertical loop will be installed. Each vertical loop is tied together in a header (Figure 17–11).

OPEN-LOOP SYSTEMS

Open-loop systems are also referred to as geothermal heat pump systems. Water is taken from the ground, heat is removed, and the water is allowed to return to the ground. The simplest is called a single-use or "pump-and-dump" system. Water is pumped from a well and dumped to a lake, pond, river, or ditch (Figure 7–12). The water is used once, only to extract heat. The other configuration pumps water from one well and returns it to another well, generally in the same aquifer, at the same depth (Figure 7–13). This open-system loop is also called an injection system. More about open loops and geothermal heat pumps is found in the next chapter. Installation is based on the local codes and requirements for the location.

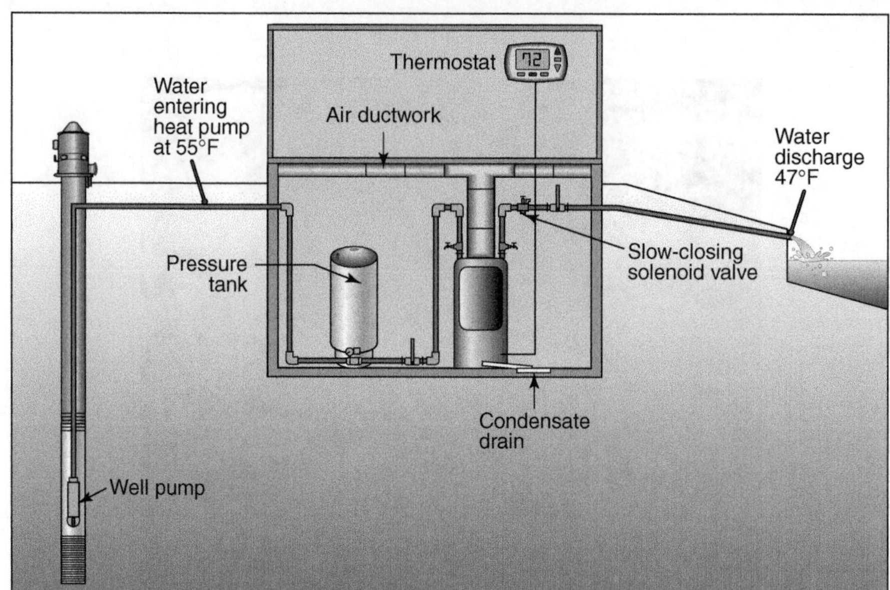

Figure 17–12
A open-loop system can draw water from a well and discharge it to a lake, pond, or ditch. (Courtesy of Delmar/Cengage Learning)

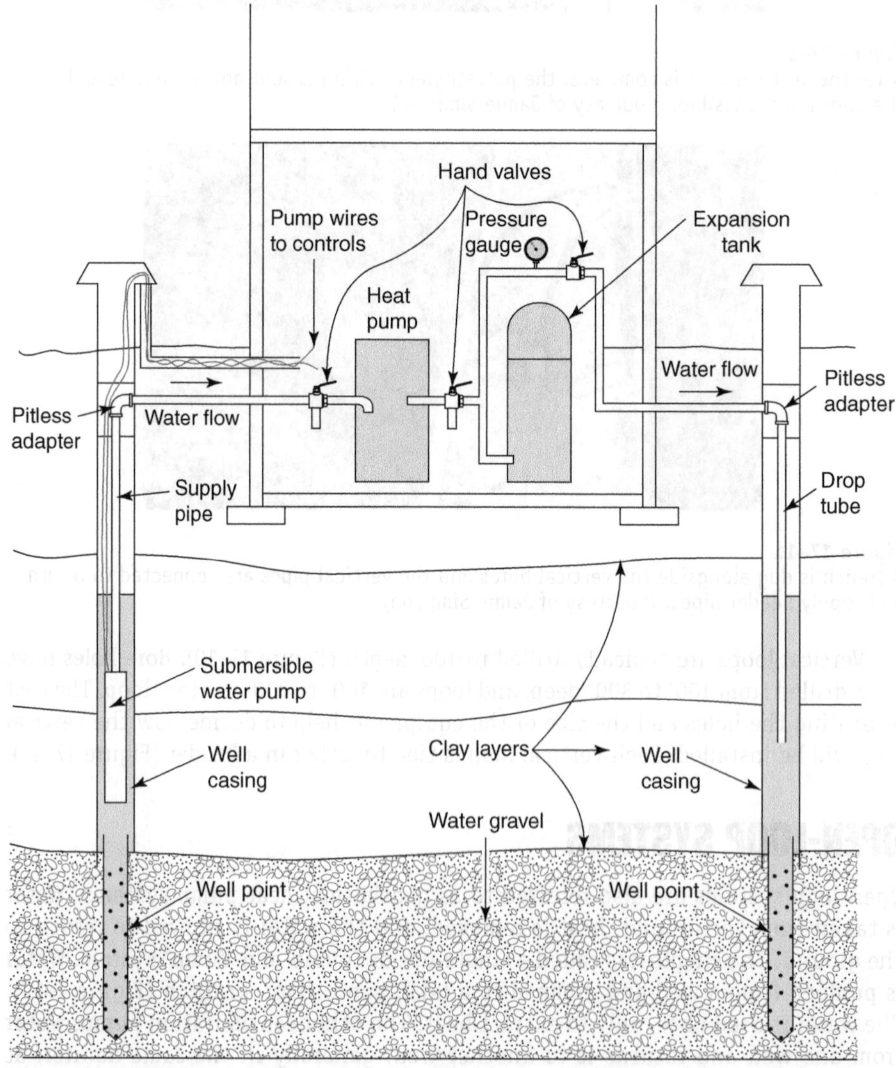

Figure 17–13
In the injection open-loop system, water can be drawn from one well and discharged to another well (as per local code requirements). (Courtesy of Delmar/Cengage Learning)

DIRECT-EXPANSION (DX) SYSTEMS

If the water loops could be eliminated and the outdoor coil could be connected directly to the ground, it should be able to absorb heat directly from the ground. This is the principle of the direct-expansion (DX) loop. The DX loop transfers heat directly from the ground to the refrigerant. To do that, oil and refrigerant must return to a common header and then to a supply and return (suction and discharge) line of the heat pump. This means that each of the loops must radiate from the header connection (Figures 17–14 and 17–15). Refrigerant (DX) loops can be installed horizontally or at a 15-degree angle from vertical (Figure 17–16). Horizontal loops are installed if there is a large area available. The angled bores are done when there is not a lot of room.

Figure 17–14a
Header connection for the DX loop. All of the common return and supply pipes are connected. (Courtesy of CoEnergies LLC, Traverse City, Michigan)

Figure 17–14b
Each is brazed to the header fitting. (Courtesy of CoEnergies LLC, Traverse City, Michigan)

Figure 17–14c
The supply and return header. (Courtesy of CoEnergies LLC, Traverse City, Michigan)

Figure 17–14d
The header cover is being positioned over the headers in preparation for backfill. (Courtesy of CoEnergies LLC, Traverse City, Michigan)

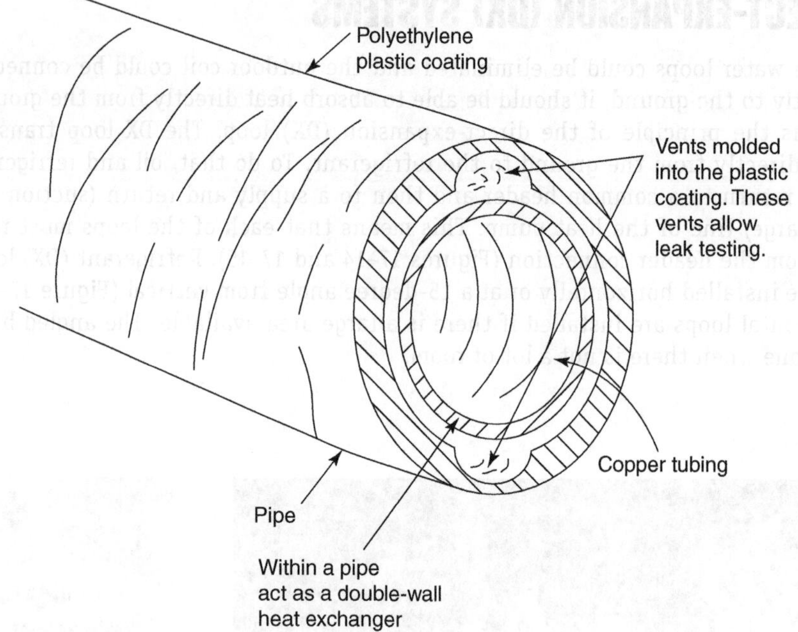

Figure 17–15
Pipe for the DX loop is coated with polyethylene. Two vent holes are formed when the coating is applied. These holes allow refrigerant to leak toward the header, in the event of a leak. (Courtesy of Delmar/Cengage Learning)

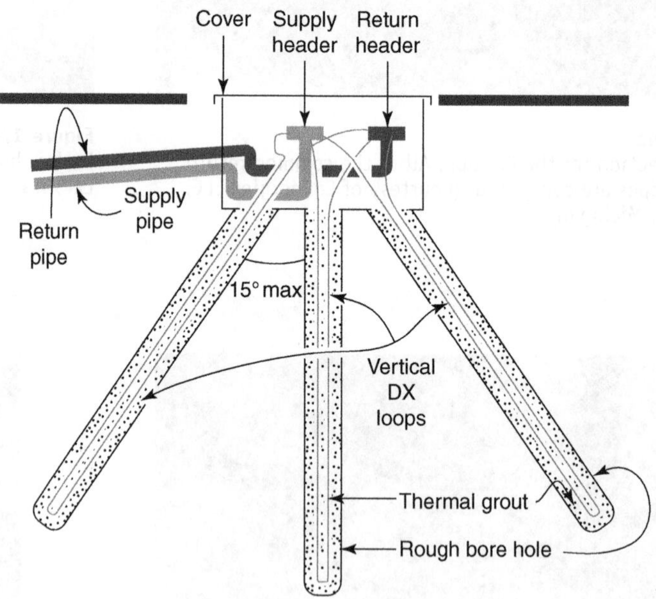

Figure 17–16
Bore holes can be made at approximately 15-degree angles to save having to make horizontal trenches. Each loop is connected to a common header and returned to the heat pump. (Courtesy of Delmar/Cengage Learning)

Angled bore holes save some of the mess with having to drill or dig. By having all of the holes in a central location, there is a minimum of surface ground that is disturbed. Drilling at an angle creates a saucer-shaped section of the earth that the heat pump uses as a source coil. In both configurations, horizontal and angled, the outdoor coil is directly in contact with the ground.

COST RELATIONSHIPS

Every installation is different. Depending on the type of design, location and equipment, installed costs are different for the same installation. Soil and land present differences in types, location, depths, and slopes. A deep vertical closed-loop system may be the only choice for a structure located on a half-acre with rocky soil, where water wells are over 400 feet deep and the ground slopes away from the structure.

When there are other choices and options, system cost is typically higher than that for conventional systems. Here are some things that impact the cost of a geothermal system:

1. Customer lifestyle, sentiments, and preferences
2. Location: soil, ground slope, vegetation, lot size, trees, other buildings, existing utilities, septic system, building perimeter drains, and so forth
3. Ground-water level and moisture content of the soil
4. Structure: size, heat loss, heat gain, air and/or water delivery system, radiant panels, ventilation, number of people, number of floors, etc.
5. Control System: simple single thermostat, multiple zones, single-wire communication system, hard-wired versus wireless, humidity control, and so forth
6. System Options: electronic air cleaning, humidification, economizer, dual-fuel back-up, and so forth
7. Labor Costs: rural, urban, residential, commercial, large company, small company, and so forth
8. Contracted Costs: well drilling, trenching/excavation, loop assembly (fusing), wetland specialist, DNR advisor, plumber, electrician, builder, and so forth
9. Licenses and Permits: well drilling, wetlands permit, building permit, mechanical permit, plumbing permit, electrical permit, and so forth

Installation of the outdoor loop system is one of the most expensive parts of a geothermal system. Pump-and-dump (single-use) systems can piggy-back on the cost of a well that had to be drilled for a new structure. If the system is being installed in an existing structure, a new well or an additional well may have to be drilled, which would increase the cost of installation. The horizontal closed-loop system requires trenching or excavation and land surface area. For new construction, some of the cost of excavation could be combined with foundation work, but the existing structure would have the added cost of trenching. Vertical closed-loop systems also need some trenching to connect the individual bores with a header, but the trenches do not have to be wide. Vertical bore drilling may require a licensed driller with specialized equipment. One well of 100 feet is typically required for each ton of cooling.

MAINTENANCE CONSIDERATIONS

Geothermal maintenance requirements and costs are well below the costs of conventional heating systems and are similar to the costs of cooling systems. If compared to air conditioning systems, heat pump maintenance requires similar processes and procedures to be followed. Heating maintenance on a heat pump is more similar to cooling systems than conventional heating systems. The most interesting aspect of geothermal systems is that if maintenance is done for the cooling system, most of the maintenance has already been done for the heating system. With a few exceptions, when the cooling system maintenance is complete, the heating system maintenance is complete.

Between the four types of geothermal systems, WSHP single use, WSHP injection, GSHP horizontal, and GSHP vertical, there are a few differences in the maintenance required, as shown in Table 17-2.

All systems have some similarities. For example, they all require:

1. Leak testing and checking
2. Fluid flow measurement and monitoring
3. Temperature measurement and monitoring
4. Control testing
5. System testing and monitoring

There is one thing that can be said for geothermal systems and all other HVAC systems: they will all require some level of maintenance to remain operating at peak efficiency.

Table 17–2 Geothermal System Characteristics

	WSHP Single-Use	WSHP Injection	GSHP Horizontal	GSHP Vertical
Mineral removal from the heat exchanger	Yes	Yes	No	No
Balancing loops	No	No	Yes	No
Pump inspection and maintenance	Limited	Limited	Yes	Yes
Screens and filters	Yes	Yes	Limited	Limited
Heat exchanger pressure-drop measurement	Yes	Yes	Yes	Yes
Solution pH	No	No	Yes	Yes
Hardness measurements	Yes	Yes	No	No
Anti-corrosion inhibitor	No	No	Yes	Yes

SUMMARY

In this chapter we have discussed the major attributes of geothermal systems. We have noted that there are three general system types: closed-loop, open-loop, and direct-expansion (DX) systems.

Closed-loop systems use polyethylene pipe to route a water-based solution through a loop of pipe that is buried in the ground. The loop can be horizontal or vertical. Horizontal loops can be straight runs of pipe or circular loops of pipe (slinky). Vertical loops have manifolds that tie together individual bore hole with loops of pipe in each.

The heat transfer solution is typically a water-based mix of propylene glycol and water. Though other solutions can be used to keep the water from freezing, propylene glycol seems to be the product that is used almost exclusively for ground-loop systems. Both horizontal and vertical loops use the same heat transfer solution.

Open-loop systems use water directly from a well. The single-use system (pump and dump) pulls water from a well, removes the heat with the heat pump, and sends the water to a lake, pond, river, or ditch. The other system uses another well to place the cooled water. This type of system is called an injection open-loop system. Water is taken from the first well and injected into the second well. Both wells are drilled to the same depth and use the same aquifer. Water in the aquifer is never depleted because of the pumping arrangement. Installation is based on the local codes and requirements for the location.

Direct-expansion (DX) systems do not use a loop or ground water. Instead, the outside coil is placed directly into the ground. This arrangement moves the refrigerant to the ground loop and the refrigerant is allowed to expand, picking up heat directly from the ground. These systems are said to be 20–25% more efficient than the ground-loop system and they take up less space. The DX system can use less ground area to do the same heat transfer work as the closed-loop system.

One system is hard to compare to another system, especially when it comes to cost. There are too many variables that could interfere with a good comparison. According to many sources, it seems that open-loop systems are the least expensive to install. The most expensive seem to be vertical loops. However, it is quite possible to have a traditionally more expensive system become less expensive after comparing the longevity of the system and cost savings over a longer period of time. Customer choices, location, structure, and equipment options all interact to make one system more preferable than another.

Maintenance expectations are reduced with the use of a geothermal system. These systems tend to have lower maintenance concerns than conventional systems over longer periods of time. With each type of system, there are also some differences. Some require more maintenance in some areas whereas others do not. We end on an assuring note: all geothermal heat pumps will need service at some point!

REVIEW QUESTIONS

1. Describe the difference between an open- and closed-loop system.
2. Describe the function and use of geothermal plastic pipe material.
3. Relate why the heat transfer solution must be protected from freezing in some installations.
4. Describe why "slinky" systems use less ground area.
5. Explain why vertical loops might be used rather than horizontal loops.

CHAPTER 18

Water-Source Heat Pumps

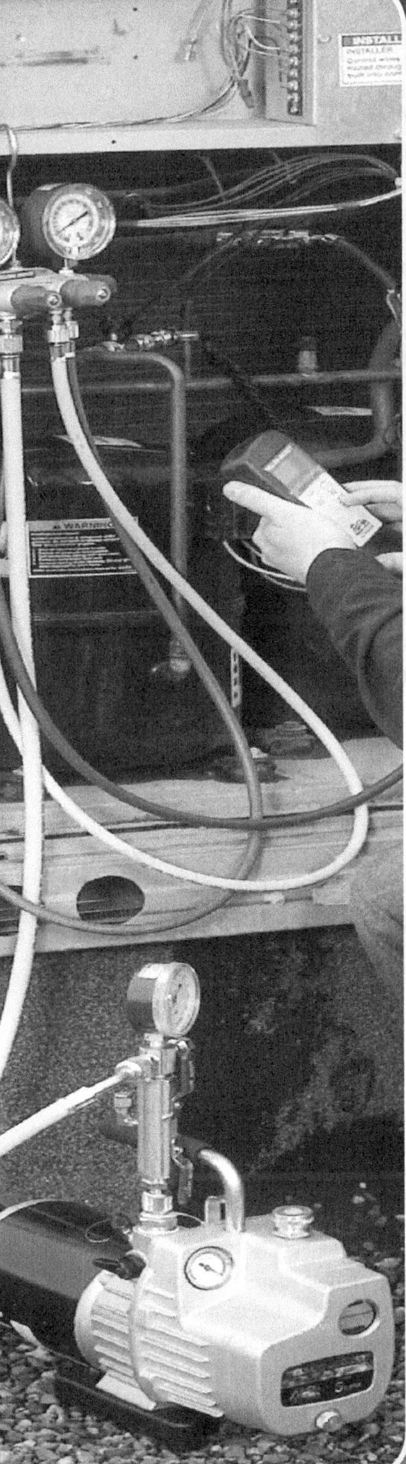

LEARNING OBJECTIVES

The student will:

- Describe how heat is absorbed and moved by a water-source heat pump (WSHP) in a ground-source application
- Describe why WSHPs are also referred to as open-loop systems
- Relate how water is used only once in a single-use system
- Describe the relationship between ground-water temperature and ground temperature
- Explain the difference between water-to-air and water-to-water systems
- Describe how cooling towers can be used with a WSHP

INTRODUCTION

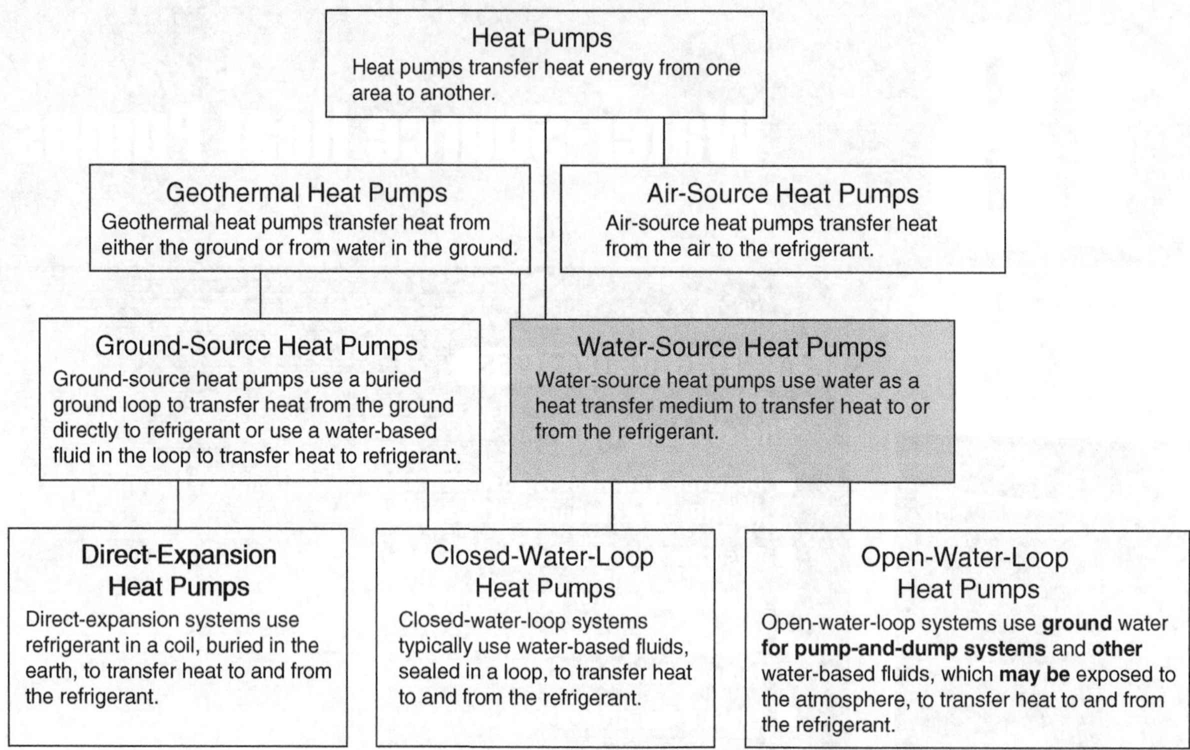
Heat Pumps
Heat pumps transfer heat energy from one area to another.
Geothermal Heat Pumps
Geothermal heat pumps transfer heat from either the ground or from water in the ground.
Air-Source Heat Pumps
Air-source heat pumps transfer heat from the air to the refrigerant.
Ground-Source Heat Pumps
Ground-source heat pumps use a buried ground loop to transfer heat from the ground directly to refrigerant or use a water-based fluid in the loop to transfer heat to refrigerant.
Water-Source Heat Pumps
Water-source heat pumps use water as a heat transfer medium to transfer heat to or from the refrigerant.
Direct-Expansion Heat Pumps
Direct-expansion systems use refrigerant in a coil, buried in the earth, to transfer heat to and from the refrigerant.
Closed-Water-Loop Heat Pumps
Closed-water-loop systems typically use water-based fluids, sealed in a loop, to transfer heat to and from the refrigerant.
Open-Water-Loop Heat Pumps
Open-water-loop systems use ground water for pump-and-dump systems and other water-based fluids, which may be exposed to the atmosphere, to transfer heat to and from the refrigerant.

Water-source heat pumps (WSHPs) were once more popular than ground-source systems. Only with the advent of plastic pipe did that change. Where water is plentiful and inexpensive, water-source heat pumps still have the advantage of overall lower installation cost.

Water temperature is related to ground temperature. Heat pumps are generally designed to take advantage of water temperatures that are in the 50°F range. These temperatures are found in the majority of the country.

Field Problem

"The floors are really cold," the homeowner exclaimed as the service technician asked the customer to describe the problem. This heat pump was a water-to-water (W2W) WSHP. The heat pump was connected to the radiant panel floors; thus the reason for the customer's description. Something was preventing the heat pump from warming the floor. As the technician headed for the system, he formed a list of possible causes.

Symptoms: Cold floors.

Possible Causes:

1. Compressor not operating
2. Contactor not pulling in
3. Pumps not operating or air locked
4. Electrical or control failures

He needed to narrow down the list. Coming to the thermostat, he checked for operation. The system light was on and there was no message to indicate that the system was not functioning. This didn't help much, because the control system might not be working properly, but at least he knew that the thermostat was set correctly and that it was calling for heat. Arriving at the unit, the compressor reported a distinct hum; now he knew the control circuit, the power supply, and the contactor could be scratched off the mental list. What was left?

The outdoor pumps were humming, and so were the indoor pumps feeding the radiant panel. Just then the compressor shut down along with the outside pumps; the system had cycled off. A quick check of temperatures at the indoor coaxial

Field Problem (Continued)

coil indicated that the coil was hot, but the radiant panel supply pipe was not. Touching the pipe in several places, the technician was able to determine where the hot water flow had stopped: at the tempering valve. After rapping the valve with a screwdriver handle, the water supply pipe immediately got hot and the heat pump cycled back on. The tempering valve was stuck in bypass and returning water from the radiant panel was being bypassed back to the panel without the benefit of heat from the heat pump. In response, the heat pump was cycling on the temperature sensor at the indoor coaxial coil. A new tempering valve and some purging slowly brought the floor back up to temperature. After checking 4 hours later, the customer reported that the system was doing fine.

WATER-SOURCE HEAT PUMPS

As noted, WSHPs were dominant before plastic pipe was developed in the late 1970s. WSHPs use ground water and the heat contained in ground water to heat occupied spaces. Water is used both as a heat source in the heating mode and as a heat sink in the cooling mode. The natural availability of water in most parts of the country made this form of geothermal heat pump popular.

Open-Loop Systems

Open-loop systems are also associated with WSHP systems. Open loop means that the water is either being taken from or discharged to a location that is at atmospheric pressure. Open-loop systems predominantly use water from a well at atmospheric pressure. Water is taken from the ground, heat is removed, and the water is allowed to return to the ground. The simplest is called a single-use or "pump-and-dump" system. Water is pumped from a well and dumped to a lake, pond, river, or ditch. The water is used only once, to extract heat (Figure 18–1).

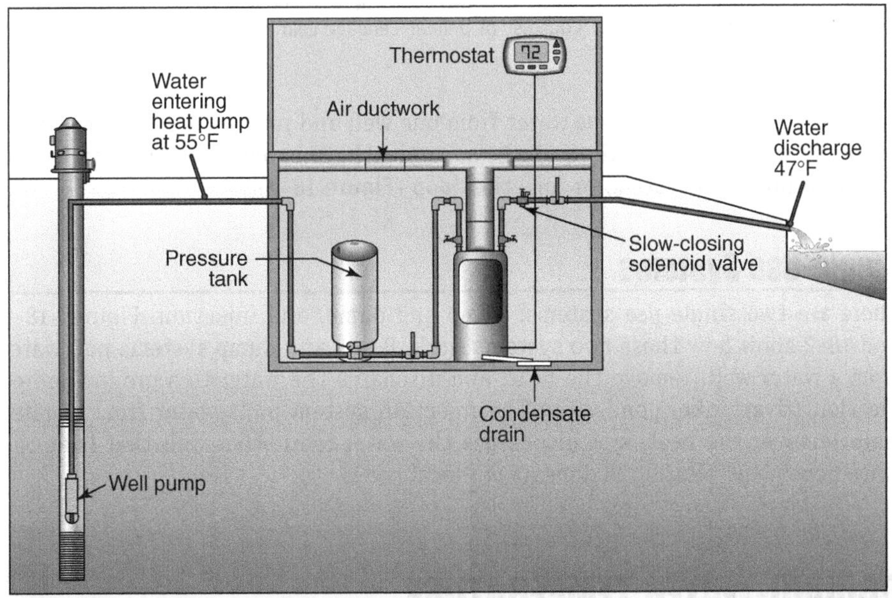

Figure 18–1
An open-loop system can draw water from a well and discharge it to a lake, pond, or ditch. (Courtesy of Delmar/Cengage Learning)

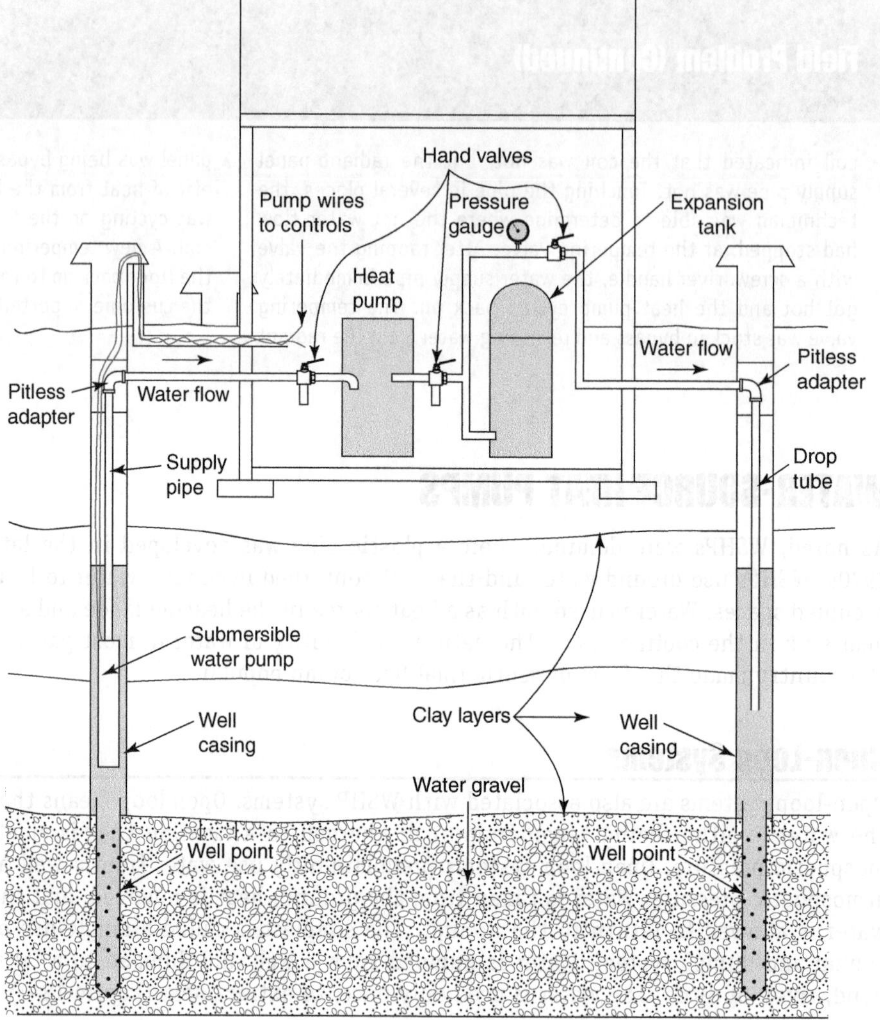

Figure 18–2
In an injection open-loop system, water is drawn from one well and discharged to another well (as per local code requirements). (Courtesy of Delmar/Cengage Learning)

The second system pumps water from one well and returns it to another well, generally in the same aquifer, at the same depth. This open-system loop is also called an injection system, or injection loop (Figure 18–2).

Single-Use Systems

There are two single-use systems: pump and dump, and injection. Figures 18–1 and 18–2 show how these two systems work. Pump-and-dump systems pull water from a water well, remove the heat, and discharge the water to a ground-source location (river, lake, pond, etc.) The injection system pulls water from a water well, removes the heat, and discharges the water to another well that is at the same depth. Installation is done as per local codes.

GROUND-WATER TEMPERATURE

Ground temperatures are related to ground-water temperatures (Figure 18–3). Where ground temperatures are high, ground-water temperatures are also high.

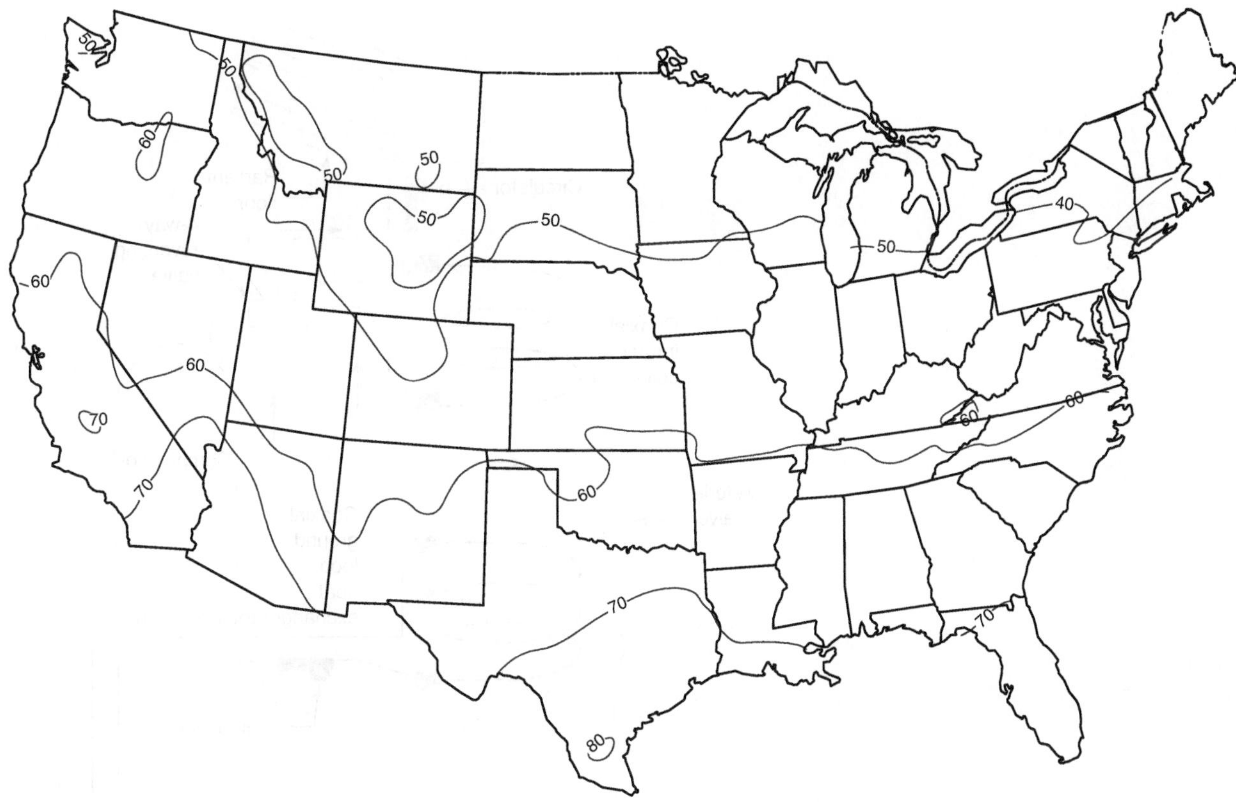

Figure 18–3
This map roughly shows the temperature gradients of ground-water temperatures at 50 to 150 feet. Notice the effect the sun has on ground-water temperatures from south to north across the country. (Courtesy of Delmar/Cengage Learning)

When ground-water temperature is high, it is easy to move heat from the ground without a heat pump. In these locations, however, there is a greater need for cooling than heating. This could present a problem in using the ground as a heat sink to move heat from the interior space to the ground, as the ground-water temperature is already higher than desired.

In most parts of the country, the ground-water temperature hovers in the 50°F range. At this temperature, ground water can be used as a heat source during the winter and put heat in the ground (heat sink) in the summer. WSHPs can take advantage of water at this temperature.

Water-to-Air Systems

The majority of WSHPs are water-to-air systems, where water acts as the source and air is being heated in the occupied space. Both air-source and water-source system indoor components are the same. An indoor coil acts as the refrigerant condenser in the heating mode. The indoor coil is a fin-tube heat exchanger and air is moved across the exchanger and into the occupied space with a blower.

Water-to-Water Systems

Water-to-water heat pumps transfer heat from a water loop that extracts heat to another water loop that is being used to transfer heat to occupied spaces. An example would be where heat is being transferred from the ground, to a water loop buried in the ground, to the refrigerant in the heat pump. The heat pump concentrates the heat and then moves it from the refrigerant to a water loop, to a radiant floor where the floor is heated, which in turn heats the air in the occupied space. The radiant floor is considered a terminal exchanger or final heat exchanger to

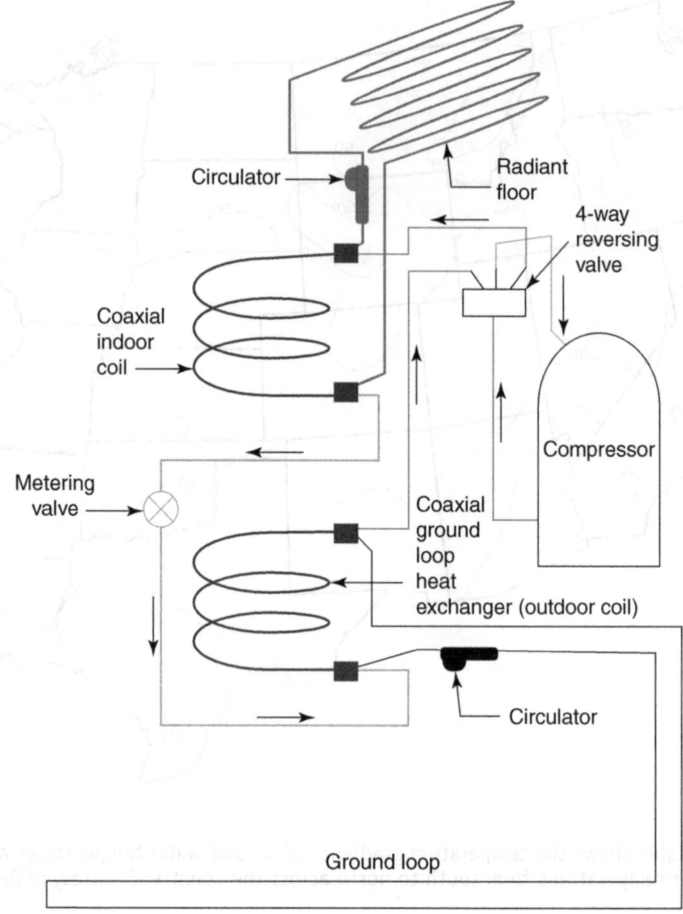

Figure 18–4
The drawing shows an example of a water-to-water heat pump. The ground loop is one of the two water-to-refrigerant exchanges. The other water-to-refrigerant exchange heats water that is sent to terminal devices used to warm occupied spaces. (Courtesy of Delmar/Cengage Learning)

heat living spaces. Water heated by the heat pump could be used to exchange heat to the occupied space using other terminal devices, such as fin-tube baseboards, fin-tube coils in air handlers, and other forms of radiant panels.

Water-to-water heat pump systems (Figure 18–4) use the concept of transferring heat to and from a water-to-refrigerant exchanger. The primary transfer is accomplished on both the high- and low-side heat exchangers transferring heat to and from water. These systems are typically used for heating. The primary heat exchanger is connected to a closed water loop with secondary heat exchangers. These secondary heat exchangers could be:

1. Fin-tube exchangers that transfer heat energy from the water to the air.
2. Radiant panels that both radiate heat and conduct heat energy to air.

When connected to radiant panels, water-to-water heat pumps are able to operate at an advantage—because the radiant panel is generally operated at lower temperatures, the water-to-water system is ideally suited. Radiant floor temperatures are purposefully kept below the 90°F recommended maximum.

Radiant Floor

A radiant floor is one type of radiant panel that can be used with a water-to-water system (Figures 18–5 and 18–6). These systems are typically used for heating. A radiant floor panel incorporates tubing that is laid in the floor. The tubing contains water that is being warmed by the heat pump and pushed through several loops.

Figure 18–5
Tubing is laid in defined patterns on the floor grade and tied to reinforcement material that holds it in place. (Courtesy of Delmar/Cengage Learning)

Figure 18–6
Concrete is poured over the tubing and the concrete is finished, forming a radiant floor panel. (Courtesy of Delmar/Cengage Learning)

Cooling Tower

In commercial and industrial building applications with large exposure to heat from the sun or heat-producing manufacturing processes, data centers, kitchens, and so forth, multiple WSHPs, connected in parallel to a water-piping loop, can be distributed throughout the building for moving heat to the places that need it. If the loop temperature gets too warm due to the lack of required building heat, a cooling tower can be applied to reduce the loop temperature (Figure 18–7). During winter operation the loop temperature can be raised with a boiler should the internal heat be insufficient to keep the loop temperature high enough to satisfy the building heating needs. Commercial and industrial applications may also be applied to ground sources with direct-expansion, closed-loop, and pond or open-loop systems, as discussed earlier.

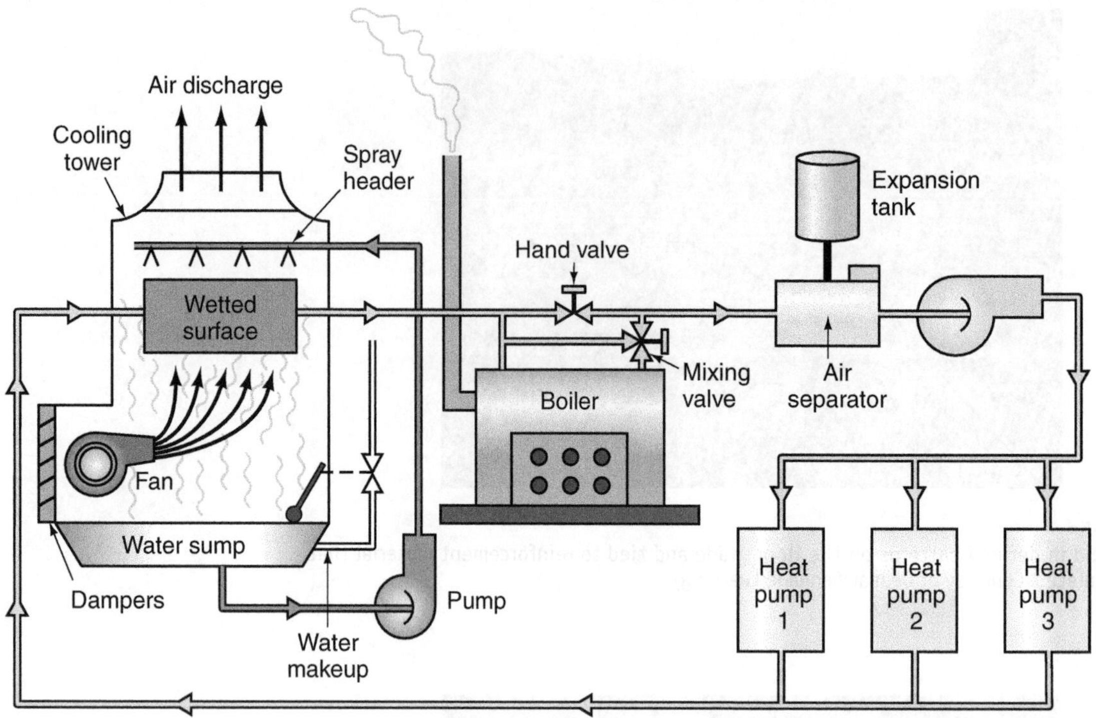

Figure 18–7
An open-loop WSHP with boiler and cooling tower to maintain the loop temperature. (Courtesy of Delmar/Cengage Learning)

SUMMARY

In this chapter we have concentrated on WSHPs. These systems operate in the same way as other heat pumps and similarly to ground-source units. The difference is in the use of water as a source. Some WSHPs are single-use systems, where water is used as a heat source or sink only once. In the case of some commercial operations, this water can be returned to the reservoir where it originated, to be used again, as with the case of cooling towers.

WSHPs are also known as open-loop systems. Water used in these systems either comes from or is discharged to an atmospheric location. This usually means that the water source is a water well. The discharge location could be a surface feature, such as a pond or lake. In other cases it could be a discharge well. Water temperature in a water well is influenced by the sun and is related to ground temperature. The temperature of ground water varies throughout the country and especially from South to North. All installations are subject to the local authority having jurisdiction.

REVIEW QUESTIONS

1. Describe how heat is absorbed and moved by a water-source heat pump (WSHP) in a ground-source application.
2. Describe why "pump and dump" WSHPs are also referred to as open-loop systems.
3. Relate how water is used only once in a single-use system.
4. Describe the relationship between ground-water temperature and ground temperature.
5. Explain the difference between water-to-air and water-to-water systems.
6. Describe how cooling towers can be used with a WSHP.

CHAPTER 19

Pumping Configurations and Flow Centers

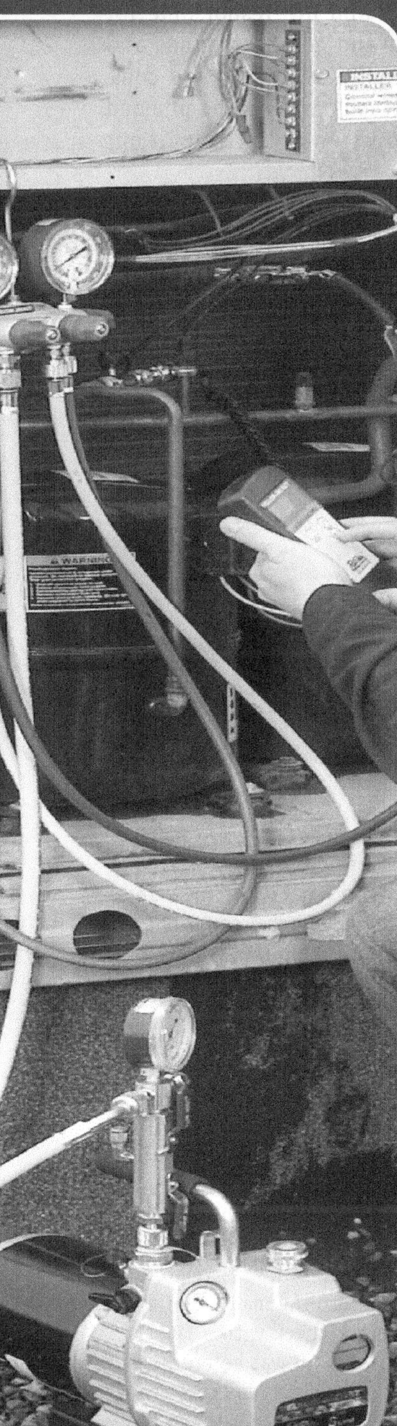

LEARNING OBJECTIVES

The student will:

- Describe the relationship between feet of head and pressure per square inch gauge (psig)
- Describe the reason for conducting a pressure-drop measurement across the coaxial heat exchanger
- Relate how temperature difference across the coaxial coil is connected to pressure drop
- Describe what causes frictional loss in a closed loop
- Explain why vertical feet of head has no effect on pumping head for closed loops, but does for open-loop systems
- Describe what components are in a flow center and the function of a flow center

INTRODUCTION

Pumping configurations are at the center of geothermal heat pumps. This is the secondary loop, which is a water-based loop that is heated or cooled and exchanges energy with the earth. Water-based solutions are used, if not water directly from the ground. Closed loops are charged with water-based antifreeze solutions. Open-loop systems can use water from a variety of sources. This chapter will discuss the physical characteristics of water and how water is moved through a loop. Friction loss, pump head pressure, water volume, and viscosity all play a role in how pumping systems are configured.

Flow centers are the heart of the geothermal loop and are responsible for moving water-based solutions through the loop, from and to the ground, and through the refrigerant-to-water heat exchanger. The characteristics and components of flow centers are discussed, as well as how flush carts can be plugged into the flow center to work in conjunction with the flow center to flush air out of the loop and charge the loop.

Flush carts are used to purge and also to charge the loop with antifreeze solution. How much antifreeze solution is needed to protect the geothermal loop is discussed. The percentage and the type are choices made by technical professionals, designers, and installers. The choice has to do with the availability, price, and characteristics of the solution while being used in a loop.

Pipe connections inside the structure can be standard plumbing-style connections, but outside, the plastic pipe connections must be fused. Fusion is the only sanctioned connection for buried applications. Fusing a pipe connection correctly requires training and certification.

Field Problem

The customer had heat, but it was operating on supplemental heat. The customer had called for service after seeing that the emergency heat light was on and had not turned off for more than 5 hours. The characteristic hum that she heard was not there and so she surmised that the geothermal unit was not running. Otherwise, the level of heat in the house was fine, but lower than the thermostat setpoint.

Symptoms:
Emergency heat light on; compressor not running; temperature lower than setpoint.

Possible Causes:
Compressor failure; electrical system malfunction; safety lockout; refrigerant system failure; loop failure.

Checking the status light-emitting diodes (LEDs), the technician noted that the system was locked out on low temperature. This condition could not occur if the compressor could not run; touching the refrigerant line at the coaxial heat exchanger confirmed that it was cold. Just then, the compressor started and almost immediately it turned off. Touching the supply and return of the water loop, the technician noticed that the supply line was cold, but only close to the heat exchanger. A quick check for pressure drop across the coil confirmed that there was no pressure drop and no flow. Moving to the flow center, the technician felt the ends of each of two pumps. They did not feel warm, nor was there any vibration that would indicate that they were on. While removing the pump relay cover, both pumps started and the compressor started as well. The system began working and ran within manufacturer pressure-drop and temperature-drop conditions until the emergency heat light turned off, shutting down the supplemental heat. Turning the pumps on and off did not make the problem show up again. A visual check of the pump relay did not reveal anything unusual—the contacts were clean and everything checked out. An electrical check did not reveal an electrical problem, except that the relay coil felt hot just after the technician had removed the cover.

The technician had run into this same type of problem once before, and he concluded that the pump relay must have been hung up. This could account for a hot relay coil. The armature of the relay had not been able to pull into the magnetic flux of the coil and the coil got hot. A quick check with a finger confirmed that the coil was cooler now than it was before. The decision was made to suggest that the relay be replaced, and the owner concurred.

PUMPING SYSTEMS

Geothermal heat pumps rely on their ability to extract heat from the ground. With the exception of direct-expansion (DX) systems, heat is removed from the ground with a secondary heat exchange loop. This loop is charged with a water-based solution that may include an antifreeze additive. Without antifreeze added, water has the following properties:

- Specific heat = 1
- Specific gravity = 1
- Viscosity 1 ssu (Saybolt Seconds Universal)= 1 cps (Centipoises – SI units)
- Weight per gallon = 8.33 pounds
- Foot of head to pressure equivalent:
  1 foot of head = .433 psig
  1 psig = 28″ w.c. (water column)/12″ = 2.33 ft.

When determining the amount of heat extracted in a ground loop, the following formula can be used:

$$\text{BTUH} = \text{Gallons/minute} \times \text{Temperature difference} \times \text{Fluid factor}$$

OR

$$\text{Temperature difference} = \text{BTUH}/(\text{GPM} \times \text{Fluid factor})$$

The fluid factor is a combination of specific heat, weight of water, and time:

$$\text{Fluid factor} = 499.8 \text{ (or 500)} = \text{Specific heat (or 1)} \times 8.33 \text{ pounds/gallon} \times 60 \text{ minutes}$$

Pressure drop through a heat exchanger is used by most manufacturers to determine the loop or heat exchanger flow rate. Using Table 19–1, if water enters the heat exchanger at 50°F and exhibits a pressure drop of 1.8 psig, the flow rate for Model XYZ will be 5 GPM. If this model uses 3 GPM/ton, the system is extracting approximately 20,000 BTUs (5/3 = 1.66 tons; 1.66 × 12,000 = 19,999 BTU).

Combining what we know and using the formula to determine temperature drop:

$$\text{Temperature difference} = \text{BTUH}/(\text{GPM} \times \text{Fluid factor})$$

$$\text{Temperature difference} = 19{,}999/(5 \times 485\ [30\%])$$

$$\text{Temperature difference} = 8.24°\text{F}$$

Table 19–1 Pressure Drop in psig through Heat Exchanger with a 30% Glycol Solution

		Loop Temperatures (at heat exchanger inlet)				
Model	**GPM**	30°F	50°F	70°F	90°F	110°F
Model XYZ	3.0	0.8	0.7	0.7	0.7	0.6
	5.0	2.0	1.8	1.7	1.6	1.5
	7.0	3.6	3.4	3.2	3.0	2.8
	9.0	5.8	5.5	5.1	4.8	4.4
	5.0	1.2	1.2	1.1	1.0	1.0

Based on actual pressure-drop chart.

Tech Tip

The fluid factor for pure water is 500. For other mixtures of propylene glycol and water, the fluid factors are: 30% = 485; 40% = 448; 50% = 425; other percentages can be looked up.

In the heating mode, the fluid temperature drop across the heat pump coaxial coil should be a maximum of 8.24°F based on the calculation above. If the coil heat transfer surfaces are fouled, the amount of heat pulled from the heat exchanger is reduced and the temperature difference will be lower. If there is a restriction in the fluid flow, causing the fluid flow rate to go down, the temperature difference will be higher.

Tech Tip

The higher the fluid flow in a heat exchanger, the lower the temperature difference. Conversely, as the flow rate through the heat exchanger goes down, fluid spends more time in the heat exchanger, allowing it to absorb or reject more heat. It is the combination of these two factors (flow and temperature) that affects heat absorption or rejection. Temperature difference alone may look alright by itself, which is why there is a need to look at other test information and compare it to the initial start-up data. The measurement of temperature difference alone cannot be used to verify proper system operation by itself.

Closed-Loop Systems

When water is pumped through a closed loop, the frictional losses of the pipe size, pipe length, fittings, and elbows or turns all add together. This frictional loss is expressed in "feet of head" (see Table 19-2). Because the loop is closed, the weight of the water in a vertical closed loop is the same weight going down as going up. For this reason there is no vertical head calculated into the total head. Pump manufacturers supply information about the capability of their pumps in charts called "pump curves." The pump curve relates to the amount of water the device can pump in GPM at a given resistance in feet of head (Figure 19-1).

A ground loop develops a frictional head through the length, diameter, and fittings (see Table 19-3). A 3-ton horizontal loop could have three circuits of 400 feet of loop piping. If each loop were connected to a header pipe and had the fewest number of fittings, it is still conceivable to have a total frictional head determined in the following way:

- Each of the three 1″-diameter pipe circuits is in a 400-ft. trench and each loop is in parallel (1,200′ of pipe)
- Each circuit is connected to a reverse return header that makes each loop equal distance from the heat pump (Figure 19-2).
- Each loop has one 180-degree bend (the equivalent of two elbows).
- Both sections of 2″ pipe header are a total of 200′ in length (supply and return combined).

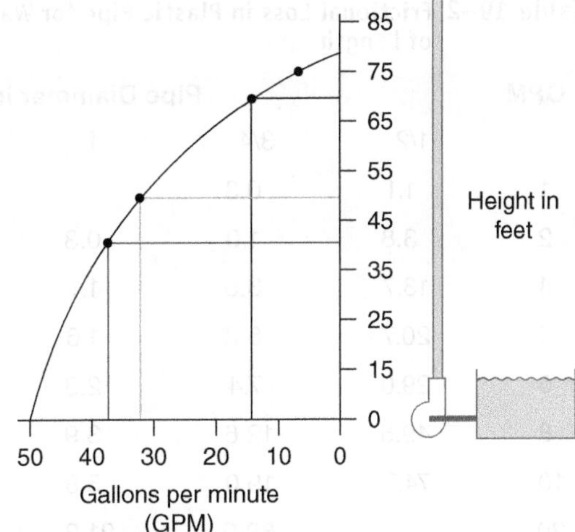

Figure 19–1
A pump by any manufacturer will have the ability to pump against a resistance measured in feet of head. Feet of head is equated to the vertical rise of water. At any given resistance in feet of head, the pump can supply a certain number of gallons per minute (GPM). The more resistance the pump works against, the lower the GPM that can be delivered. At the maximum feet of head, the pump cannot supply water and simply runs to push against the maximum friction—in this example, 0 gallons of water at 78 feet of head. (Courtesy of Delmar/Cengage Learning)

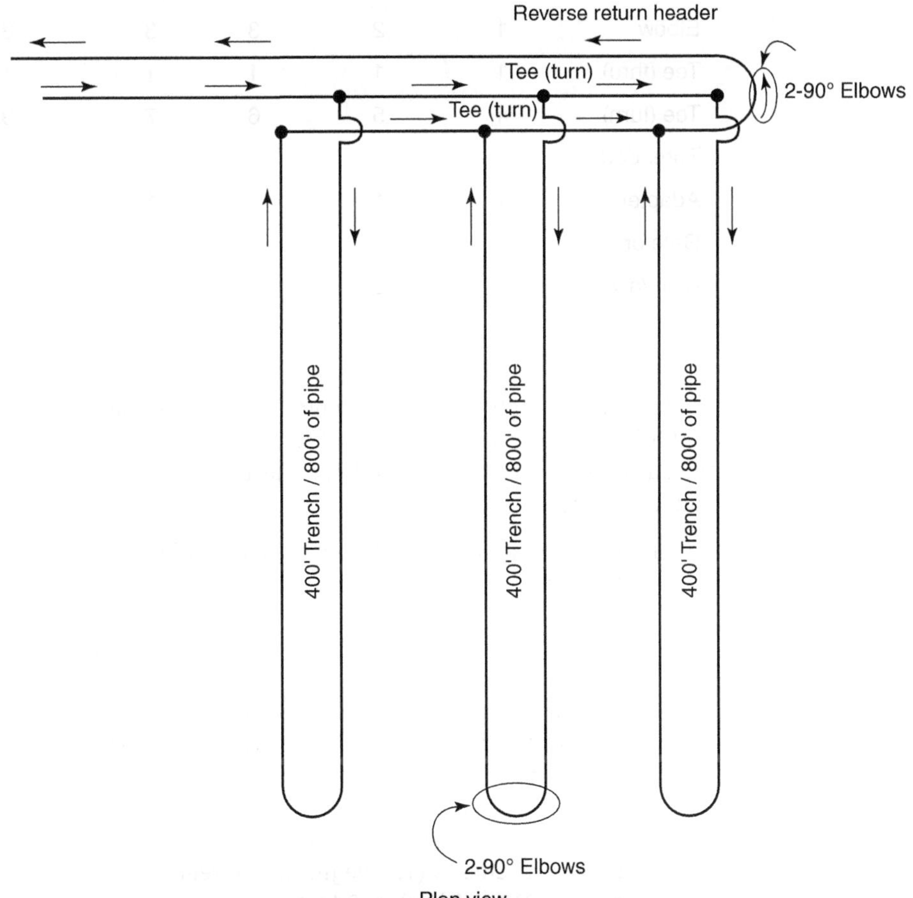

Figure 19–2
An example of a loop where each circuit has the same length because of the reverse return header. Water flows the same distance from the heat pump through the loop and back to the heat pump. (Courtesy of Delmar/Cengage Learning)

Table 19–2 Frictional Loss in Plastic Pipe for Water in Feet of Head per 100 Feet of Length

GPM	Pipe Diameter in Nominal Inches					
	1/2	3/4	1	1 1/4	1 1/2	2
1	1.1	0.3				
2	3.8	1.0	0.3	0.1		
4	13.7	3.5	1.1	0.3	0.1	
5	20.7	5.3	1.6	0.4	0.2	
6	29.0	7.4	2.3	0.6	0.3	
8	49.5	12.6	3.9	1.0	0.5	0.1
10	74.7	19.0	5.9	1.6	0.7	0.2
20		68.6	21.2	5.6	2.6	0.8
30				11.8	5.6	1.7

Table 19–3 Frictional Loss for Pipe Fittings, Plastic (in feet)

Fitting	Pipe Diameter in Nominal Inches					
	1/2	3/4	1	1 1/4	1 1/2	2
Elbow	1	2	3	3	3	4
Tee (thru)	1	1	1	1	1	1
Tee (turn)	4	5	6	7	9	11
Threaded Adapter	1	1	1	1	1	1
Gate or Ball Valve	2	3	4	5	6	7

Let's go through a sizing example. A 3-ton heat pump was designed to pump 2.66 GPM/ton or a total of 7.98 GPM. Water being pumped in this example loop would develop a frictional total for the loop and be calculated in basically the following way:

1. One circuit that is 800′ long and 1″ in diameter is multiplied by the friction loss from the friction chart read for 2.66 GPM (3 loops with equal flow, so 8 GPM ÷ 3 loops = 2.66 GPM in each loop), which would be approximately 1 foot of head for every 100 feet of pipe: 1 × 8 = 8 feet of head.
2. 200′ of 2″ pipe would be figured in the same manner: 0.1 × 2 = 0.2.
3. The calculation uses the loop with the highest frictional loss. It is the circuit with the highest frictional loss because water flowed through the fittings as follows (Figure 19–3):
   a. 2 – 2″ tee (turn) = 22 feet
   b. 2 – 1″ elbows (180 degrees) = 6 feet
   c. 2 – 2″ tee (thru) = 2 feet
   d. 2 – 2″ elbows (180 degrees) = 8 feet
   e. Total frictional feet from fittings: 22 + 6 + 2 + 8 = 38 feet.
4. All frictions added together: 8 + 0.2 + 38 = 46.2 feet of frictional head.

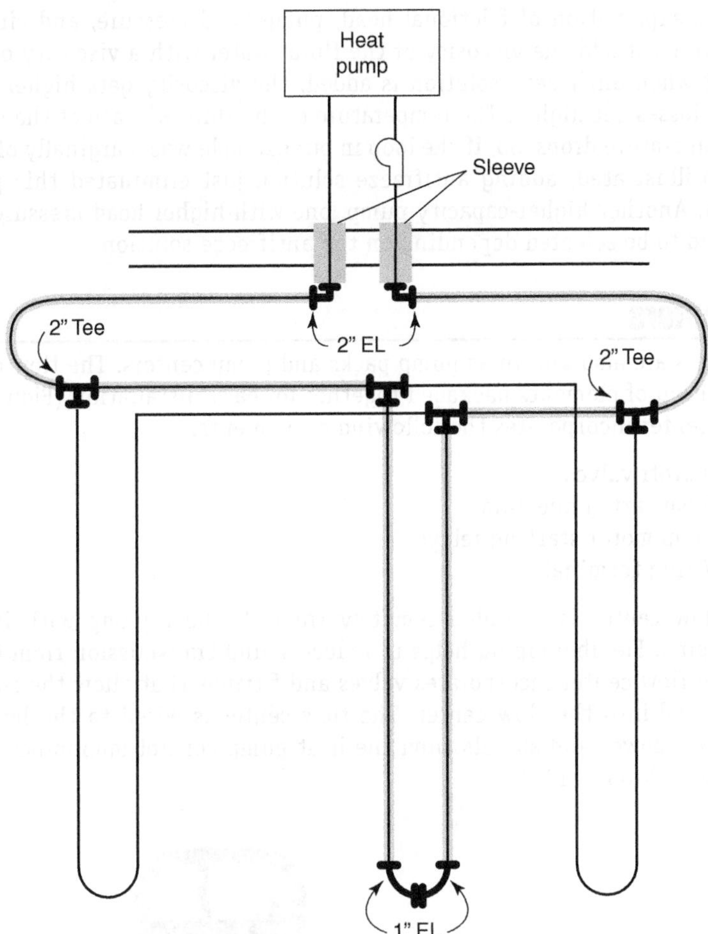

Figure 19–3
This is a simplistic example of a typical loop design and shows the loop with the highest frictional resistance because of fittings. The example is a reverse return loop that tends to be self-balancing. (Courtesy of Delmar/Cengage Learning)

Tech Tip

Adding parallel loops is one method of reducing overall pressure drop in a closed-loop system. Remember that pump head equals system head.

Each of the three circuits of the ground loop would exert no more than 46.2 feet of frictional head pressure. If the pressure drop of the heat exchanger were a maximum of 4 psig (equivalent to 9.24 feet of head), the pump would have to overcome a total of 55.44 feet of head for the loop and heat exchanger. A circulating pump would have to deliver a minimum of 8 GPM with a 55.44 head pressure. Using the pump curve (Figure 19-1), the pump illustrated would deliver approximately 11 GPM with a resistance of 55 feet of head. This would be sufficient to produce the 8 GPM needed for the loop.

Tech Tip

If the pump selection is close or between two pumps, the larger pump is chosen.

To this explanation of frictional head, pump head pressure, and circulation or flow, we must add the viscosity of the fluid. Water with a viscosity of 1 is no issue, but when antifreeze solution is added, the viscosity gets higher and the frictional losses get higher. The temperature of the fluid will affect the viscosity as the temperature drops. So, if the loop in our example was marginally okay with the pump illustrated, adding antifreeze solution just eliminated this pump as an option. Another higher-capacity pump, one with higher head pressure rating, would need to be selected depending on the antifreeze solution.

Flow Centers

Flow centers are also known as pump packs and pump centers. The flow center is a combination of elements packaged together for easy installation (Figure 19–4). The flow center incorporates the following components:

1. Shutoff valves
2. Flush cart connectors
3. Pump motor-starting relays
4. Wiring terminals

The flow center is installed remotely from the heat pump with interconnected piping. Flexible piping helps to reduce sound transmission from the heat pump. The flow center incorporates valves and fittings that allow the flush cart to be plugged into the flow center. The flow center is wired to the heat pump and receives power and signals from the heat pump control microprocessor. See Figures 19–5 through 19–7.

Figure 19–4
A flow center for a geothermal loop is designed to house the pumps, valves, and controls necessary to operate a geothermal heat pump. (Courtesy of WaterFurnace International)

Figure 19–5
The flush cart can be used for flushing and charging the geothermal closed loop. This flush cart is designed to connect to the flow center. (Courtesy of WaterFurnace International)

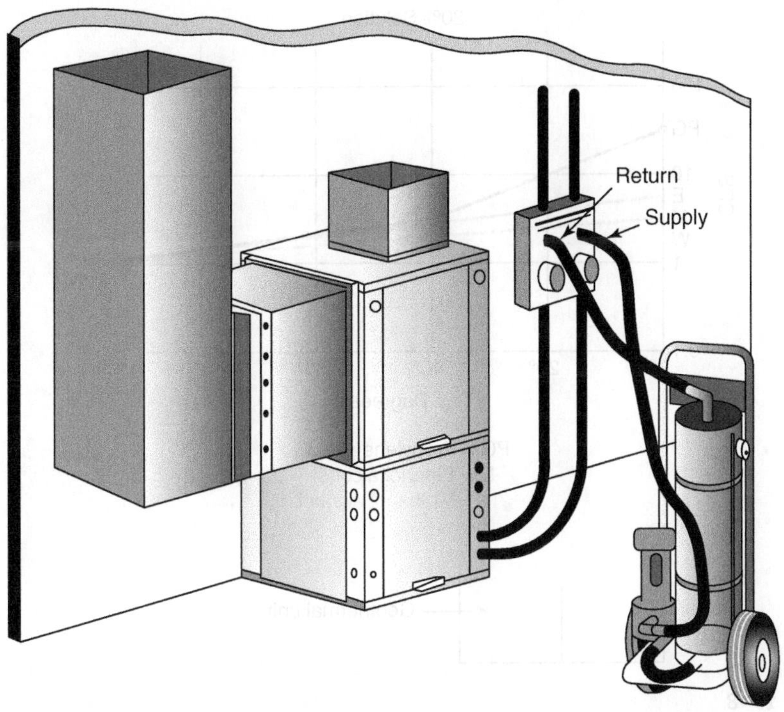

Figure 19–6
The flush cart is plugged into the flow center. The flow center is remote from the heat pump with interconnecting lines to the geothermal loop and to the heat pump. (Courtesy of WaterFurnace International)

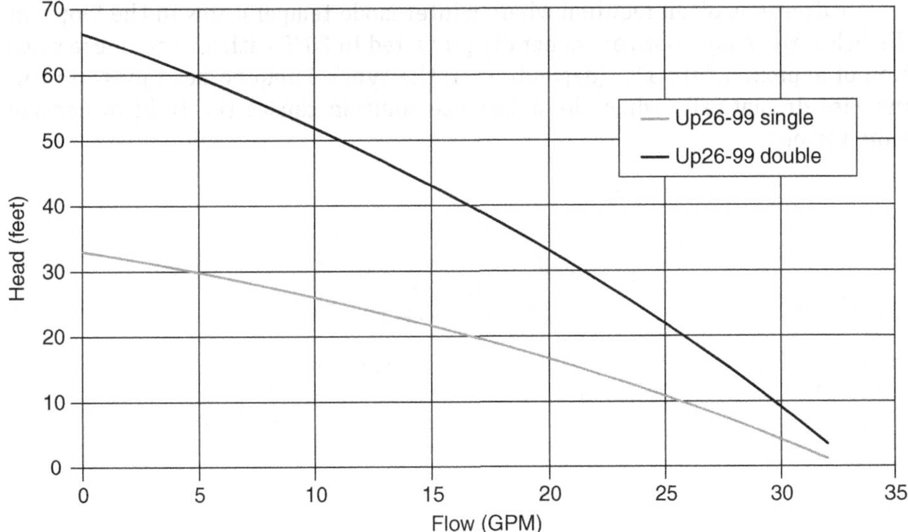

Figure 19–7
The pump curve for the flush cart, fitted with one or two pumps, has the capability of producing a large static head and greater flow volumes for flushing and charging. (Courtesy of WaterFurnace International)

Antifreeze in the Closed-Loop System

There are many types of antifreeze solutions that can be used (Figure 19–8):

1. Propylene glycol
2. Methanol alcohol
3. Ethylene glycol

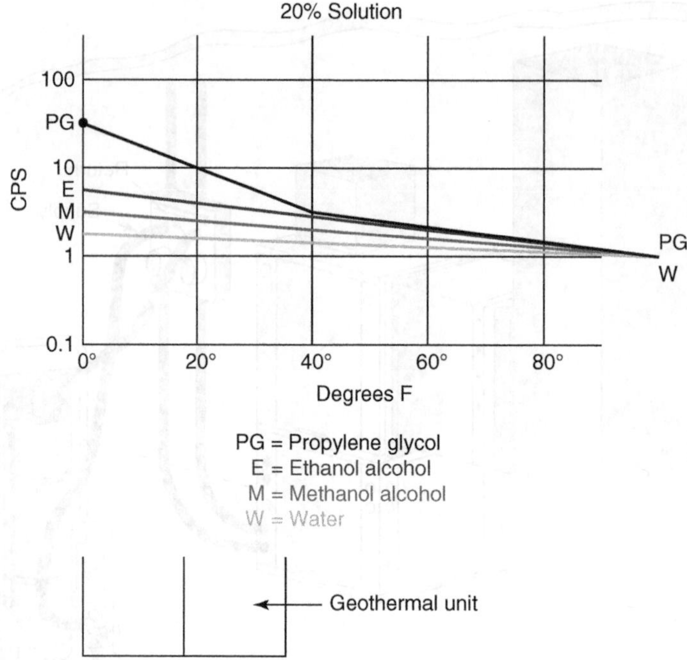

Figure 19–8
As the temperature goes down, a 20% solution of antifreeze affects the viscosity of the water-based solution. Each substance added increases the viscosity and makes it harder for the pump to push the water through a loop. Propylene glycol becomes very viscous below 20°F. Methanol and ethanol alcohol are less viscous at the same temperature. (Courtesy of Delmar/Cengage Learning)

Antifreeze is often required where winter-mode temperatures in the loop can dip below freezing. Loops are generally protected to 20°F with an antifreeze solution of approximately 30% (depending on the type). Pumping head pressure can increase dramatically when the antifreeze solution causes the fluid to become more viscous.

Tech Tip

The equipment manufacturer will suggest the type of antifreeze to be used, and this will often depend on the configuration of the manufacturer's heat exchanger. The contracting company may also have a preferred antifreeze that is used in loops it installs. The technician should be aware that there are many different types and be sure to check the system tags or documents to determine what type of antifreeze product is being used. Do not add water and antifreeze solution without being sure of the type of solution in the closed-loop system.

Antifreeze concentrations are determined by using a hydrometer (Figure 19–9). In some cases, the hydrometer is calibrated specifically for the antifreeze being used. If the hydrometer is designed for the antifreeze solution used, the hydrometer is usually read directly for the percentage of solution or the amount of freeze protection. Otherwise, a chart needs to be used and the hydrometer is read normally for specific gravity. Water has a specific gravity of 1. The specific gravity decreases when antifreeze is mixed into the solution.

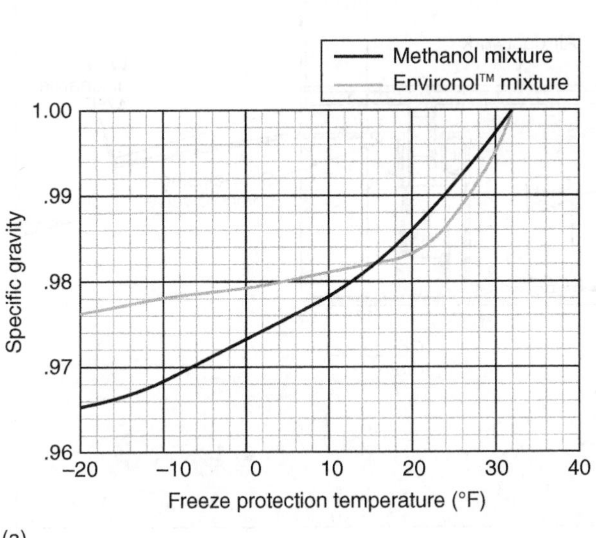

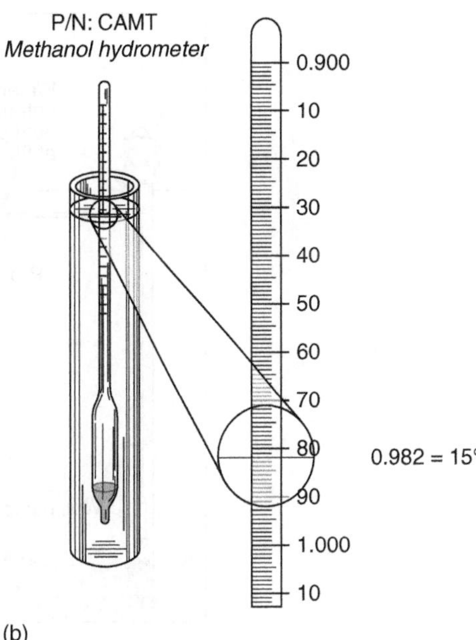

Figure 19–9
This methanol-calibrated hydrometer shows a reading that corresponds to the graph and indicates a level of freeze protection close to 15°F. (Courtesy of WaterFurnace International)

Tech Tip

All systems that use an antifreeze solution can be protected to a particular low temperature by using a solution chart. At low-temperature percentages, the antifreeze solution protects the water from becoming higher in viscosity and hard to pump. However, burst protection typically is a much lower temperature. At these low temperatures, the solution becomes slushy, but does not freeze solid. The burst-protection temperature ensures that the system does not freeze and burst pipes (loops) as temperatures go extremely low.

Open-Loop Systems

Many of the same requirements for closed-loop systems apply to open-loop systems. Some differences are:

1. Water friction is calculated to include vertical head pressure and water pumping height (Figure 19–10).
2. Tanks (holding tanks) are used to minimize the start and stop of the pump.
3. Either constant-pressure or constant-volume systems are used in conjunction with geothermal installations where water is used for both drinking and heating.

Piping frictional losses are calculated in the same fashion as for a closed-loop system, with the exception that geothermal plumbing may be integral to the building drinking water system. It should be noted that when drinking water (potable water) systems are involved, it is suggested that a well driller and plumber be part of the installation team. Problems with the building water system should be referred to both of these team members to ensure that adequate water volume and pressures are provided.

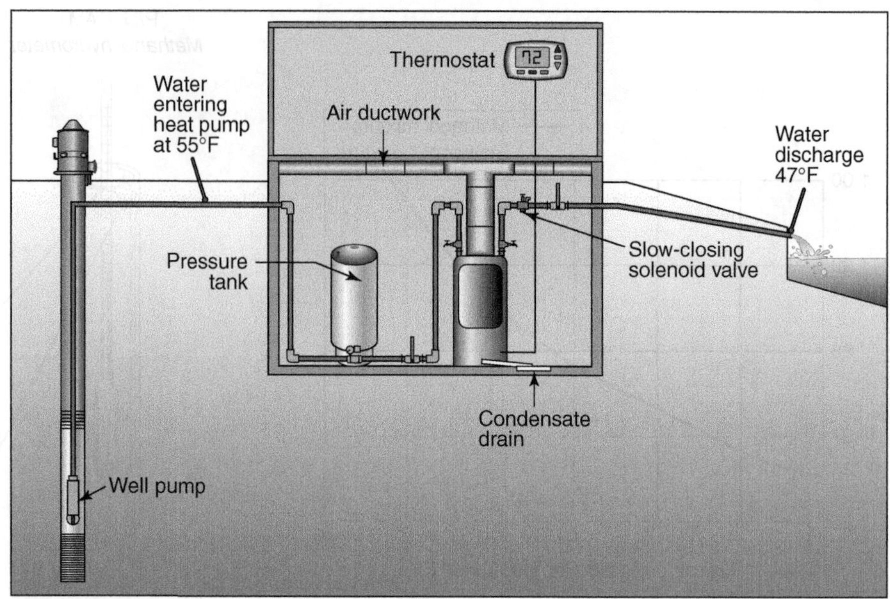

Figure 19–10
Water well systems include the calculation of vertical feet of head. (Courtesy of Delmar/Cengage Learning)

Water well information that should be included with the installation data for open-loop geothermal systems includes:

1. Water Well
   a. Depth—total depth from ground level to screen depth
   b. Pump depth—the depth from the ground to the pump (submersible pump applications)
   c. Water-table depth—depth from ground level to the height of the water in the well
   d. Total GPM—draw-down capacity of the well, also known as the replenish rate
   e. Well draw-down height—the draw-down height of water in the well (with pump operating)
2. Water Usage
   a. Domestic water usage in GPM
   b. Geothermal water usage in GPM

Water can be supplied in two ways to the building and a geothermal system: (1) constant pressure, or (2) constant volume. The constant-pressure system is based on the standard expansion (pressure) tank and pressure controls. As the water is being used and the pressure goes down, the pressure control senses the pressure drop and turns on the well pump to replenish the water. Pressure controls can be set so that the pressure range is only a few pounds when coupled with a large expansion tank. This type of system still requires the well pump to start and stop. If the frequency of pump starts is high, the well pump will wear out sooner.

Another solution is the constant-volume system. This system does not require a large expansion tank. The well pump is variable speed and operates nearly constantly while any part of the system requires water. As the water consumption goes up, the well pump is operated at a higher speed and delivers a larger volume of water. As water usage drops, the well pump throttles back and delivers less volume. This system also has a soft-start feature that reduces the stress of starting and reduces pump wear.

PIPING

The International Ground Source Heat Pump Association (IGSHPA) certifies installers in the technique of fusing plastic pipe for ground loops. Fused joints are the only joints that can be buried. Fusing is a heating process that brings the pipe connection to a semi-fluid condition. While in this semi-fluid condition, the two parts to be joined are pressed together and become one contiguous piece of pipe. When cooled, the connection performs as if it were molded in that condition.

Outside the structure and in the ground, all pipe joined connections are fused. The installer must be certified and familiar with working in field conditions that are cramped and awkward. The joining technique must be followed precisely or a good joint will not be formed. Improperly fused joints will leak and must be cut out of the loop and a new joint fused. It is very important to be sure that all joints are not leaking before the loop field is backfilled. If a loop is suspected of leaking, it must be dug up, leak-checked, repaired, and reburied—a cost that no one wants to pay for.

Inside the structure, piping connections can be fused or mechanical. Typically, where the ground loop enters the structure, pipe connections revert to mechanical connections. Piping is insulated to keep from sweating during operation and is secured to keep the pipe relatively straight, or flexible where it could transmit sound and vibration (Figure 19-11).

Geothermal piping includes:

1. Piping and pumps from and to the ground loop and geothermal unit
2. Condensate drain

Figure 19–11
This flow center is piped with hard pipe and fittings between the flow center and the loop connections. Between the flow center and the heat pump, the piping is flexible. (Courtesy of Jamie Simpson)

Loop Penetration

A possible leak situation exists where the ground loop enters the structure. As pipe heats and cools, it expands and contracts. This movement can loosen connections as the pipe moves through the below-ground wall. If the pipe is foamed in or grouted in place, the pipe could loosen the seal and the seal will leak. An alternative method is shown in Figure 19-12. In this example, the hole

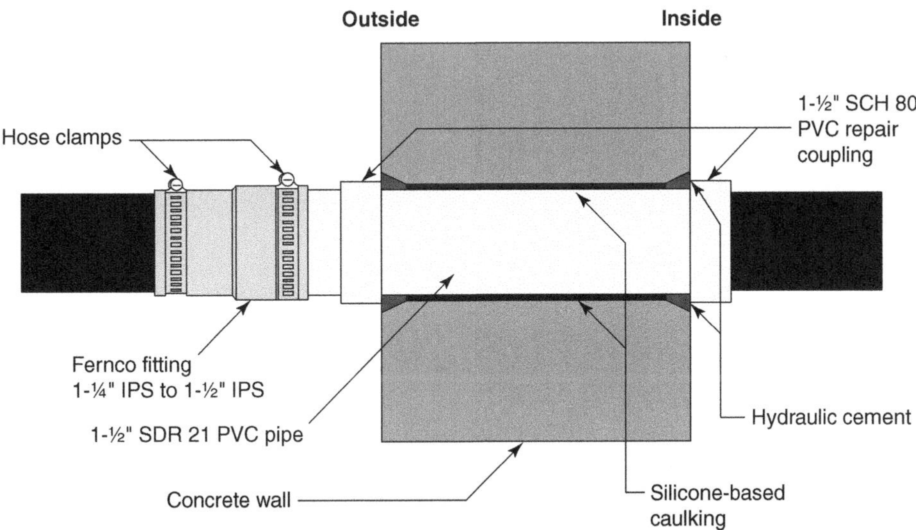

Figure 19–12
An alternative method to seal loop pipe extending through a below-grade wall. (Courtesy of WaterFurnace International)

Safety Tip

Working in open trenches can be very hazardous and as such, OSHA has very stringent rules regarding shoring, sloping, and other aspects of trench safety. These regulations are found in OSHA 29 CFR 1926.650.

is bored and then fitted with a PVC pipe and a half of a coupling. The pipe is slipped through the wall and grouted with hydraulic cement. The other half of the PVC coupling is cemented on the outside of the wall where the PVC extends into the ground. Where the PVC ends, a rubber fitting is clamped over both the loop pipe and the PVC coupling. The rubber fitting allows the pipe to move while maintaining a leak-proof connection. The hydro-cement maintains a solid seal to the PVC pipe.

SUMMARY

In this chapter we have discussed pumping systems as they apply to closed- and open-loop systems. Because geothermal loops use a water-based solution, we started with the basic characteristics of water and how to determine the amount of fluid flow in a loop using a pressure-drop determination across the refrigerant-to-water heat exchanger.

Closed loops have frictional forces that need to be overcome by the pump in order to move a water-based solution through the loop. We reviewed how the friction is determined and what causes it. Frictional head pressure is also known as feet of head pressure. Pump manufacturers rate their pumps based on the feet of head, which determines the amount of water volume in gallons per minute (GPM) that the pump is able to deliver.

Flow centers, or pump packs, circulate water through a closed loop. These devices are typically remote from the heat pump and incorporate pumps, fittings, valves, and controls into a single system that can be mounted independently from the heat pump. Flush carts can plug into the flow center, which makes it convenient to flush and charge the geothermal loop.

The flush cart can also be involved in charging the system with antifreeze solution. There are many different types of solutions; one type is selected over the others by the manufacturer, the installer, or by local ordinance. Typically, the solution is approximately 30% antifreeze, which protects the geothermal loop from freezing down to a temperature of approximately 20°F (this is dependent on location, use, and conditions).

Open-loop systems use a water well. Because the loops are open, vertical lift or feet of head needs to be incorporated into the calculation for a water well pump. The selection and the installation of a water well pump will typically involve a water well driller. The installation of a plumbing system for a structure also involves a plumber. The water well system and the delivery of water to the heat pump may be handled by professionals other than the HVAC installer/technician.

Piping connections for the loop are fused. Fusion is the only technique that is approved for buried connections. To be able to fuse plastic pipe, the installer needs to be certified by IGSHPA. The reason it is so important to be able to make leak-free connections is because it costs too much to dig up the buried loop to repair the leak.

Where loops penetrate the below-grade wall, there is a possibility that the pipe, through its own action, could cause the wall to leak. When the pipe expands and contracts through cooling and heating modes, the pipe could loosen the sealing material. An alternative method is discussed to eliminate this potential leak.

REVIEW QUESTIONS

1. Describe the relationship between feet of head and pressure per square inch gauge (psig).
2. Describe the reason for conducting a pressure-drop measurement across the coaxial heat exchanger.
3. Relate how temperature difference across the coaxial coil, pressure difference, and flow are related.
4. Describe what causes frictional loss in a closed loop.
5. Explain why vertical feet of head is not important for closed loops, but is for open-loop systems.
6. Describe what components are in a flow center and the function of a flow center.

CHAPTER 20

Geothermal Sequence of Operation and System Checks

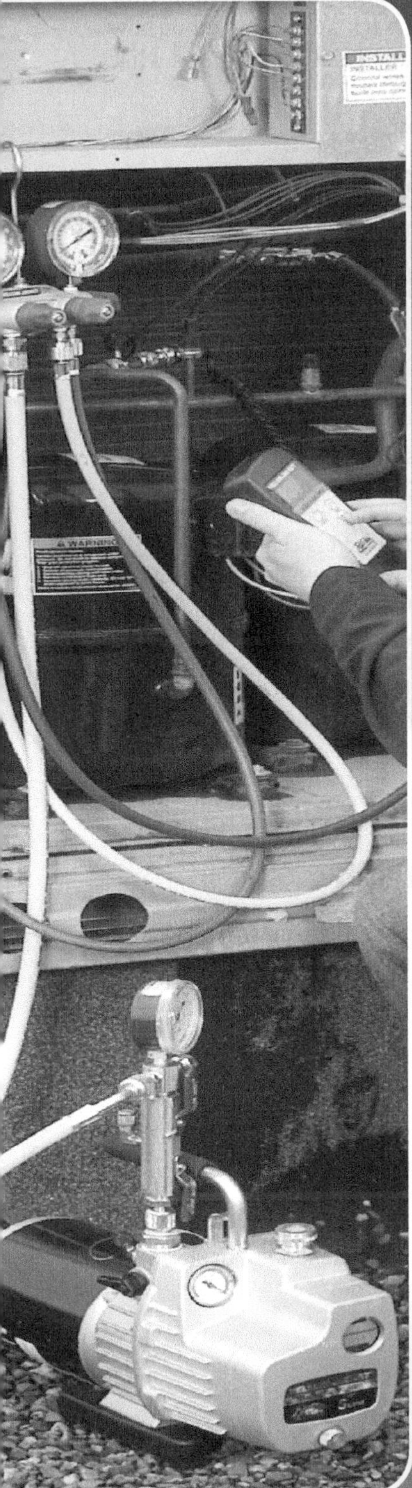

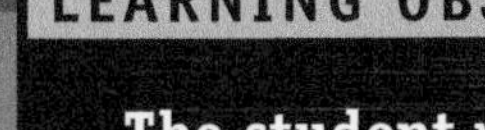

LEARNING OBJECTIVES

The student will:

- Describe why a defrost system is not necessary for a geothermal heat pump
- Describe the construction of a coaxial heat exchanger
- Describe the construction of a shell-and-coil heat exchanger
- Explain the concept of counter-flow heat exchange
- Describe the difference between open- and closed-loop water-source heat pumps (WSHPs)
- Explain how flow through the water-source heat exchanger is affected by pressure drop
- Explain the concept of "approach" temperature
- Describe what happens to electric resistance strip heat on a third-stage call for heat (use the wiring diagram in this chapter)

INTRODUCTION

Ground-source, or geothermal, heat pumps (GSHPs) operate in the same fashion as air-source heat pumps (ASHPs). There is a cooling and a heating mode that uses a reversing valve to switch modes. There are also differences because the GSHP is not required to pull heat from outside air that is constantly changing temperature. Because the ground is nearly a constant temperature year-round, the GSHP can function at higher efficiency levels.

The unique systems of a GSHP will be discussed as they apply to water-source heat pumps (WSHPs) that transfer heat from and to the ground using water as a transfer medium. Heat exchangers, water-based system components, and flow rates will also be discussed. The chapter will conclude with a description of how a WSHP operates and a wiring diagram. What happens on a call for heat and what is sensed will be described.

Field Problem

Summer outside temperatures had recently risen and the service technician was receiving more service calls. The system was a geothermal ground-loop GSHP. This particular customer complained that the system was short-cycling and not keeping the house cool.

Symptoms:
Compressor short-cycling; blower runs all of the time in thermostat auto position.

Possible Causes:
Low-pressure control or high-pressure control is opening; air blockage or dirty filter; fouled coaxial coil or blocked water flow outside; refrigerant leak causing low-pressure trip; loop pumps not operating to specification.

The technician began a series of sensory checks by touching, listening, looking, and smelling. There were no unusual smells. There were no unusual noises, except for the short-cycling of the compressor. There was nothing out of place or unusual, except for the flashing light-emitting diode (LED) display. Checking the manufacturer's legend, the technician noted that the flashing code indicated a pressure trip. Touching the loop pipes, the technician noticed a larger-than-usual temperature difference between the supply and return piping to the heat exchanger. Checking the pressure drop across the coaxial coil, the technician measured a very small pressure drop. Checking with a clamp-on ammeter, he found that one of the two loop pumps was not running. Disconnecting the power and checking with an ohmmeter, he verified that one of the loop pumps had an open circuit. After replacing the pump and checking the pressure drop across the outdoor coil, he confirmed that the system was operating correctly.

THINGS UNIQUE TO GEOTHERMAL SYSTEMS

Geothermal systems are distinguished from ASHPs in significant ways. This is predominantly because they transfer heat from and to the ground and not the air. Table 20–1 compares geothermal systems and ASHPs.

Table 20–1 Comparison of GSHPs and ASHPs

Component/System	ASHP	GSHP
Compressor location	Outside	Indoor
Heat exchanger	Finned tube air-to-refrigerant heat exchanger located outside	Coaxial water-to-refrigerant heat exchanger located indoors
Fluid-moving devices	Fan	Pump
Defrost system	Yes	No
Freeze protection	No	Yes

The most notable difference is that the ASHP requires a defrost system, but not freeze protection because it is required to transfer heat from air that is below freezing. The GSHP removes heat from a constant heat source, one that changes very little when compared to outside air temperatures. However, heat absorbed by the refrigerant from the water could leave the water temperature below freezing, hence the need for freeze protection.

COAXIAL HEAT EXCHANGER

The water-to-refrigerant heat exchanger of the water-loop geothermal heat pump is typically a coaxial heat exchanger. Coaxial means that one tube is located inside another tube (also called a tube-within-a-tube heat exchanger). The inner tube typically carries water or a water-based solution (water and an antifreeze solution) and the outer tube surrounds the water tube carrying the refrigerant. See Figures 20–1 and 20–2.

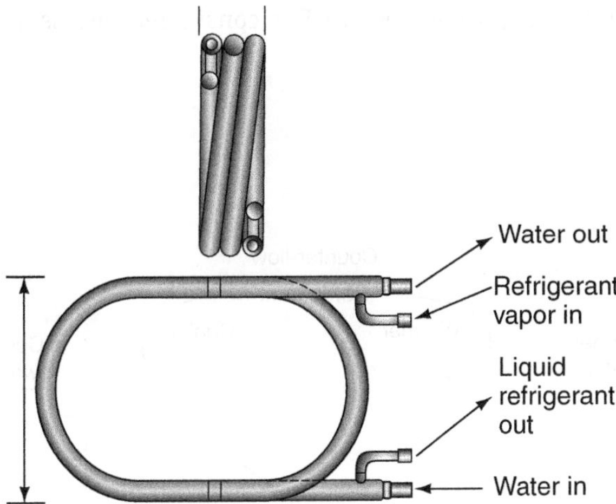

Figure 20–1
The coaxial (tube-in-tube) heat exchanger is typically coiled to take up less room. Refrigerant moves through the outer tube and water is in the inner tube. (Courtesy of Noranda Metal Industries, Inc.)

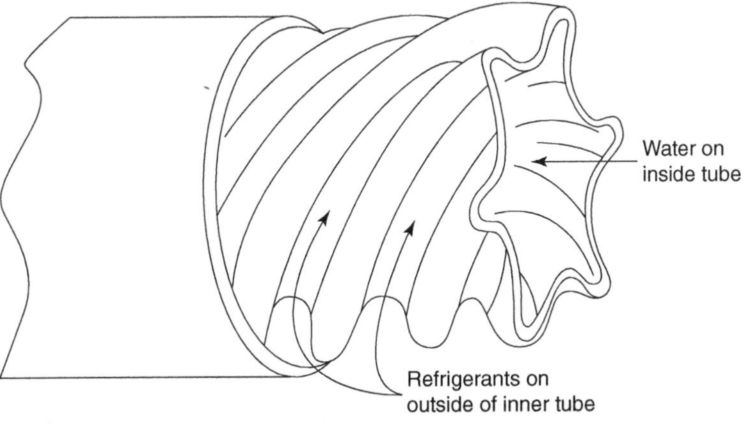

Figure 20–2
Cutaway of a coaxial heat exchanger showing the corrugated inner tube and the outer tube. The inner tube is shaped to increase surface area and turbulent flow of fluids. (Courtesy of Delmar/Cengage Learning)

Most coaxial heat exchangers are constructed with a steel outer tube and a cupronickel inner tube. The inner tube is formed (or twisted) to increase the surface area of the tube and to produce turbulence in both the refrigerant and water. The turbulence and the increased area of the inner tube both work to maximize heat exchange. The flow in most of these heat exchangers is counterflow. This means that the water-based solution moves in the opposite direction of the refrigerant. During the heating mode, cooler refrigerant enters the heat exchanger and gains heat throughout its travel. It is warmer when it leaves the coaxial heat exchanger. Ground-temperature water enters the exchanger at the point where the cooler refrigerant is exiting and loses heat throughout the coaxial heat exchanger, leaving the heat exchanger at a much cooler temperature.

The reason for counter-flowing the water and refrigerant is to achieve a continual temperature difference throughout the length of the coaxial heat exchanger (Figure 20–3). This is to maximize the amount of heat transfer by maintaining the largest difference in temperature between the water solution and the refrigerant. The greater the difference in temperature between fluids, the greater the heat transfer will be. A temperature difference between the refrigerant and the water of approximately 10°F is considered necessary for good heat transfer.

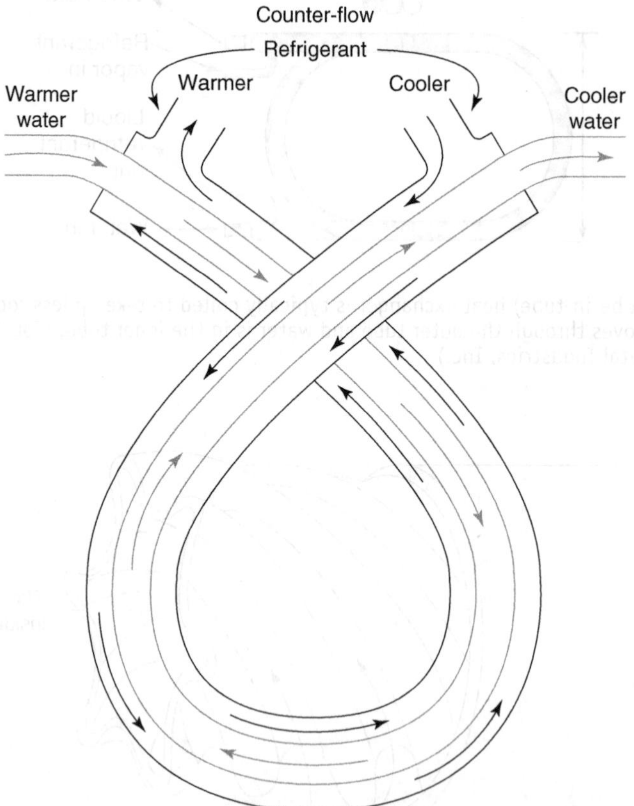

Figure 20–3
Moving water solution and refrigerant in opposite directions is called counter-flow. This type of flow attempts to maintain a constant temperature difference between the water and refrigerant to maximize heat flow. Refrigerant on the outside tube gives off heat to the ambient air, and with the coldest water meeting condensed refrigerant out, maximum subcooling is accomplished. (Courtesy of Delmar/Cengage Learning)

SHELL-AND-COIL HEAT EXCHANGER

Shell-and-coil heat exchangers use an outer tank to contain the inner tube (Figure 20–4). Water solution flows through the tube while refrigerant moves through the shell or tank and around the tube.

Shell-and-coil heat exchangers are not easily cleaned. Just like tube-in-tube (coaxial) heat exchangers, they are sealed so that cleaning can only be done by passing a cleaning solution through the exchanger. Physical cleaning with a brush is nearly impossible. The process of passing a solution through the heat exchanger is referred to as chemical cleaning.

REGULATING WATER FLOW RATES

For geothermal water-loop systems (GSHPs), proper flow rates must be established and maintained for the system to operate at maximum efficiency. Both open- and closed-loop systems use different strategies to regulate flow.

Open-Loop Systems

Water in an open-loop geothermal system must be pumped from the ground and supplied to the geothermal unit to transfer heat. A submersible pump (Figure 20–5) at the bottom of the water well is used to lift water and pressurize a holding tank (also known as an expansion tank; Figure 20–6). This may be the same water well that is used to supply drinking water for the structure. As the water is used, the pressure in the tank decreases. The decrease in pressure is sensed by the pressure switch, which operates the water well pump. The pump is energized and pumps water to the expansion tank, increasing the pressure in the tank until the switch

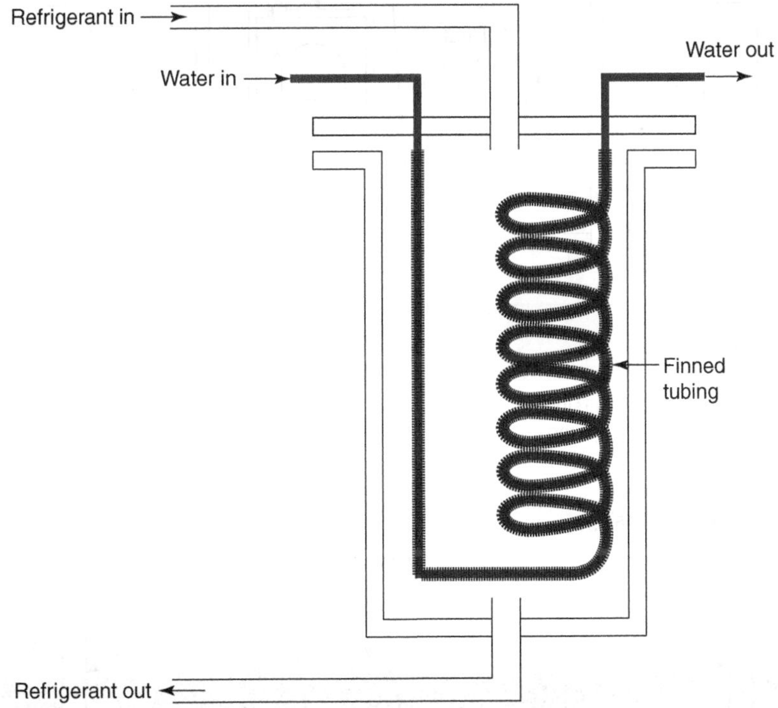

Figure 20–4
Shell-and-coil heat exchangers are similar to tube-in-tube exchangers. The inner coil is filled with water and the shell contains refrigerant. (Courtesy of Delmar/Cengage Learning)

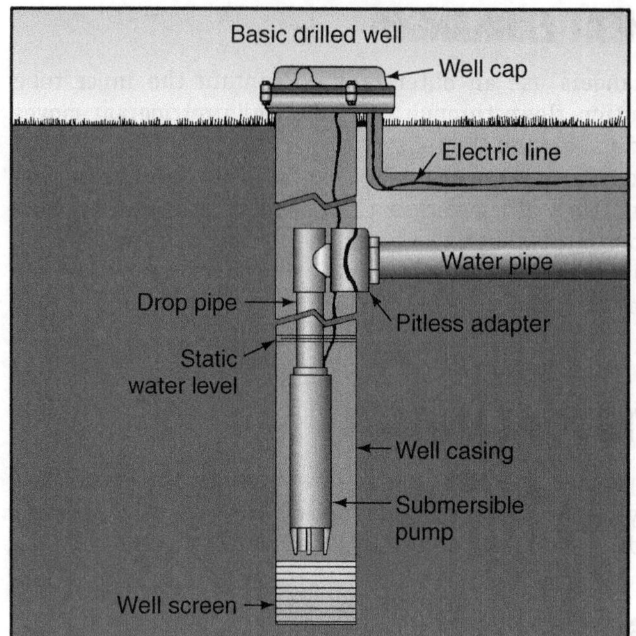

Figure 20–5
The basic construction of a drilled water well, showing the submersible pump and connections from the well to the structure. (Courtesy of Delmar/Cengage Learning)

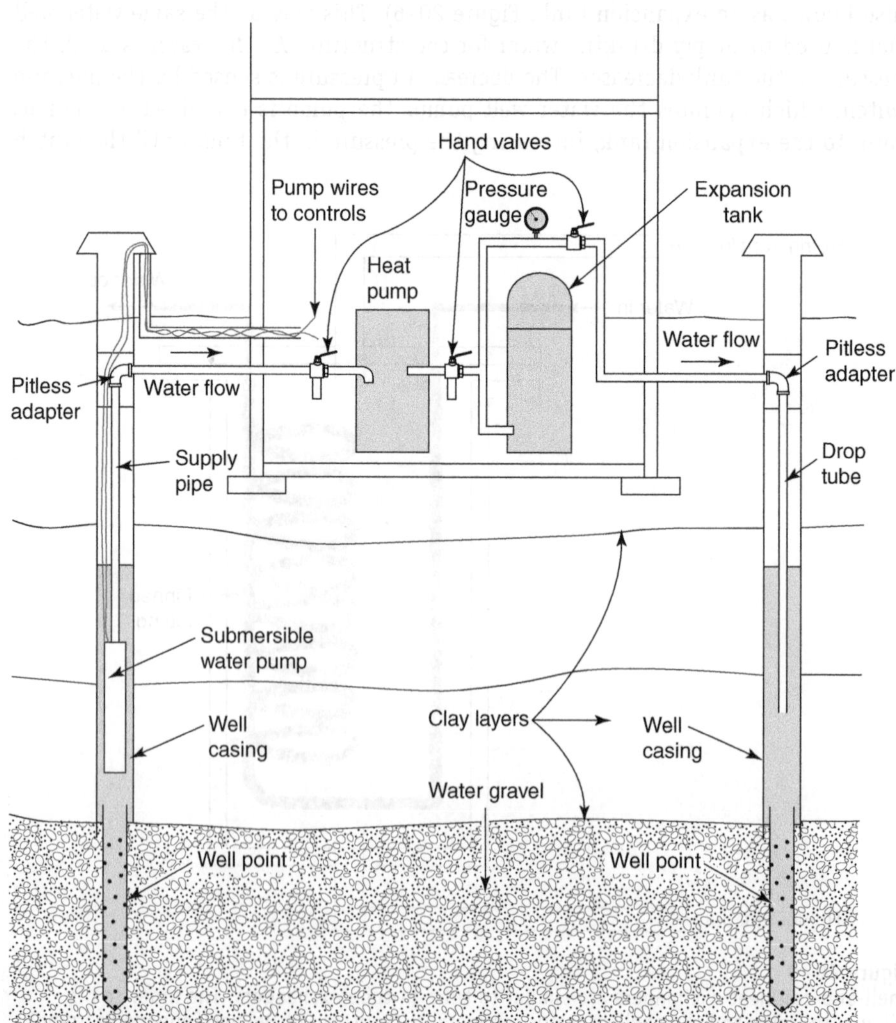

Figure 20–6
The water well supplies water to a pressurized holding tank. Water drawn from the tank moves through the heat pump and is either put back down another well or drained to a surface location. Consult local codes.(Courtesy of Delmar/Cengage Learning)

pressure is satisfied and the water well pump turns off. The expansion tank's main function is to reduce cycling of the well pump.

Water is regulated by restricting the flow, using a ball or balancing valve, to the number of gallons per minute (GPM) as recommended by the manufacturer. On a call for heating (or cooling), water flow is initiated by opening a motorized valve. Pressure taps on either side of the balancing valve allow for pressure readings that can be converted to flow rates given the manufacturer's flow characteristics of that valve.

Tech Tip

Some manufacturers suggest that a simple ball valve may be used to regulate water to the heat pump instead of a water regulator. If there is sediment in the water or the water quality is poor, a ball valve may be a better choice than a water regulator.

Closed-Loop Systems

Water is regulated or controlled by the design of the loop and selection of the pump. The system may or may not have a balancing valve. The system is designed to deliver the GPM as required by the manufacturer.

FLOW RATES

To ensure the proper flow of water (or water-based antifreeze solution) through the heat exchanger, manufacturers require the technician to follow certain procedures. Be sure to read, understand, and follow the manufacturer's recommendations to adjust water flow.

The following steps describe how to make the water flow adjustment:

1. Connect the pressure and temperature gauges to the P/T (pressure/temperature, also called Pete's plug) connectors at the unit (Figures 20-7 and 20-8).
2. Set the thermostat to energize the blower motor. Check for blower operation.
3. Set the thermostat control setpoint to place the system in the cooling mode. Check for compressor operation. (Note: Some systems have a time delay on start. Allow sufficient time for the compressor to operate in steady state.)
4. Ensure that the water control valve or loop pump(s) is operating.
5. Insert the thermometer into the Pete's plug.
6. Measure the water temperature entering and leaving the heat exchanger.
7. Insert the pressure gauge into the Pete's plug—both supply and return.
8. Measure the pressure difference from the inlet to the outlet of the heat exchanger.
9. Determine the water flow rate by comparing the pressure drop through the heat exchanger to the manufacturer's chart.
10. Compare the operation of the system to the manufacturer's cooling operation table.

Refer to the manufacturer's flow rate charts for examples and a complete description of how the manufacturer determines flow rates in the field.

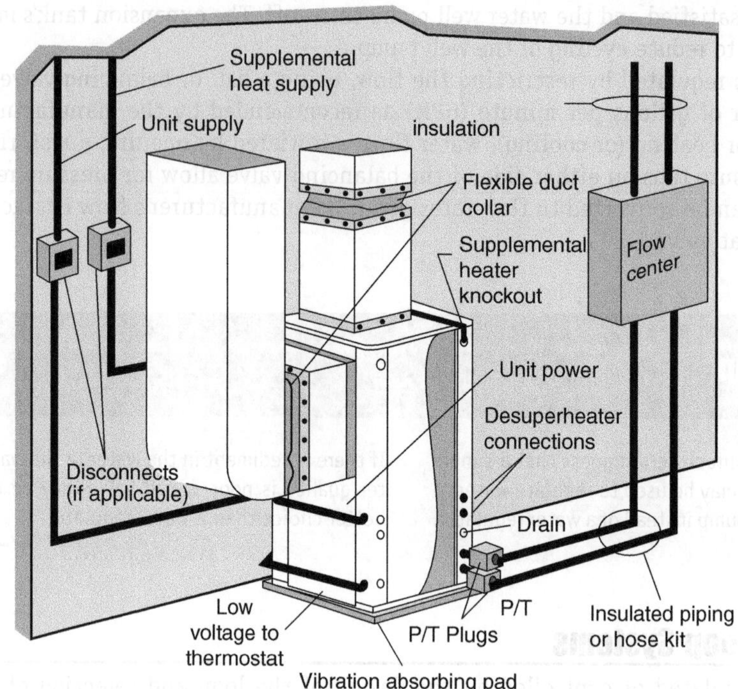

Figure 20–7
Closed-loop GSHP application. P/T (pressure/temperature) access fittings are located close to the unit at the supply and return connections of the ground loop. (Courtesy of WaterFurnace International)

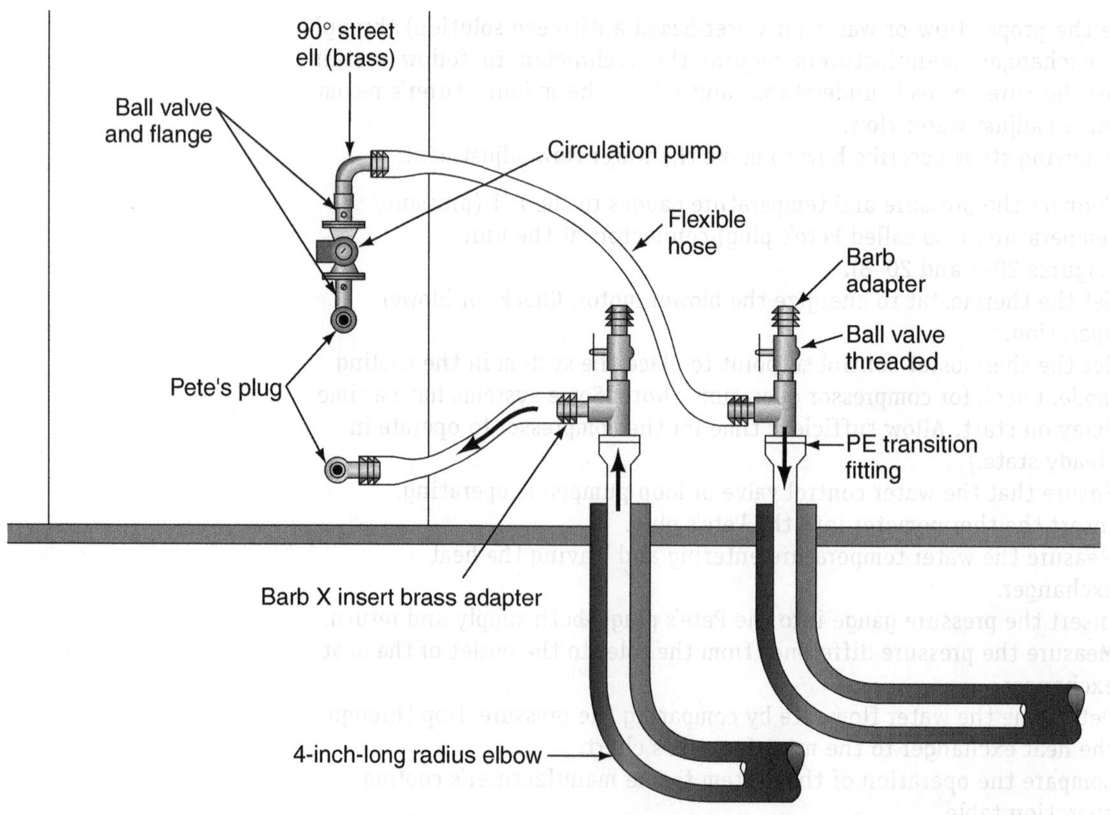

Figure 20–8a
The P/T connection (also sometimes referred to as a Pete's plug) is used to measure the pressure and temperature of water flowing through the heat exchanger. (Courtesy of Delmar/Cengage Learning)

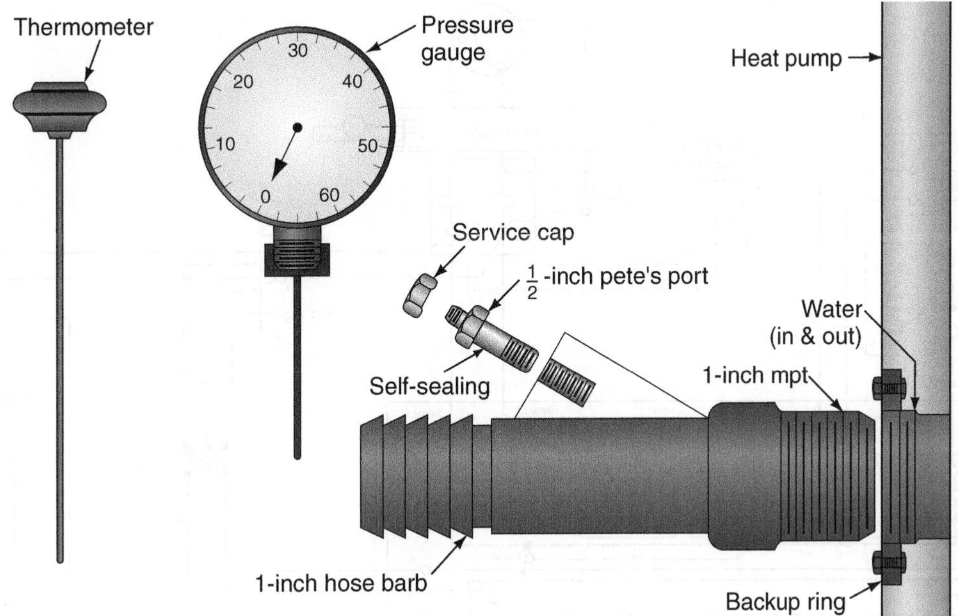

Figure 20–8b
Pressure and temperature sending adapters allow the pressure and temperature probes to access water conditions inside the piping while keeping air out of the system. (Courtesy of Delmar/Cengage Learning)

APPROACH TEMPERATURES

"Approach," as applied in HVAC, has to do with the difference in temperature between the refrigerant (in a refrigeration heat exchanger) and the temperature of the medium (water or air) being heated or cooled. It is generally considered to be a sign of good heat exchange when the temperature of the medium is very close to or "approaches" the temperature of the refrigerant after proper flow is verified. This temperature difference is measured and provides a good indicator of how heat is being transferred between a geothermal system and the ground loop.

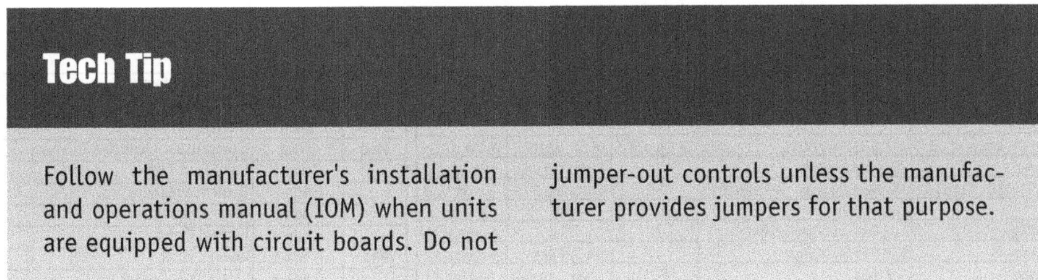

Tech Tip

Follow the manufacturer's installation and operations manual (IOM) when units are equipped with circuit boards. Do not jumper-out controls unless the manufacturer provides jumpers for that purpose.

ELECTRICAL SYSTEM OPERATION

The following is a normal sequence of operation for a geothermal heat pump (with domestic water heating) in the heating mode. The reader should read the text description and try to read the wiring schematic in Figure 20–9 at the same time. Read until an operation is completed and then review the electrical schematic. Move back and forth until the entire description and schematic have been read. The description is of a dual-capacity geothermal heat pump and will also include the operation of domestic hot water (DHW) using the WaterFurnace print, courtesy of WaterFurnace International.

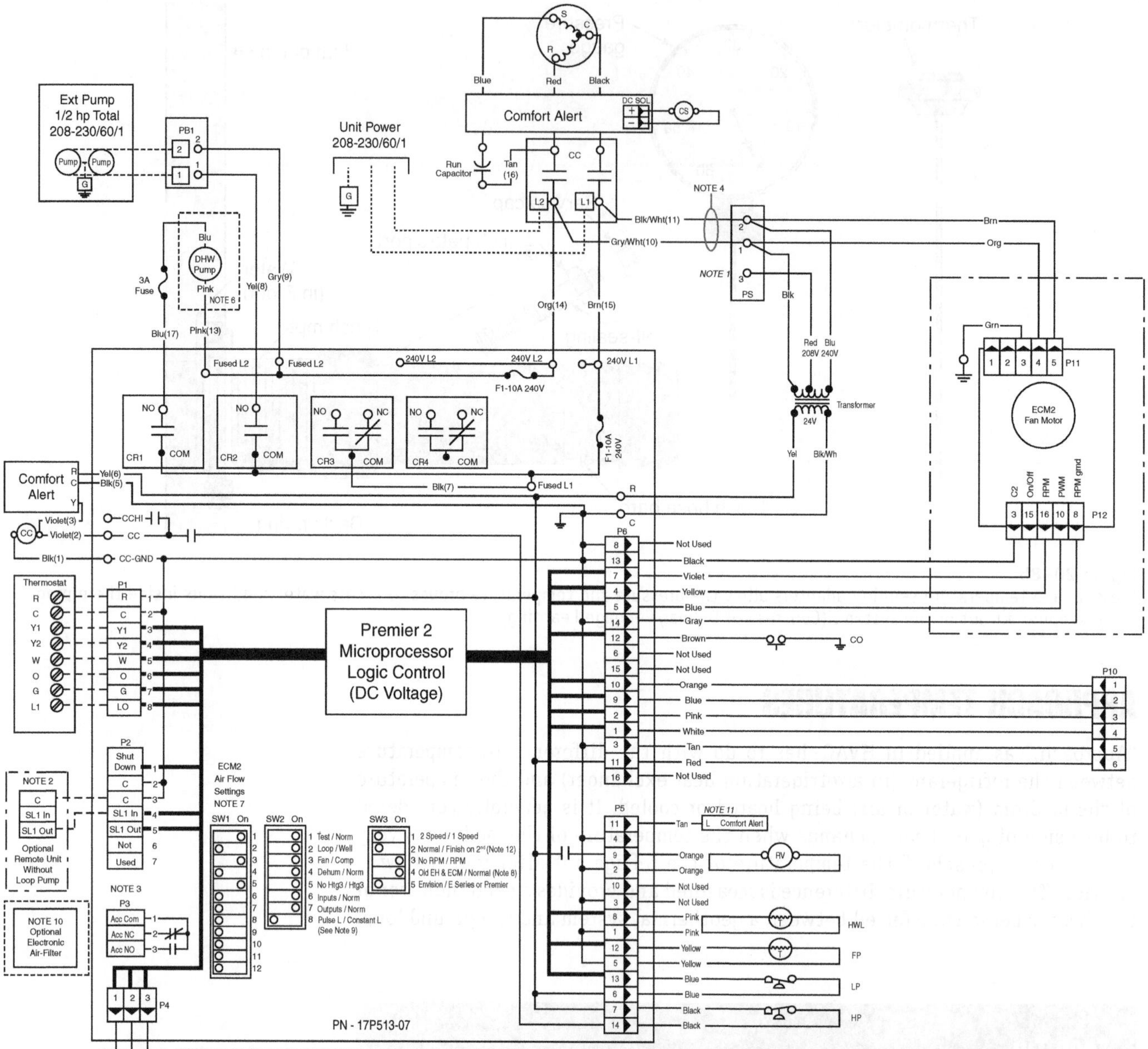

PN - 17P513-07

LED	Normal Display Mode	Diagnostic Modes: Current Fault Status	Inputs	Outputs	Outputs 2
	Field Selection Dips - #1 On, #6 On, #7 On	#1 Off, #6 On, #7 On	#6 Off, #7 On	#6 Off, #7 Off	#6 Off, #7 Off
Drain	Drain pan overflow Lockout	Drain pan overflow	Y1	Compressor Lo	Blower Lo
Water Flow	FP thermistor (loop<15°F, well<30°F) Lockout	FP thermistor (loop<15°F, well<30°F)	Y2	Compressor Hi	Blower Med
High Press	High Pressure	High Pressure	O	RV	Blower Hi
Low Press / CA	Low Pressure / Comfort Alert	Low Pressure / Comfort Alert	G	FAN	Aux Heat #1
Air Flow	ECM2 RPM < 100 rpm Lockout	ECM2 RPM < 100 rpm	W	DHW Pump	Aux Heat #2
Status	Microprocessor malfunction*	Not Used	SL1	Loop Pump(s)	AuxHeat #3
DHW Limit	HWL thermistor > 130°F	HWL thermistor > 130°F	--	--	Aux Heat #4
DHW off	DHW pump switch off	DHW pump switch off	--	--	--

*Green LED not flashing

Comfort Alert Status		
LED	Flash Code	Description
Green	Solid	Module Has Power
Red	Solid	Compressor Overload Trip
	Code 1	Long Run Time
	Code 2	System Pressure Trip
	Code 3	Short Cycling
	Code 4	Locked Rotor
Yellow	Code 5	Open Circuit
	Code 6	Open Start Circuit
	Code 7	Open Run Circuit
	Code 8	Welded Contactor
	Code 9	Low Voltage

97P774-02 8/1/08

(continued)

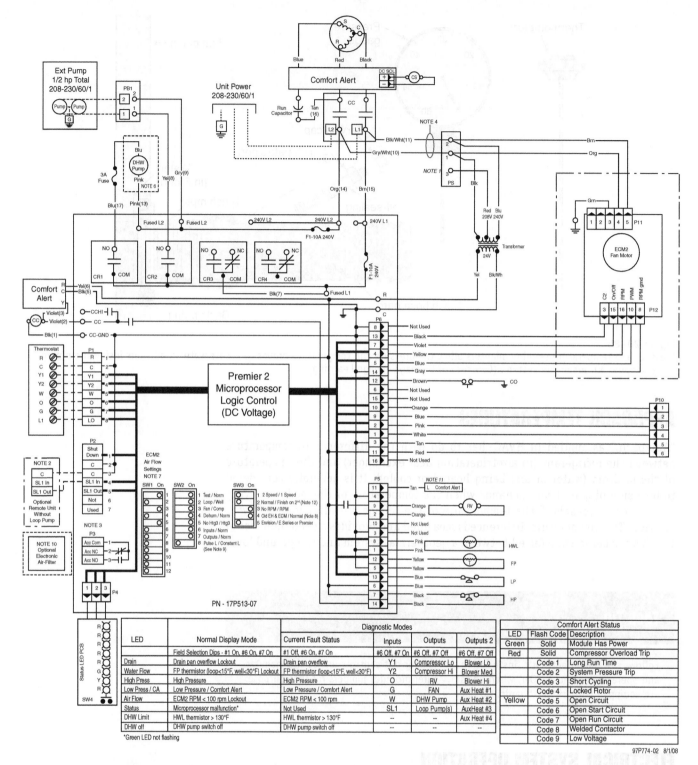

Figure 20–9a
Dual-speed geothermal system wiring. (Courtesy of WaterFurnace International)

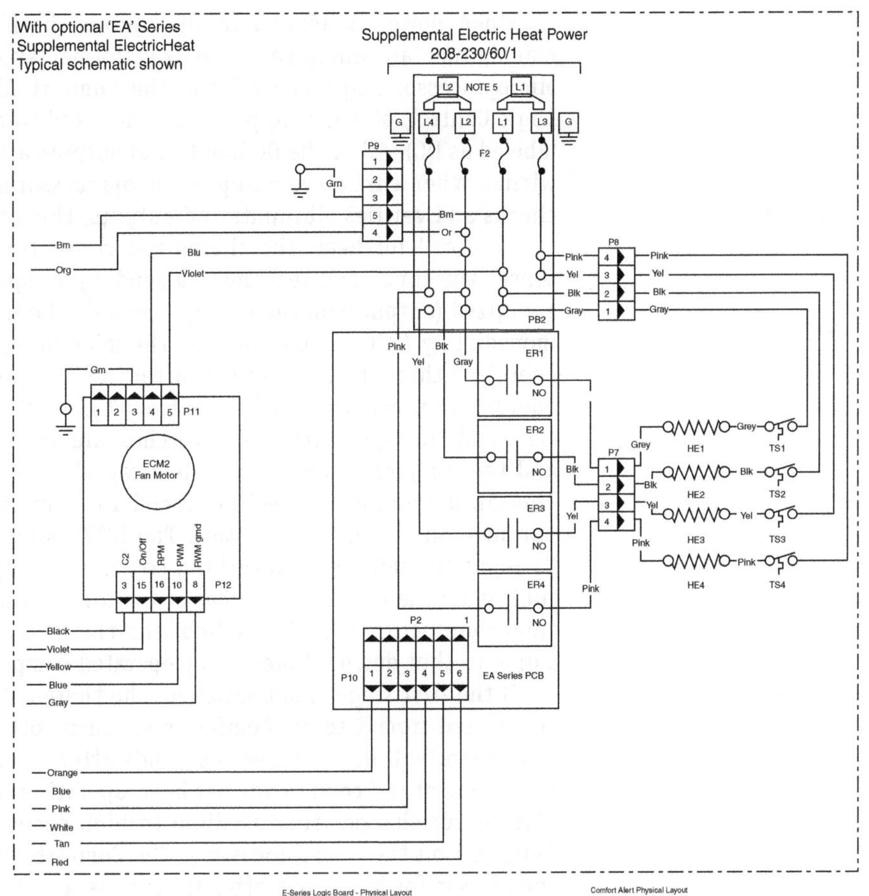

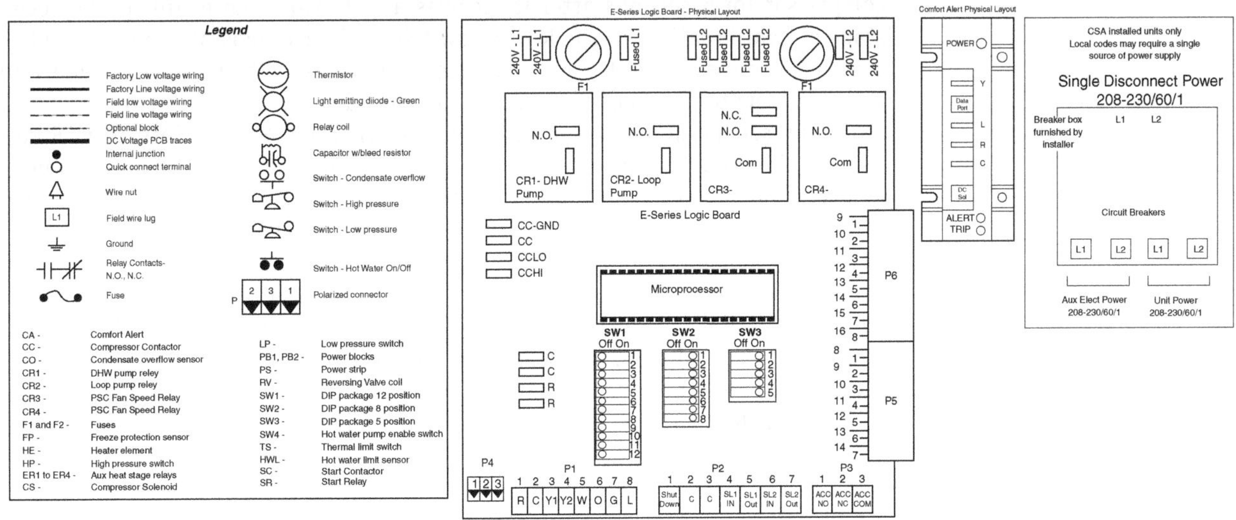

NOTES

1. Switch Blue and Red wires for 208V operation.
2. Connection of remote unit that does not have a ;oop pump for slave operation.
3. 24V Accessory relay (see SW2 -3 for description of operation).
4. The blk/wh and gray/wh wires are removed when Aux Heat is installed.
5. Buss lugs L1 and L2 can be removed and dual power wire sets connected directly to box lugs L1, L2 and L3, L4.
6. DHW pump only in models with hot water generation option.
7. Air Flow Configuration Example: SW1 configured for dip 1 as low, dip 3 as medium, and dip 5 as high speed ECM2 fan.
8. SW3-4 should be in the OFF position when using ECM motor and 17P501A01 electric heat board and should be ON when using ECM2 with 17P514A01 electric heat board.
9. SW2-8 must be in the OFF position for pulsed "L" lockout signal and in the ON position for constant "L" lockout signal.
10. When optional electronic air-filter is installed, power for the electronic air-filter is provided by P2-2 and 24 VAC.
11. Comfort Alert fault output to Premier Control Board.
12. This switch allows the unit to down stage with t-stat when off and finish with second stage when on. Finish on second stage reduces stages changing in recip dual capacity compressors and should be ON for unzoned Dual Cap E Series or Premier 2 Speed.

97P774-02 8/1/08

Figure 20–9b (continued)
Dual-speed geothermal system wiring. (Courtesy of WaterFurnace International)

When power is applied to the system, both the ECM2 fan motor and the transformer are energized. The transformer supplies power to the Premier 2 Microprocessor Logic Control and the Comfort Alert module (Note: Within the Logic Control, alternating power is converted to direct current; this is why it is labeled as DC). All of the DC inputs and outputs are labeled by larger, gray-colored wiring. When first powered-up, a microprocessor may delay all operations until it checks and verifies all inputs and outputs; this could take several minutes.

On a call for heat, the thermostat closes from R to Y1. The microprocessor senses the input and responds by sending a signal. The reversing valve (RV) is energized (output 9 on the microprocessor). The fan motor is started on low speed immediately (PSC ON; output 7 on the microprocessor); the ground-loop pump is energized through CR2 5 seconds after the Y1 input is received. The compressor is energized through the comfort control (CC), which closes CC contacts. Low-pressure (LP) and high-pressure (HP) switches are closed. Comfort Alert operates CS, and the compressor operates on low capacity for 10 seconds after the Y1 input. The fan is switched to medium speed 15 seconds after the Y1 input (ECM only; output 4 on the microprocessor). The DHW pump is cycled 30 seconds after the Y1 input through CR1. Everything continues to operate unless there is an interruption by the pressure switches or input from the high water limit (HWL) or the freeze protector (FP). If satisfied, the thermostat will open R to Y1, sending a signal to shut down all previously operated components.

If the first stage is not satisfied, the thermostat will close from R to Y2 while still closed from R to Y1. Comfort Alert turns off CS and the second stage of the compressor will be activated 5 seconds after receiving a Y2 input as long as the first stage of the compressor has been operating for a minimum of 1 minute. The ECM blower changes from medium to high speed 15 seconds after the Y2 input (output 5 on the microprocessor). The Comfort Alert will delay the second-stage compressor for 5 seconds after it receives a "Y2" from the board. If the second stage of heat is satisfied, the R-to-Y2 switch in the thermostat is opened and the system continues on first-stage heat after CS is operated by the Comfort Alert and the blower motor changes back to medium speed.

If the second stage is not satisfied, the thermostat will close from R to W while still closed from R to Y2 and R to Y1. The domestic water pump continues to be de-energized, which directs all heat to satisfy the thermostat. The first stage of resistance heat is energized 10 seconds after "W" input, and with continuous third-stage demand, the additional stages of resistance heat engage sequentially every 5 minutes until all stages of electric resistance heat are operating (outputs 10, 9, 2, 1, and 3, 11 from the microprocessor to the EA PCB). Relays ER1, ER2, ER3, and ER4 are closed in sequence. (Note: The ECM2 blower continues to operate at high speed.) When the third stage of heat is satisfied, the thermostat will open R to W and the system will revert to second-stage operation.

SUMMARY

In this chapter we have discussed the operation and system checks for a geothermal heat pump. The chapter began with those characteristics unique to geothermal systems and ended with a description of how the system operates using a manufacturer's electrical diagram. The outside air coil is not present. Instead, there is a water coil that connects the system to the ground by using ground water or a ground-water loop. Either a coaxial or a shell-and-coil heat exchanger is used as the outside heat exchanger. Water is received directly from the ground (open loop) or circulated in the ground loop (closed loop). Water is controlled or regulated through the outdoor water coil by a simple manual ball valve, water regulator, pumps, or a combination of devices to control the flow.

Correct operation of a functioning geothermal heat pump was provided through a description of operation with an accompanying electrical diagram. It is important to understand the correct operation of a heat pump. With that understanding, the technician is better able to diagnose and troubleshoot a malfunctioning unit.

REVIEW QUESTIONS

1. Describe why a defrost system is not necessary for a geothermal heat pump.
2. Describe the construction of a coaxial heat exchanger.
3. Describe the construction of a shell-and-coil heat exchanger.
4. Explain the concept of counter-flow heat exchange.
5. Describe the difference between open- and closed-loop water-source heat pumps (WSHPs).
6. Explain the concept of "approach" temperature.
7. Describe what happens to resistance heat on a third-stage call for heat (use the wiring diagram in this chapter).

CHAPTER 21

Geothermal Installation

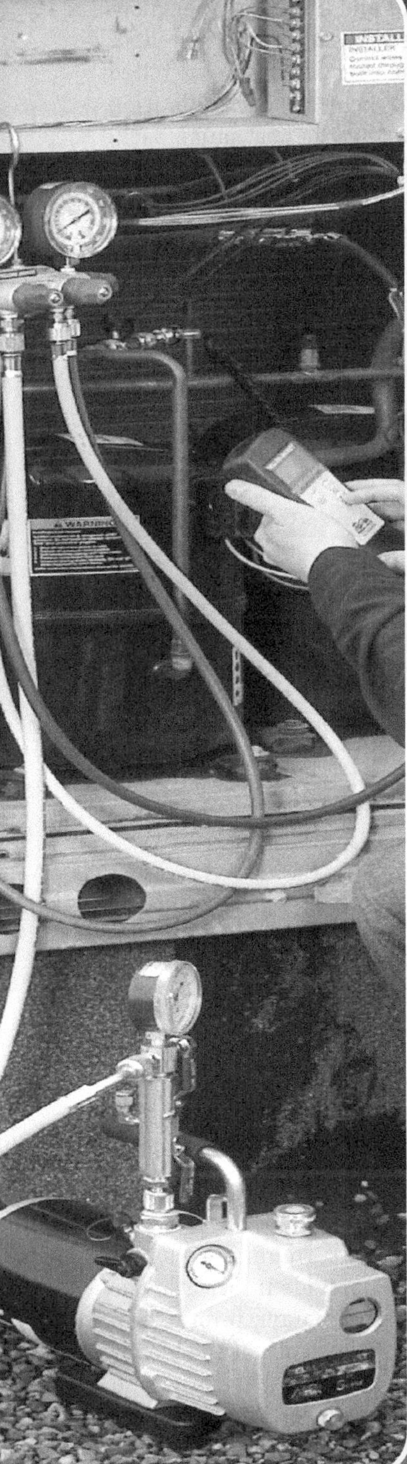

LEARNING OBJECTIVES

The student will:

- Describe what surface considerations affect geothermal loop installation
- Describe the difference between closed-loop and open-loop installation
- Relate the importance of flushing and pressurizing the ground loop
- Describe the fluid-charging operation for a ground loop
- Describe the reason why thermal balance point is of lesser concern with geothermal heat pumps

INTRODUCTION

Geothermal heat pump installation sometimes requires a large area of ground or surface area to install a horizontal loop. Vertical loops take up a smaller footprint, but can be very deep. Well water open-loop systems may use the same well for the heat pump and for domestic water in the building. This chapter explores the issues involved in installing the ground loop. A vertical loop installation will be used for discussion regarding closed loops. Open-loop installation is also discussed, including well sizing, in terms of gallons per minute.

After the ground loop is installed, the loop needs to be flushed and charged. There are two chapters' sections covering the flushing and charging operation. At the same time the ground loop is going in, the heat pump can be installed. Because all of the rest of the components, including the compressor, are installed indoors, a suitable location and careful installation will help to reduce vibration and sound transmission. Supplemental heat is often part of the heat pump package. This heat is available if the heat pump cannot maintain indoor temperature conditions and acts as a back-up source of heat.

The chapter will conclude with a discussion regarding dehumidification and psychrometrics. It is important to be able to remove moisture from, as well as lowering the temperature of, the indoor air.

Field Problem

The geothermal system had been operating on supplemental heat, according to the customer. The emergency heat light had come on and stayed on, he reported. The customer had not noticed the light until yesterday and was unsure if it had been on very long.

Symptoms:
Emergency heat light on at the thermostat.

Possible Causes:
Heat pump locked out on a safety; compressor failed or mechanical problem; contactor stuck or coil is burned out; control circuit problems on the first stage of heating; no flow at loop.

The technician checked the control board for other light-emitting diode (LED) indicators. A yellow LED was flashing five times before pausing. The legend noted that five flashes indicated an open circuit. The technician read the voltage at the terminal strip in the control panel from Y1 to common—it read 24 volts, indicating that the thermostat was calling for the compressor to be on. Checking for low voltage at the contactor, the meter indicated that 24 volts was being applied to the coil. Checking across the contactor contacts gave a voltage reading of 240 volts, indicating that contactor was open. There was something wrong with the contactor and it needed to be replaced, but was it the coil or the mechanical action of the contactor? Turning the power off, the technician verified that the contactor was the problem and what was wrong. Placing the meter in ohms, the technician attempted to check the continuity of the coil—no continuity. The contactor coil was bad.

After reporting the problem to the customer and getting the customer's okay to make the replacement, the technician replaced the contactor and verified the operation of the heat pump.

SURFACE CONSIDERATIONS

The installation of a geothermal system requires the ground surrounding a building to be evaluated (see Figure 21-1). Design programs can provide information as to the size of the geothermal loop area. To use the design program, information about the soil strata and the temperature of the ground at the depth of the loop is important. Geothermal loops can be open or closed. Closed loops can be horizontal or vertical. Each of these water-based loop designs requires a certain amount of ground area. Generally, open-loop systems require the least amount of ground surface area. Required amounts of ground surface area are as follows:

- Single-use (one well) and injection (two wells) units can be employed to remove or reject heat to ground water; these require the least amount of surface area.
- Vertical closed-loop designs require minimal ground surface area.
- Horizontal closed-loop designs use the most land surface area.

Wherever the loop is installed, there can be no obstructions. Buried electrical lines, sewer connections, foundation drains, water wells (not used for open loops), and other underground obstructions must be identified and worked around when installing geothermal systems. Septic fields must be at a distance from the geothermal loop, as determined by local code. Trees, bushes, and gardens may also need to be considered during the planning process and guarded from damage during the installation phase.

It is very simple to say that the loop goes into the ground, but it can be very difficult to find a convenient place to put the loop. A complete review and measurement of the loop location is needed prior to installation.

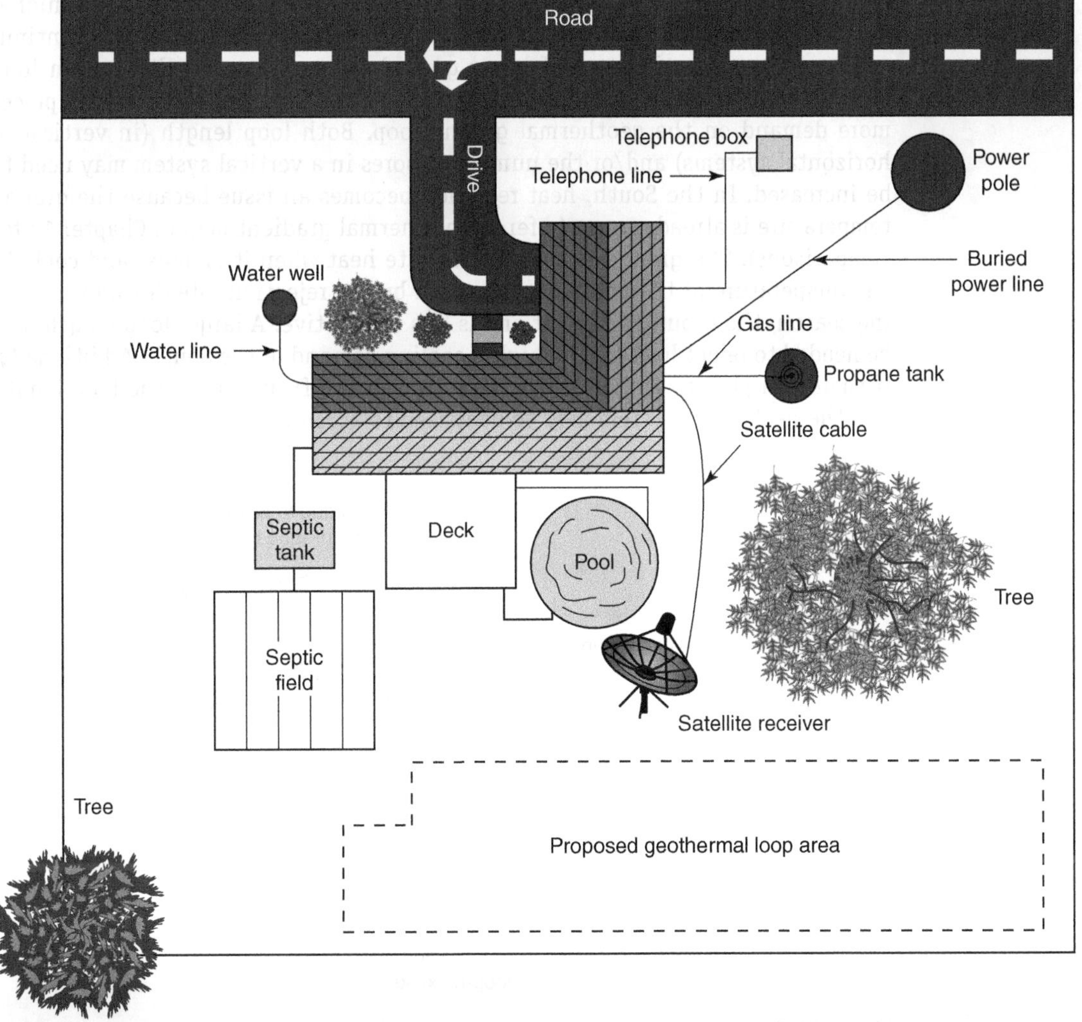

Figure 21–1
The location for a geothermal loop must consider all of the obstacles and the amount of surface area that is available for the loop. (Courtesy of Delmar/Cengage Learning)

Safety Tip

Working in open trenches can be very hazardous and as such, OSHA has very stringent rules regarding shoring, sloping, and other aspects of trench safety. These regulations are found in OSHA 29 CFR 1926.650.

Closed-Loop Piping and Installation

Horizontal trenching is required for the installation of a horizontal loop. This type of trenching is dug to 6 feet of depth or more and removes a lot of soil. There needs to be room to dig the trench and to place the soil being removed. Other things to consider are:

1. Trenches should not have sharp bends and pipe laid in the trench should not bend to a sharp angle.
2. Backfilling of the trench will prevent or remove sharp rocks from touching the pipe. All such sharp objects must be removed from the backfill.
3. The first layer of backfill should be sifted soil, sand, or pea stone. The first layer may be up to 6 inches thick.
4. All air pockets must be removed around the loop pipe. All voids must be compressed and filled for continuous contact with the pipe. Water may need to be used to settle the backfill.
5. Return bends must be backfilled with care. This may require that the bend is backfilled by hand to prevent kinking (Figure 21–2).

Vertical and horizontal loops require correct spacing as suggested by the manufacturer or loop designer. For horizontal loops, oversizing the field is common for heating in the northern part of the country and cooling in the southern part of the country. In the North, the field may need to be larger due to the higher heating demands. The amount of heat that can be effectively taken on a continuous basis decreases because of lower ground temperatures resulting from long periods of heat removal. Additionally, the greater heating requirement places more demand on the geothermal ground loop. Both loop length (in vertical or horizontal systems) and/or the number of bores in a vertical system may need to be increased. In the South, heat rejection becomes an issue because the ground temperature is already warm (refer to the thermal gradient map in Chapter 18 for comparisons). The ground is able to dissipate heat when it is moist and cool. As the temperature of the ground is increased by the rejection rate during the cooling season, the ground dries out and is less conductive. A larger loop length may be needed to reject heat for the higher cooling demand of the South. Additionally, if there is a plan to add on to the structure in the future, both the loop length and the heat pump system may be oversized to accommodate building additions.

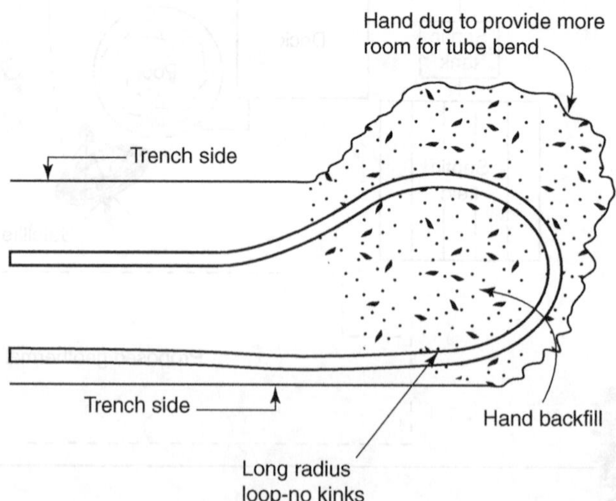

Figure 21–2
Hand digging and backfilling may be required where ground loops turn in the trench. The tubing must not make severe bends. Care must be taken to eliminate kinks. (Courtesy of Delmar/Cengage Learning)

Vertical loops are less susceptible to the effects of northern or southern climates. This is because the vertical loop taps the ground heat at a depth that is less affected by geographical location. Loop depth is generally around 100 feet and can go as deep as 600 feet. The vertical bore is made by a boring or well-drilling machine (Figure 21–3). The vertical loop piping is supplied with a fused U-bend and rebar is taped to the U-bend to help drop the pipe to the bottom of the bore (Figure 21–4).

Figure 21–3
A well driller bores vertical holes in the same fashion as a water well is drilled. Sediment must be moved away from the bore hole. In this picture, the loop is ready to insert in the bore hole. Re-rod is taped to the loop U-bend as a weight to help drop it to the bottom of the bore. (Courtesy of Jamie Simpson)

Figure 21–4
The U-bend is fused to the bottom of the vertical loop pipes. (Courtesy of Jamie Simpson)

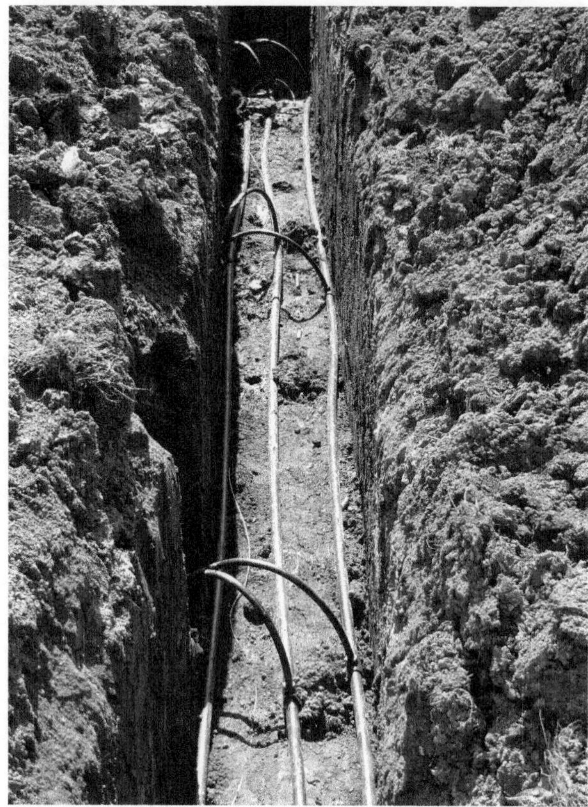

Figure 21–5
A trench alongside each vertical bore hole connects the vertical loop to a main header pipe. The connections are fused using approved fusing techniques. (Courtesy of Jamie Simpson)

Figure 21–6
The trench extends from the vertical bore field to the structure. (Courtesy of Jamie Simpson)

After the loop pipe is dropped into the vertical bore, the bore is grouted with bentonite or another heat transfer material that will prevent voids around the loop pipe and will seal the bore and make good contact with the surrounding ground. Each vertical well must be connected by digging a trench (Figure 21–5). The trench also extends to the building structure, where the piping is passed through a basement wall (Figure 21–6).

Where pipes pass through below-grade walls, special precautions must be made to ensure that the piping is protected and seals to the wall. The wall is drilled and the pipes are passed through (Figure 21–7). The pipes can be fitted with a rubber sleeve that allows the pipes to move and expand. The pipe can also be foamed in place (Figure 21–8).

After the ground loop is installed and before backfilling, the loop is pressure tested to check for leaks and flushed. Any leaks are repaired or sections of pipe replaced before the pipes are buried. The test pressure is 1.5 times the working pressure. If the working pressure is 50 pounds, the minimum test pressure would be 75 psig. The length of time that the loop must hold this pressure is generally 30 minutes or more. If there is a loss of pressure, it will require a visual check for leaks.

Ground-loop installation requires training. Loop contractors typically have certified installers/technicians trained by the International Ground Source Heat Pump Association (IGSHPA) or through an IGSHPA-sanctioned training program. This training is usually required to be able to fuse loop pipe in the field—there is no other acceptable underground joining practice. This type of training requires the certificate holder to perform a heat-fused connection under the observation of a certified trainer. Certified technicians may also be required to attend training on an annual basis.

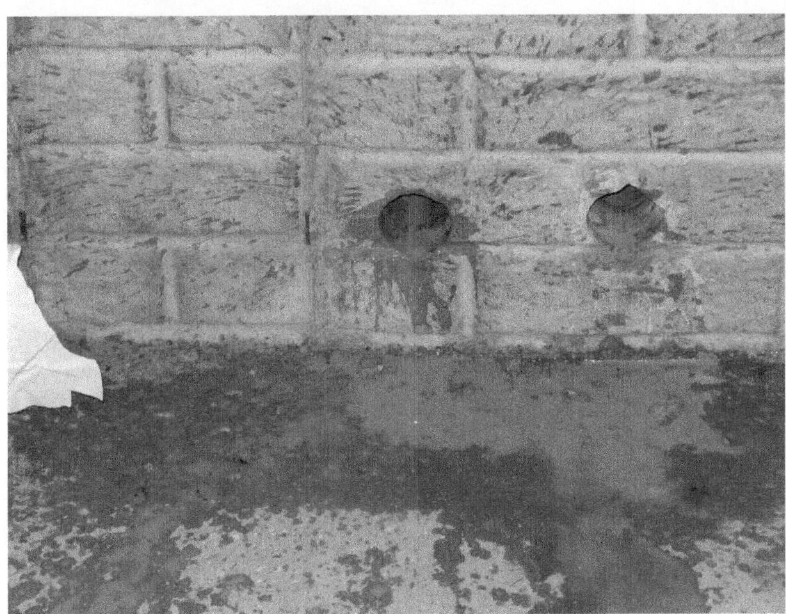

Figure 21–7
Holes are drilled through the below-grade wall. (Courtesy of Jamie Simpson)

Figure 21–8
Pipes passed through the wall act as sleeves for the loop piping. (Courtesy of Jamie Simpson)

Local and state requirements may need to be met wherever geothermal loops are installed. Some states require well drillers to perform any excavations or boring activity below a certain depth from the ground surface. Be sure to know all of the regulations that apply before installing a ground loop.

Figure 21–9
If undocumented, the vertical loop location could be lost. Tracer wire that is foamed in along with the pipe lies in the trench and will act as a locator for the bore holes and interconnecting loop piping. Drawings and measurements are also important documentation for the loop location. (Courtesy of Jamie Simpson)

The ground loop must be documented and in some cases marked at the surface. Corners of the loop can be marked with steel rods that can be found with a metal detector. Tracer or location wire can also be used to mark the location. Tracer wire is laid in the trench with the loop during installation. This provides a record of the location of the loop and prevents the loop from being damaged by accidental excavation. The loop diagram must be placed with the rest of the system documentation and left with the owner (Figure 21–9). The other documents that should be in the same packet are the heat pump manufacturer's installation manual and the owner's manual.

Open-Loop Piping

Open-loop geothermal systems are generally connected to the water well system. When a geothermal system is connected to the drinking water system, the well and its associated components must be sized to handle the normal water use and the geothermal demand.

Most water-source heat pumps (WSHPs) use from 1.5 to 3 gallons of water per minute (GPM) per ton. If the maximum use of 3 GPM were used for a 3-ton system, it would use a total of 9 GPM, or 540 gallons per hour (GPH). This amount of water usage would be added to the house estimate. Using the Peak Demand Estimator in Table 21–1, a home with two bathrooms would require an additional 14 GPM or 840 GPH. The total amount of water needed would be 13 GPM or 1,380 GPH. This example shows that the typical water usage would nearly double if a WSHP was installed. For existing homes, a second well is sometimes drilled for the exclusive use of the heat pump.

When the usage of a well increases, the pump, expansion tank, and the well diameter also increase. The size of the well also considers several other factors, such as water flow in the aquifer, well capacity or pump down, and the well recharge rate. All of these are determined by a qualified water-well driller. It is important to have a knowledgeable and experienced well driller who can assist

Table 21-1 **Peak Demand Estimator**

Peak Demand Estimator			Number of Bathrooms			
Fixtures	**Usage in GPM**	**Maximum in Gallons**	**1**	**1 1/2**	**2–2 1/2**	**3–4**
Shower/Tub	5	35	35	35	53	70
Lavatory	4	2	2	4	6	8
Toilet	4	5	5	10	15	20
Kitchen Sink	5	3	3	3	3	3
Washer	5	35	–	18	18	18
Dishwasher	2	14	–	–	3	3
7- Minute Peak			45	70	98	122
Minimum	**GPM**		**7**	**10**	**14**	**17**
Pump Size	**GPH**		**420**	**600**	**840**	**1,020**

with sizing the water well and specifying the components needed to satisfy the total water requirement of the structure.

As with a closed-loop system, if there is any plan to expand the building or if there are extreme circumstances (northern or southern location), the size of the water well may be increased or oversized to accommodate the building expansion or the other design requirements.

Flushing the Loop

After the ground loop is installed and before connecting it to the heat pump, the ground loop is flushed. Flushing removes all of the dirt and debris that may have gotten into the piping. During installation, all caution is taken to keep dirt out of the lines. Tape and pipe plugs are used until final connections are made to the headers. Even with all of this care, dirt will make its way into the piping and will need to be flushed. It is also possible that some part of the pipe or a stone will accidentally fall into the pipe. Flushing is where the pipe is cleaned and readied for charging.

In large systems, several pipes are brought back to the mechanical room. Each of these will have valves to isolate a part of the loop from the rest of the loop. In small systems, there may only be one set of supply and return pipes. In order to flush the loop, a high velocity must be developed to move the debris to a holding tank. Pre-filters and a final dirt filter are used to remove sediment from the flush water.

The flushing operation continues until the water returning to the drum is clear. Depending on the size of the system and the amount of dirt, the flushing operation could take a few minutes to several hours. After this process is complete, the loop is pressurized and checked for leaks.

Tech Tip

Pressurizing with compressed air is another way to test for leaks. Pressurization should be done before flushing; if a leak is found, it can be repaired prior to putting water into the loop. Pressurization is typically 1.5 times the working pressure of the loop and held for 30 minutes. Air pressurization requires that all joints and connections are tested with soap bubble solution.

Charging the Loop

After the heat pump is set and connected to the loop and the loop is flushed and leak checked, the loop is charged with water and glycol. In order to remove all air bubbles, a minimum velocity of 2 ft/sec. must be developed in each piping section and maintained for a minimum of 15 minutes. This requires a pump that can provide a large volume of water. Typically, a 1.5-horsepower pump is used on a charging cart (also known as a flush cart and sometimes used for the flushing operation as well as charging).

The charging cart is able to displace the water used for flushing with purified water used for the charge. Some systems are charged with distilled or purified water. This water has the minerals removed that might deposit on heat exchanger surfaces. The charge water may also be pre-mixed with an antifreeze solution. The charging procedure continues for the minimum amount of time or until the charge water runs clear and without air bubbles.

Tech Tip

The antifreeze solution type and amount will depend on the application and location. Be sure that the correct type of antifreeze is used and the correct percentage of solution is applied to provide freeze protection. Propylene glycol, ethylene glycol, and ethanol alcohol can be used. Protection temperatures may be as low as 0°F. The type of antifreeze and the level of freeze protection should always be placed on a tag on the unit, so that it is identifiable and can be checked during future maintenance procedures.

After the air is eliminated, the charging cart is used to apply a final working pressure in the system. Static pressure in most ground loops can be as high as 75 psig. Because of heating and cooling of the loop, the loop static pressure will fluctuate between seasons as the system adds or removes heat.

INDOOR COMPONENT INSTALLATION

All of the other components of the geothermal heat pump are generally placed within the structure. The only portion outside is underground—the geothermal loop. Because the compressor and pumps are located inside, it is important to locate these devices where sound and vibration are reduced or eliminated. Piping to these devices must not extend into walkways or create headroom obstructions. System components should be easily accessible and clearances for service maintained. The manufacturer will supply minimum clearance requirements. If possible, increase the amount of clearance so that servicing can be done with minimal effort.

Sound Transmission and Vibration

The heat pump should be located in a mechanical room that can be isolated from the rest of the living space. The mechanical room can have soundproofing applied to the walls and ceilings to aid in the suppression of sound from any mechanical device. When the heat pump is installed as a free-standing unit, a vibration absorption pad should be used and the unit placed on the pad. There is no need to bolt the system to the floor. The pad can be made of 2″ extruded polystyrene and is similar to the vibration pads used for air conditioning units that are placed outside on the ground. The better the vibration pad, the quieter

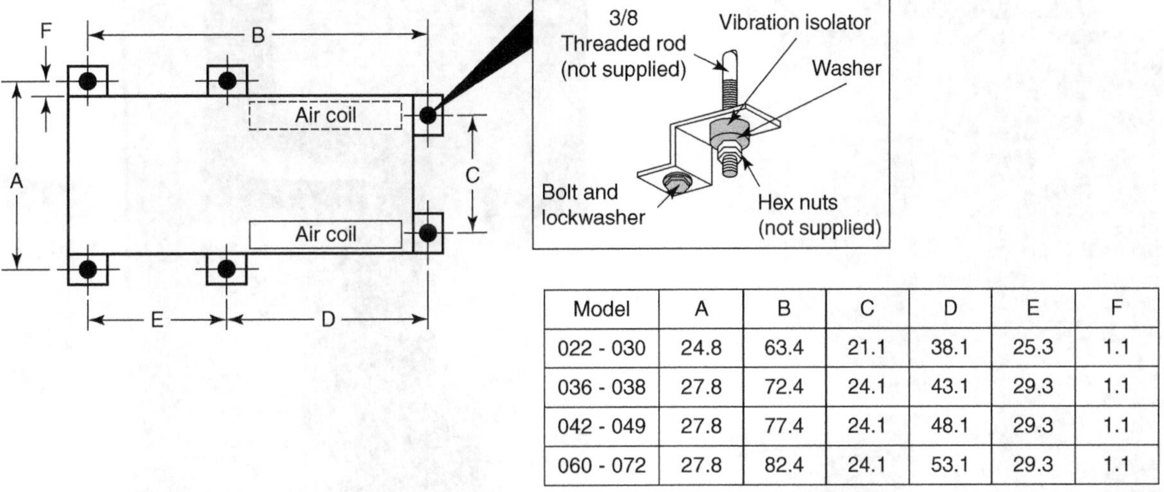

Model	A	B	C	D	E	F
022 - 030	24.8	63.4	21.1	38.1	25.3	1.1
036 - 038	27.8	72.4	24.1	43.1	29.3	1.1
042 - 049	27.8	77.4	24.1	48.1	29.3	1.1
060 - 072	27.8	82.4	24.1	53.1	29.3	1.1

Figure 21–10
The mounting location and configuration of the vibration isolator used for the hanger rod in a horizontal heat pump installation. (Courtesy of WaterFurnace International)

the system will be. Units that are installed horizontally in an attic are typically installed with a secondary condensation pan. The vibration elimination pad is installed below the pan and between the pan and the material used to support the heat pump. This material could be a plywood shelf that is suspended by hanger rods. The hanger rods should also have vibration-elimination material applied (Figure 21–10).

Pipe connections to the heat pump should be slack (Figure 21–11). Allowance for extra pipe will allow the pipe to absorb vibration and eliminate transmitted noise (Figure 21–12). Pipe connections are not strained in this way and will survive expansion and contraction during heating and cooling cycles.

Ductwork connections to the heat pump must also have vibration and sound protection. Flexible boot connectors are used to eliminate sound transmission through sheet-metal ducts. The duct is also lined with insulation to reduce sound transmission from the blower. All of these things are done to isolate the heat pump from the rest of the structure in an effort to reduce sound and vibration transmission. The result is a very quiet system that the customer will not notice.

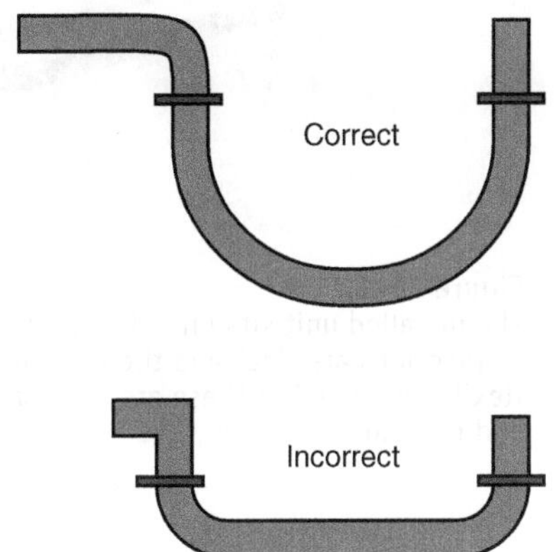

Figure 21–11
Pipe connections should have allowance or slack. Piping pulled tight will transmit sound and vibration. Tight connections could also tear the pipe connection apart during operation. (Courtesy of Delmar/Cengage Learning)

Tech Tip

The vibration elimination base can be glued to the floor and glued to the bottom of the heat pump. Using a mastic or glue will keep the base from "walking" or vibrating out from under the heat pump. Mounting tape (double-sided tape) can also be used to keep the heat pump in the middle of the vibration pad.

Tech Tip

All piping must also be protected from damage and condensation. Piping that could be damaged should be protected. Any pipe that could cause water condensation should be insulated.

Figure 21–12
The installed unit sits on a double-insulated pad. The loop connections are slack and the ducting connection uses a flexible boot. All of these are necessary to reduce vibration and transmitted sound. (Courtesy of Jamie Simpson)

Figure 21–13
This pump pack is shown installed on the stud wall of the mechanical room. A plywood wall mount was applied to the wall and the pump pack was screwed to the wall mount. (Courtesy of Jamie Simpson)

Pump Packs

Closed-loop systems have pumps that circulate the water-based fluid through the loop and heat exchanger. The pumps are packaged so that they can be placed remotely and away from the heat pump. In some installations, the pump packs are placed where the loop enters through the below-grade wall (Figure 21–13).

Pump packs are field wired to the heat pump, where they obtain their power and signal for operation. Sometimes these packs contain more than one pump. Pumps are generally piped in parallel and operate at the same time. The technician must check to see that both pumps are operating when installed. The operation can be checked with a clamp-around ammeter. Flow sensors and temperature sensors are typically mounted in the heat pump.

SUPPLEMENTAL HEAT

Supplemental heat can take many forms. Electricity, propane, oil, and natural gas can all be used to supply supplemental heat needs. When a fuel other than electricity is used, it is referred to as a dual-fuel system. Most geothermal heat pumps have an integral electric strip heat, and supplemental heating is typically done with electricity. When this is wired with the heat pump, two disconnects and two separate electrical branch circuits are used. In this way the refrigerant system can be isolated from the supplemental heat for service (Figure 21–14).

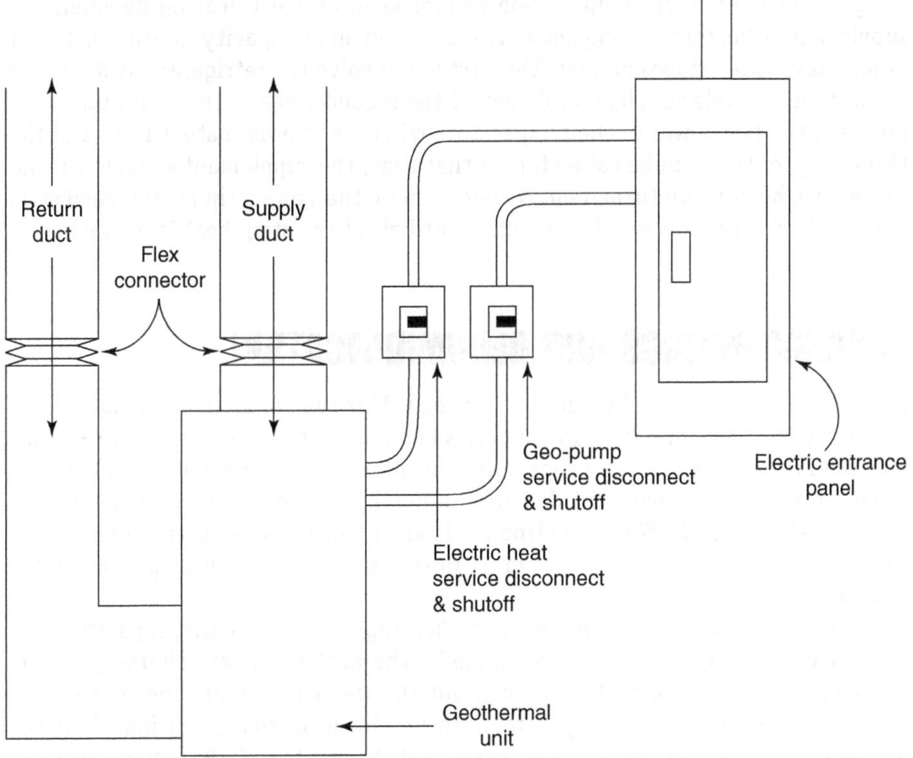

Figure 21–14
The geothermal heat pump and supplemental electrical heat have separate electrical runs with their own disconnect. It is wired this way to allow for electrical isolation and service of either the geo-pump or the electric heat system. (Courtesy of Delmar/Cengage Learning)

Tech Tip

Both the geothermal refrigeration system and the supplemental electric heat have separate electrical requirements. The geothermal unit may house both heating systems, but they are wired separately and each requires the right size of electrical conductor, overcurrent protection, and service disconnect. The conductor may also be encased in electrical conduit if the conductor is exposed. In all cases, be sure that all electrical wiring and components are sized and installed per the NEC. In many cases, the geothermal refrigeration system uses less amperage than the supplemental electric heat. This means that larger conductor sizes will be used for supplemental heat than for the refrigeration system.

Geothermal heating systems are sized using the BIN method, which averages the last 30 years worth of climatic data for a location. By using this method, data can be derived to estimate the amount of energy needed for heating and cooling 80–90% of the time. At those times when the outdoor temperature drops below the estimate, supplemental heat is engaged to augment the amount of heat the heat pump can supply. Designing and installing the heat pump system with this level of capacity reduces the size of the heat pump because the system does not need to meet the most extreme temperature requirements. These extremes occur infrequently, but when they do occur, the supplemental heating system is available when needed.

Unlike air-source heat pumps, geothermal systems do not have outdoor thermostats or settings for thermal balance point. The thermal balance point is by design. When the heat pump system cannot keep up with heating demand, the supplemental heaters are engaged. Two-speed or dual-capacity geothermal heat pumps have three stages of heat. The first two involve the refrigerant system. The third stage controls supplemental heat. If the second stage of the heat pump cannot keep the house warm, the temperature will drop approximately 1.5°F and the third stage of heat will be called for. At that time, the supplemental electric resistance strip heaters are turned on. If there is more than one strip heater, each strip is turned on with a delay relay or timer until all of the strip heat is operating.

PSYCHROMETRICS AND DEHUMIDIFICATION

Most heat pumps are sized to the cooling load. This type of sizing generally works to satisfy the cooling needs and supplies enough heat during the winter to satisfy the majority of heating needs. Supplemental heat is provided when the heat pump cannot keep up with the demand. In northern climates, this sizing method does not always work. Where cooling loads are a third of the heating need, heating becomes the priority and sizing estimates shift toward an emphasis on the heating season.

When the system is sized more for the heating load, the cooling capacity tends to be oversized. When in the cooling mode, the heat pump will short-cycle. The air temperature will be cooled quickly, but the water in the air does not have a chance to condense on the evaporator surface. This leads to a condition where the dry bulb (DB) temperature setpoint is met, but the wet bulb (WB) temperature has not dropped because moisture has not been removed from the air. Indoor conditions can feel cool and clammy because the relative humidity stays high.

To remove moisture from the indoor air, air needs to move slowly over the surface of the evaporator in order for the air to drop in temperature and for water to condense. When dew point is reached, the air needs to stay in contact with the evaporator to condense moisture from the air. The longer the air stays in contact with the evaporator, the more the moisture has a chance to condense, and the drier the leaving air will be.

Slowing the blower speed helps moisture to condense. This is one of the methods the technician can employ to partially remedy conditions where moisture is not being removed in an oversized system. However, slowing the blower too much will cause the coil to freeze moisture and the coil will become iced over. Once moisture freezes on the coil, it compounds the problem by further reducing the amount of air moving across the coil and further dropping the temperature of the coil. Both of these will cause the coil to freeze to a block of ice very quickly, and the building owner will be complaining because there is no cooling.

Two-stage or dual-capacity heat pumps can be used to satisfy the need for higher heat capacity during the winter and lower capacity for cooling in the summer. These systems are sized to supply cooling while running the compressor at lower capacity. System sizing involves using the smaller compressor capacity specifications to match summer cooling needs. When the system operates, the smaller capacity is used exclusively for cooling. These systems also use electronically commuted motors (ECMs) for the blower, which can be operated at lower blower speeds. Both lower capacity and lower blower speed work well to control both temperature and moisture.

SUMMARY

In this chapter we have discussed surface conditions where the geothermal loop is installed. The location of the loop, vertical bores, or wells has to do with the number of obstructions encountered. Locations of trees, buried obstructions, and surface-use features are important to know and must be worked around. When the loop is being installed, there needs to be enough room to work and move soil. After the loop is installed, it is important to have documentation that identifies the location of the loop.

Open loops require less surface space but have other requirements. Many times an additional well needs to be drilled for the exclusive use of the heat pump. Existing water wells may not be big enough to handle the water demand of both the house and the heat pump.

The loop requires flushing to ensure that dirt and debris that tend to get into the loop pipe during the installation process are removed. Before flushing, the loop is pressurized and checked for leaks. Any leak is repaired before flushing.

After the loop is installed, leak checked, and flushed, it is ready to connect to the heat pump. After making the connection, the loop and heat pump heat exchanger can be charged with water and antifreeze solution and given a final static pressure. This pressure tends to fluctuate throughout the year and depending on the season in which the installation is made, the static pressure could be slightly higher or lower.

All of the rest of the heat pump components are placed indoors, usually in the mechanical room. This means that the heat pump, circulation pumps, and controls need to operate quietly. Vibration elimination and sound-proofing help to reduce noise and sound transmission to the living space.

Supplemental heat is typically part of the heat pump package. Unlike air-source heat pumps, geothermal supplemental heat is only used when indoor conditions cannot be met by the geothermal system. Supplemental heat is controlled by the thermostat when space temperature drops below the setpoint of the heat pump.

REVIEW QUESTIONS

1. Describe what surface considerations affect geothermal loop installation.
2. Describe the difference between closed-loop and open-loop installation.
3. Relate the importance of flushing and pressurizing the ground loop.
4. Describe the fluid-charging operation for a ground loop.
5. Describe the reason why thermal balance point is of lesser concern with geothermal heat pumps.

CHAPTER 22

Geothermal Scheduled Maintenance

LEARNING OBJECTIVES

The student will:

- Describe what makes a geo-system different to maintain
- Describe the checks performed on the contactor
- Relate why voltages and amperages are recorded
- Describe what to listen for when checking a pump
- Describe the function of the indoor coil during the heating mode
- Explain why the outdoor coil may have to be checked more often for an open-loop system
- Describe what a desuperheater is and what needs to be maintained

INTRODUCTION

Unlike some HVAC systems, geo-source heat pumps (GSHPs) have reduced requirements for maintenance. That doesn't mean that maintenance should be delayed or not done; it means that these systems have reduced complexity and are easy to maintain.

This chapter will cover the essential maintenance that should be done for geothermal systems. General maintenance for mechanical and electrical components will be covered. One focal point is the outdoor coil, also known as a source coil. "Source" is a reference to the heat source, and the coil pulls heat from this source—the ground. Maintenance of the source coil is discussed along with the desuperheater, a similar coil to the source coil that can supply domestic hot water (DHW) to the home.

Field Problem

In the heating mode, the customer was complaining about the supplemental heat coming on all the time. The system was using ground water from a well and returning it to a pond.

There didn't seem to be enough temperature rise through the air coil. Checking the manufacturer's operational data, the technician determined that the head pressure was too low

Symptoms:

1. Compressor short cycling on LP freeze
2. Supplemental heat on

Possible Causes:

1. Plugged coaxial coil
2. Low water flow
3. Problems with the well pump for the open loop
4. Mechanical refrigeration problems

Starting with his first thought, the next step would be to check the pressure drop on the water side of the source loop (outdoor coil). The pump flow was marked on the data sheet from the installer as being 7 gallons per minute (GPM). Checking the manufacturer's data sheet under the 50°F column, the pressure drop should be 2.1 psi. With that in mind, the technician connected pressure gauges on the supply and discharge of the source coil and determined the pressure difference by subtracting the two readings—3.4 psi! He concluded that the strainer was plugged. The technician isolated the strainer, cleaned it, and put the system back in operation. He then did a system check for proper operation.

OVERVIEW

Geothermal heat pump maintenance is different than that for an air-source system, but only on the outdoor side of the system. The indoor-air side of a water-to-air geothermal heat pump is the same. The indoor unit will be configured in the same fashion as an air-source heat pump (ASHP), and like any air delivery system, the same requirements for airflow will be required by manufacturer specifications and referenced by performance sheets. From this perspective, all of the scheduled maintenance activities that are required for the indoor unit of an ASHP would be the same for a geothermal heat pump air delivery system.

It is from the perspective that the water side of the GSHP is different and may require additional knowledge and skill that the following sections are presented. For information about maintenance on the air side, the reader is reminded that general information and guidance for quality maintenance are to be found in the ACCA/QM standards described in Chapter 12 of this book.

Electrical

Pumps are electrically operated by line voltage components with low-voltage control. This means that they are controlled with electromechanical relays or contactors. Electrical maintenance means that:

1. All electrical connections are checked for tightness and corrosion.
2. Measure the voltage to the pump while it is operating and ensure that the voltage is not appreciably lower than the nameplate rating on the

pump. A ±10% difference would be acceptable. Voltage readings should be taken at the service disconnect.

3. Inspect and test the condition of the contactor or relay. This should include:
   a. Visually check the condition of the contacts. Look for pitted or burned contacts. Excessively deteriorated contacts should be replaced.
   b. Electrically check the condition of the contacts. Voltage drop across the contact (power on) should not be more than 1 volt.
   c. Check for smooth closing of the contactor or relay. The action should be smooth and quick, without hang-ups or other noise (a single click should be heard).
   d. Check the coil for discoloration and signs of getting too warm. Discoloration or melting of the wires or coil cover indicates that it needs to be replaced or that there is some other cause of heating (loose electrical connections).
   e. Check for proper control voltage. Verify that the control voltage transformer primary is wired for proper line voltage input (i.e. 208V vs. 230V).
4. Inspect all electrical connections and wire for loose connections, discolored wire, corrosion, and tightness at each terminal. This should be done with the power off.
5. Check the microfarad rating of all capacitors.
6. Measure the amp draw for all motors, pumps, and compressor(s). Compare the amp draw against the manufacturer's data tables.

All voltage and amperage measurements are to be recorded and a copy left in the customer information packet. This copy should be used for the next scheduled maintenance visit. Data and measurements should be compared to identify any trend that is occurring with the unit. If, for example, amperage draw is getting higher with each visit, this would mean that more extensive checking for the cause of higher amperage is needed.

MECHANICAL

The only mechanical device that is part of the outdoor unit on a GSHP is the pumping mechanism. Sometimes referred to as pump packs, they move the fluid through the outdoor coil (sometimes referred to as the source coil) and the loops of a ground-source (closed-loop) heat pump. If the system is a water-source (open-loop) system, the pump(s) may be injection and/or water well pumps.

Most of these pumps have enclosed impellers and mechanical drive components. When checking pumps, the following steps are suggested:

1. Ensure that the pump is operating by starting and stopping the pump. If possible, look for shaft rotation. If not visible, check for water pressure indicating pumping action.
2. Check for signs of water leaks. Look for dried residue indicating old leaks. Look for active leaks. If leaking, repair the leak. Inspect for the source of dried residue and determine if repair is necessary. Clean all residue from the pump.
3. Listen for pump sounds that could indicate trouble. In addition to the normal sounds of a pump, high-pitch sounds or intermittent sounds could mean:
   a. Pump bearings are worn.
   b. Pump impeller has lost vanes, or is out of balance (dirty).
   c. There is air in the pump housing (cavitation).

   Investigate and repair sources of noise.

The interconnection mechanical connections to the pump and piping should also be checked for kinks, leaks, and loose fittings.

WATER COIL

The water coil of a GSHP is also referred to as the source coil. If the geo-source unit is a water-to-water system, the indoor coil is referred to as the load coil. The source coil is typically a coaxial heat exchanger (Figures 22–1 and 22–2).

In a closed loop, the water is contained; no new water is introduced. In this system, the water quality should not be as much of an issue as in an open loop. Water in a closed-loop system can be treated and is usually a solution of propylene glycol and water. Water in an open-loop system is constantly new. Minerals in the water can deposit on the inside of the heat exchanger. Though heat exchangers in both systems should be checked, an open-loop coaxial heat exchanger needs to be cleaned to maintain efficiency.

Maintenance on the source coil might include:

1. Measure and record the temperature difference (TD) of the water entering and leaving the coil.

Tech Tip

It is recommended that gauge manifolds NOT be attached to a heat pump system during routine or scheduled maintenance. Attaching gauges or opening the refrigerant lines for any reason runs the risk of contaminating the system and losing refrigerant. Take measurements of temperatures externally and only connect gauges when a problem is suspected.

2. Measure the pressure drop of the water from the inlet to the outlet of the coil.
3. Inspect and clean any screen or filter in the water line.
4. Test the expansion tank for air pressure and operation.

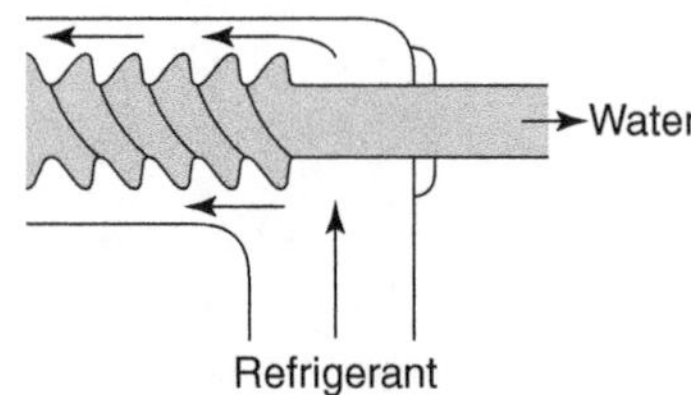

Figure 22–1
The coaxial heat exchanger is typically made of copper or cupronickel material. Refrigerant and water solution are separated by a heat exchange surface. The refrigerant flows opposite to the water, called opposed or counter-flow. (Courtesy of Delmar/Cengage Learning)

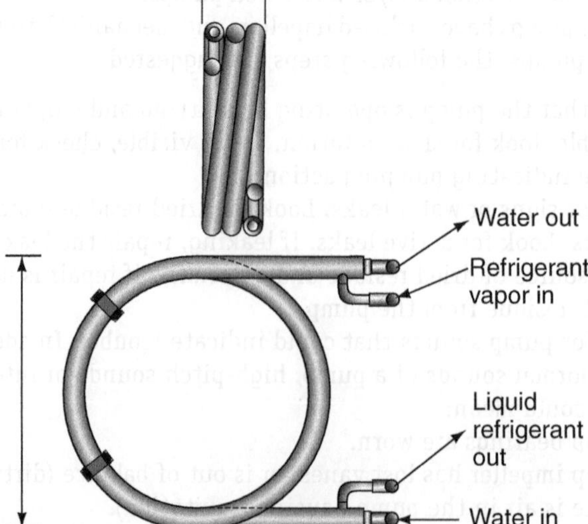

Figure 22–2
Coaxial heat exchanger coiled and ready for installation. Notice that refrigerant tubes are connected perpendicular to the water tubes. (Courtesy of Noranda Metal Industries, Inc.)

Note

Temperature differences and pressure-drop measurements should be compared with the manufacturer's data and the initial start-up data. Any differences could mean that the heat exchanger needs to be cleaned.

5. Operate the pressure-reducing valve (if installed) to ensure that the valve is operating and maintaining control pressure on the outlet of the valve.
6. Check the operating pressure on the coil. There should be a positive pressure (higher than atmospheric pressure).
7. Check the pH of the water in a closed-loop system. pH should be close to 7, which is neutral.

An open-loop system has some additional maintenance checks:

1. Air should be purged from the well head. If air exists, the piping should be checked for air leaks.
2. Maintain pressure on the water well system. Loss of pressure will allow air into the system.
3. Check the water pressure control valve in the discharge line to ensure that it is operating and maintaining pressure on the source coil.
4. If the water has a high mineral content, higher than 120 parts per million (PPM), the coil should be checked regularly and cleaned using standard coil-cleaning procedures. The procedure and products for cleaning should be those recommended for copper or cupronickel heat exchangers.

Tech Tip

Minerals in the water of an open-loop system will come out of solution (precipitate from solution) when there is a large temperature difference and the water changes temperature drastically. If a water system has high mineral content, water flow can be increased to prevent large changes in temperature and large amounts of mineral deposit. Increased water flow will reduce the amount of cleaning that may be necessary in an installation that has high mineral content.

CLEAN AND ADJUST

As for any HVAC system, the cabinet of the unit, pumps, and other exposed equipment should be cleaned/dusted. Cleaning is a sign of professionalism. Clean off any dried residue that may indicate a leak and check for leaks. If the leak has stopped, the next maintenance check will not find residue at that location. Potential leak points or suspicions that could not be confirmed should be recorded and checked during the next maintenance check.

If the coaxial heat exchanger needs to be cleaned, a solution pump is connected to the coaxial heat exchanger so that the supply of the pump is connected to the outlet of the heat exchanger (Figure 22-3). Cleaning solution is pumped in reverse of the normal flow of water. This action will help loosen mineral scale and

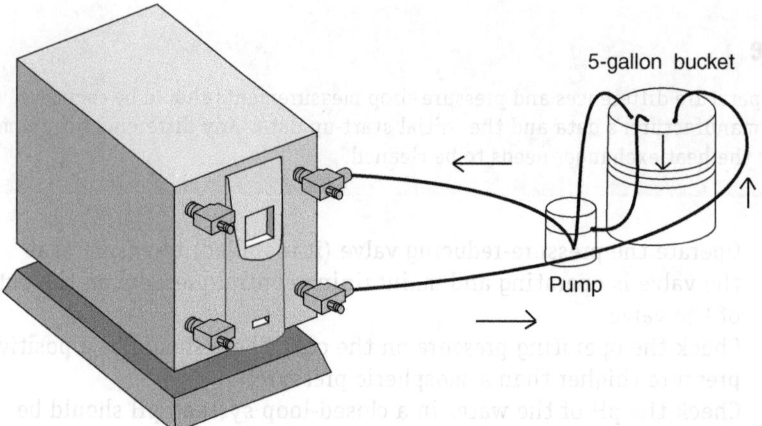

Figure 22–3
The cleaning solution pump is connected to a geothermal system to clean the coaxial heat exchanger. (Courtesy of WaterFurnace International)

clean the heat exchanger faster than hooking the pump to flow through the heat exchanger in the normal direction. The cleaning solution should be recommended for copper or cupronickel heat exchangers. Steps in the process are as follows:

1. Drain the heat exchanger.
2. Disconnect the water lines to the heat exchanger.
3. Connect the solution pump discharge to the heat exchanger outlet.
4. Connect a hose from the heat exchanger inlet to the solution bucket (large).
5. Place the solution pump in a smaller pail that is filled halfway with clear water.
6. Prime the solution pump and continue to add clear water until the lines are full and water is returning to the large bucket.
7. Connect a siphon hose between the large bucket and pail and fill with clear water. Operate the pump to ensure that the siphon hose is functioning.
8. Add cleaning/acid solution as per chemical company directions. Inhibited products are recommended to protect the metal.
9. Continue circulation until scale has been dissolved.
10. Pull the siphon hose from the large bucket and pail and add clear water while the solution pump operates.
11. Continue adding clear water until clear water returns to the large bucket.
12. Dispose of the chemical solution in the large bucket.
13. Flush the heat exchanger one more time with clear water and dispose of the flush water collected in the large bucket.
14. Remove the solution pump.
15. Reconnect the heat exchanger inlet and outlet to the water system.
16. Pressurize the heat exchanger and purge air.
17. Operate the heat pump and check temperature difference and pressure drop. Compare readings with the manufacturer's data.

Some GSHPs have a feature that can heat domestic hot water (DHW) as well as heat occupied space. The device used is called a desuperheater, and it is another refrigerant-to-water heat exchanger (Figure 22–4). It looks like a small version of the coaxial source coil. A slight difference is that this coil has a double wall between the water and the refrigerant. If a leak should develop, the water and the refrigerant will not mix.

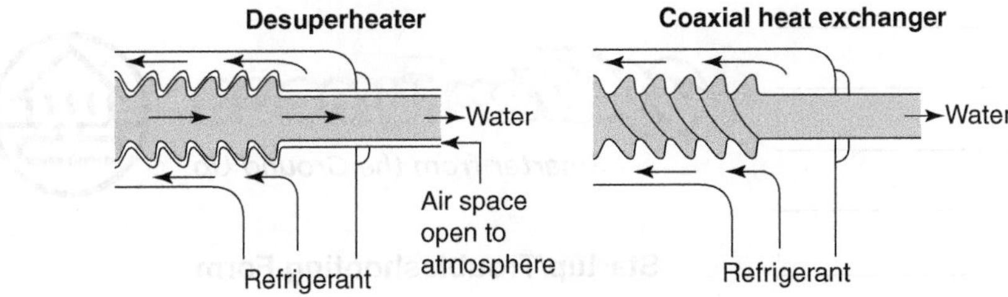

Figure 22–4
The desuperheater is a double-wall coaxial heat exchanger to heat domestic hot water (DHW). This double-wall design helps ensure that the DHW system cannot be contaminated by refrigerant should a leak occur, and also prevents the refrigeration system from being contaminated by water should a leak occur. If either water or refrigerant leak, the leak will be conducted through the air space between the double heat exchanger. The leak will go to atmosphere. A water leak will be visible. A refrigerant leak can be detected with an electronic leak detector. (Courtesy of Delmar/Cengage Learning)

If the heat pump has one of these devices, maintenance for the desuperheater is as follows:

1. Measure and record the amperage (as previously noted) of the DWH circulator.
2. Measure and record the inlet and outlet temperature of the water connections.
3. Check the operation of the temperature limit switch. This could be a thermistor and should conform to manufacturer specifications for resistance.
4. Check and clean the coil as necessary (use the same cleaning procedure as for the source coil).

Final adjustments or checks on adjustments should be done at this time. Pump flow rates, pressure-reducing valves, pressure-control valves, and balancing valves should already have been checked and their settings recorded.

Dealer: ______________________

Phone #: ______________ Date: ______________

Problem: ______________________

Model #: ______________________

Serial #: ______________________

Startup/Troubleshooting Form

Cooling cyle analysis

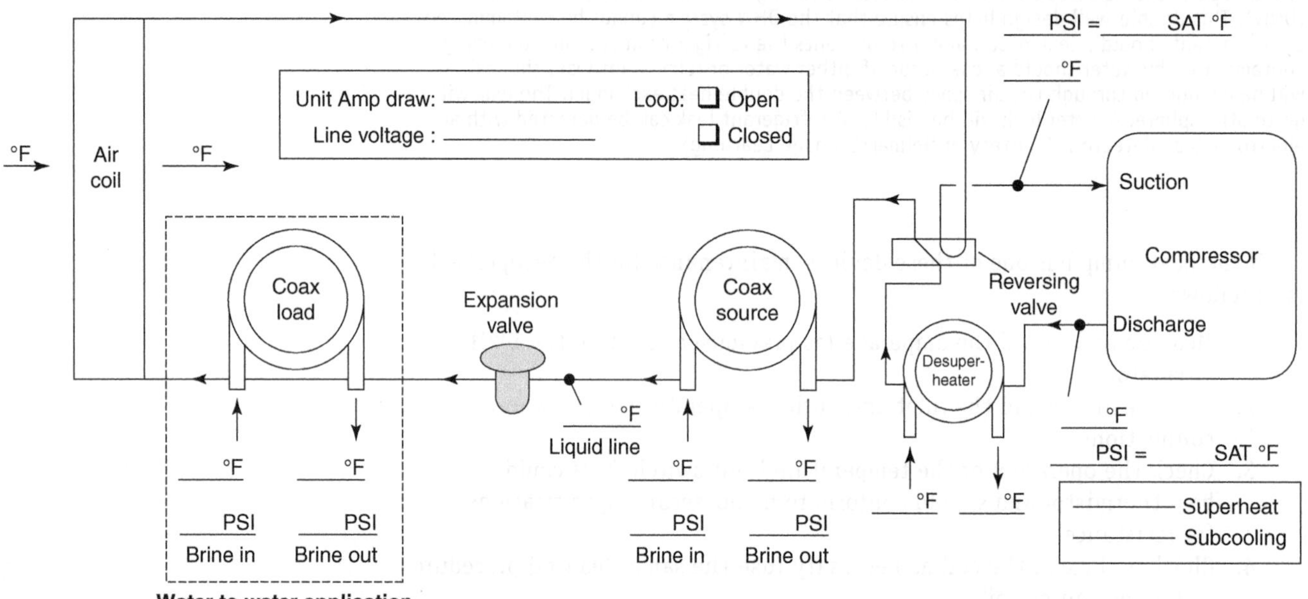

Heat of extraction/rejection = GPM × 500 (485 for water/antifreeze) × ΔT
Note: Do not hook up pressure gauges unless there appears to be a performance problem.

Heating Cycle analysis

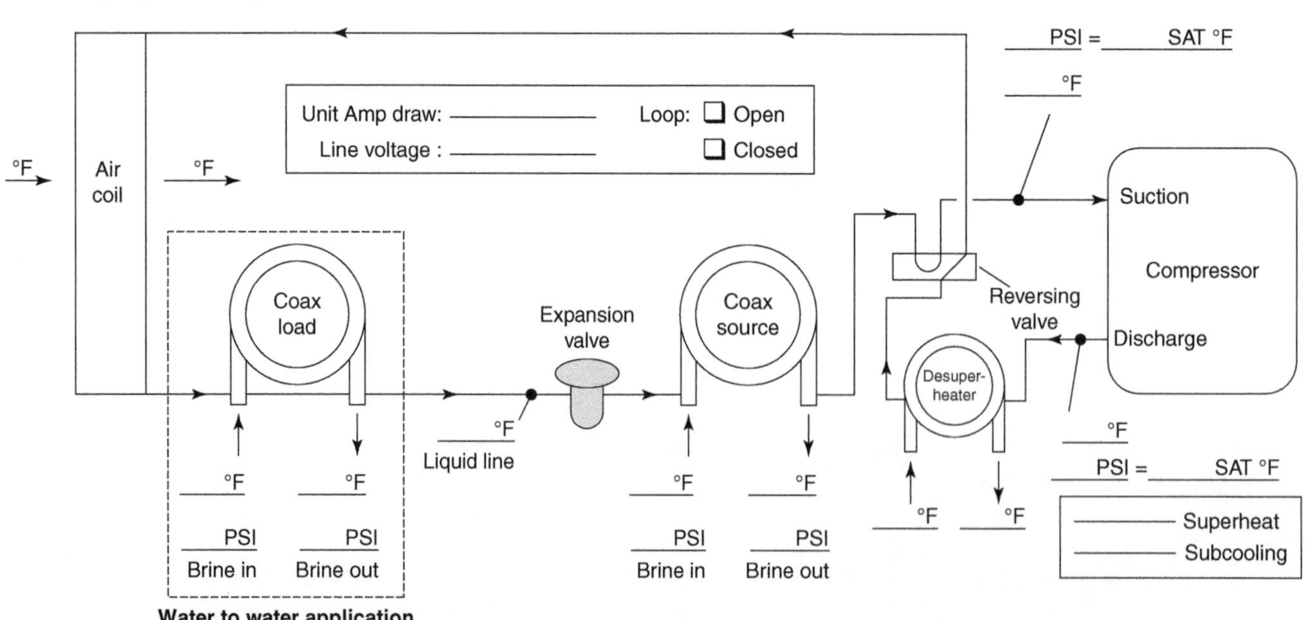

Figure 22–5
Example of the type of form used to record the operation of a geothermal heat pump system. The form can be used for scheduled maintenance as well. (Courtesy of WaterFurnace International)

1. JOB INFORMATION

WFI Model #: ______	Job Name: ______	Loop: Open / Closed
WFI Serial #: ______	Install. Date: ______	Desuperheater: Y / N

2. FLOW RATE IN GPM

	SOURCE COAX COOLING	SOURCE COAX HEATING	SOURCE COAX (WATER TO WATER) COOLING	SOURCE COAX (WATER TO WATER) HEATING
WATER IN Pressure:	a. ____ PSI	a. ____ PSI	a. ____ PSI	a. ____ PSI
WATER OUT Pressure:	b. ____ PSI	b. ____ PSI	b. ____ PSI	b. ____ PSI
Pressure Drop = a − b	c. ____ PSI	c. ____ PSI	c. ____ PSI	c. ____ PSI
Look up flow rate in table:	d. ____ GPM	d. ____ GPM	d. ____ GPM	d. ____ GPM

3. TEMPERATURE RISE / DROP ACROSS COAXIAL HEAT EXCHANGER*

	COOLING	HEATING
WATER IN Temperature:	e. ____ °F	e. ____ °F
WATER OUT Temperature:	f. ____ °F	f. ____ °F
Temperature Difference:	g. ____ °F	g. ____ °F

*Steps 3 - 9 should be conducted with the desuperheater disconnected.

4. TEMPERATURE RISE/DROP ACROSS AIR COIL

	COOLING	HEATING
SUPPLY AIR Temperature:	h. ____ °F	h. ____ °F
RETURN AIR Temperature:	i. ____ °F	i. ____ °F
Temperature Difference:	j. ____ °F	j. ____ °F

LOAD COAX (WATER TO WATER)

	COOLING	HEATING
LWT:	h. ____ °F	h. ____ °F
EWT:	i. ____ °F	i. ____ °F
	j. ____ °F	j. ____ °F

5. HEAT OF REJECTION (HR) / HEAT OF EXTRACTION (HE)

	COOLING (HR)	HEATING (HE)
Brine Factor*:	k. ______	
HR / HE = d × g × k	l. ____ BTU / HR	l. ____ BTU / HR

*Use 500 for pure water, 485 for methanol or Environol™. (This constant is derived by multiplying the weight of one gallon of water (8.34) times the minutes in one hour (60) times the specific heat of the fluid. Water has a specific heat of 1.0

STEPS 6 - 9 NEED ONLY BE COMPLETED IF A PROBLEM IS SUSPECTED

6. WATTS

	COOLING	HEATING
Volts:	m. ____ VOLTS	m. ____ VOLTS
Total Amps (Camp. + Fan)*:	n. ____ AMPS	n. ____ AMPS
Watts = m × n × 0.85	o. ____ WATTS	o. ____ WATTS

*If there is only one source of power for the compressor and fan, Amp draw can be measured at the source wiring connections.

7. CAPACITY

	COOLING	HEATING
Cooling Capacity = l − (o × 3.413) Heating Capacity = l + (o × 3.413)	p. ____ BTU / HR	p. ____ BTU / HR

8. EFFICIENCY

	COOLING	HEATING
Cooling EER = p / o Heating COP = p / (o × 3.413)	q. ____ BTU / W	q. ____ BTU / BTU

9. SUPERHEAT (S.H.) / SUBCOOLING (S.C.)

	COOLING	HEATING
Suction Pressure:	r. ____ PSI	r. ____ PSI
Suction Saturation Temperature:	s. ____ DEG. F	s. ____ DEG. F
Suction Line Temperature:	t. ____ DEG. F	t. ____ DEG. F
S.H. = t − s	u. ____ DEG. F	u. ____ DEG. F
Head Pressure:	v. ____ PSI	v. ____ PSI
High Pressure Saturation Temperature:	w. ____ DEG. F	w. ____ DEG. F
Liquid Line Temperature*:	x. ____ DEG. F	x. ____ DEG. F
S.C. = w − x	y. ____ DEG. F	y. ____ DEG. F

*Liquid line is between the coax and the expansion device in the cooling mode; between the air coil and the expansion device in the heating mode.

Figure 22-5 (Continued)

SUMMARY

In this chapter we have discussed the types of scheduled maintenance that should be performed for the outside unit of a geo-source heat pump. Even though we call this the outside unit, it may be mounted inside. It is simply a reference to the outside coil (also called the source coil).

Electrical systems that relate to the outside unit of the GSHP are confined to a set of pumps that circulate fluid in a closed loop or pumps that move ground water through the heat exchanger of an open-loop system. All of the electrical checks are typical of electrical checks done on other heat pumps and HVAC electrical systems.

Mechanical checks of the pump(s) and connection piping are done to look for evidence of leaks or prior leaks. Dried residue is evidence of a prior leak. Those areas of prior and potential leaks are checked and repaired if necessary. The residue is removed so that the next maintenance check can rule out the old leak as a current leak.

The water coil (source coil) may need more frequent maintenance if it is connected to an open-loop system. Open-loop systems use well water and the continual introduction of new, high-mineral-content water has the tendency to build mineral deposit in the coil. Maintenance steps can be taken to reduce the mineral deposit by increasing water flow. Coils may also need to be cleaned on a regular basis. The process used for cleaning the coil was discussed.

Some GHSPs heat domestic hot water with a device called a desuperheater. This is another refrigerant-to-water heat exchanger that looks like the coaxial source coil. The same cleaning and maintenance procedures are used for this coil as for the source coil.

Figure 22–5 is an example of a start-up and troubleshooting form that can also be used for scheduled maintenance. The form provides a place for the technician to record the current operation of the heat pump.

REVIEW QUESTIONS

1. Describe what makes a geo-system different to maintain.
2. Describe the checks performed on the contactor.
3. Relate why voltages and amperages are recorded.
4. Describe what to listen for when checking a pump.
5. Describe the function of the indoor coil during the heating mode.
6. Explain why the outdoor coil may have to be checked more often for an open-loop system.
7. Describe what a desuperheater is and what needs to be maintained.

CHAPTER 23

Troubleshooting GSHPs

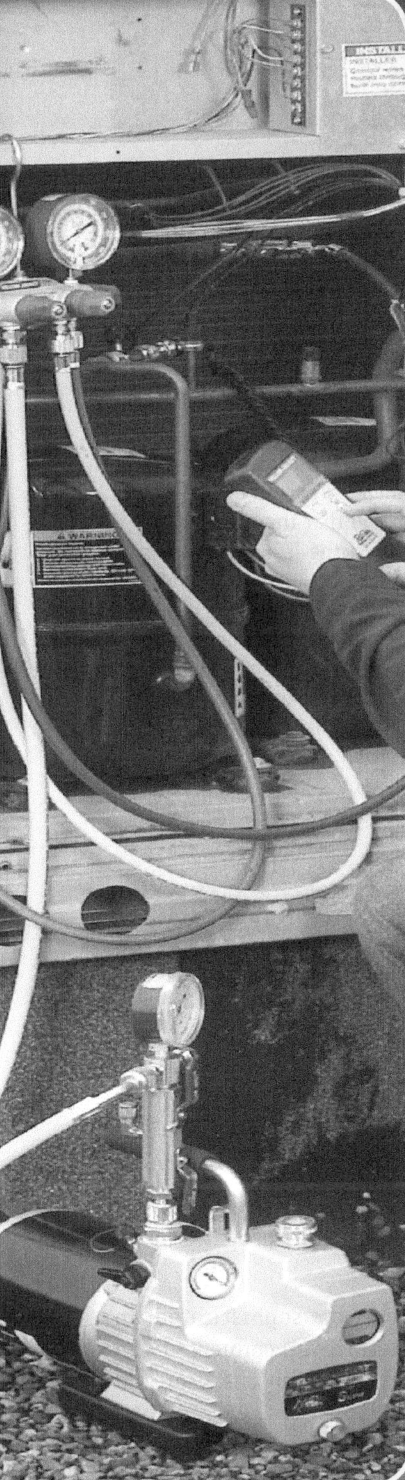

LEARNING OBJECTIVES

The student will:

- Describe user-level troubleshooting
- Describe how a thermostat might send a message for service directly to the service company
- Relate what observational issues are part of owner maintenance
- Describe the relationship between electronic modules and the microprocessor
- Explain how electronic modules and microprocessors are used in troubleshooting
- Describe the difference between mechanical and electrical troubleshooting

INTRODUCTION

This chapter will explore a few scenarios where troubleshooting is used to determine the cause of a set of symptoms. The reader is encouraged to get involved with troubleshooting. When new at this, the learner will ask questions and require explanations of how something works or how a particular technician approaches a troubleshooting situation. One thing to keep in mind is that all technicians do not use a uniform methodology. Troubleshooting requires a thorough understanding of the mechanical/electrical systems and mechanical/electrical sequence of operation.

FIELD PROBLEM

The system was 16 years old, and with the exception of air filter changes, had not had a single service or maintenance procedure done since installation. The customer was under the impression that geothermal systems did not require maintenance. This request for service was because the system was not keeping up with the heating needs. Indoor air temperature was at the setpoint and the supplemental heaters were coming on intermittently. The system was a geothermal closed loop with supplemental electric heat, installed in a basement.

Symptoms: Intermittent supplemental heat; outside air temperatures not excessively low; indoor air temperature at the setpoint.

Possible Causes: Mechanical refrigeration problem; loss of loop circulation; refrigerant-to-water exchanger fouling.

Temperature drop across the coaxial coil was low, only 3°F. Loop pressure drop showed that there was no fouling, according to the manufacturer's pressure-drop chart. The loop seemed to be pressurized and the loop pumps were running. Temperature measurements of the refrigeration lines also indicated that the high side was not hot enough and the low side was not cool enough. After checking all of the other possible causes, the technician connected a gauge manifold to the refrigeration system to measure saturation and confirmed that both the high-side and low-side pressures were low, superheat was high, and subcooling was low. Knowing that one of the possible causes could be a leak, the technician pulled out the electronic leak detector and started a methodic check of the refrigeration system, starting with all of the mechanical connections that could vibrate loose. The detector picked up a leak at the expansion valve body. The valve needed to be replaced. Before discussing this problem with the customer, the technician had learned from previous experience that he should continue the leak-check process for the entire system. As he pulled back the insulation on the coaxial heat exchanger, a shroud of rust fell and the electronic leak detector signaled that there were more leaks. From visual inspection along with the leak detector, it became obvious that the coaxial heat exchanger was rusting in the damp basement. Pin holes perforated a large portion of the heat exchanger, and it became more obvious that the extent of refrigerant leaks was more significant than just the metering device. The technician spoke to the customer and provided several service scenarios along with the possibility of a new unit. The customer would make a decision within the next few days and the technician's service company would be in touch.

Tech Tip

If a leak is suspected in the water-to-refrigerant coil, the technician should isolate and drain the coil. Then using an electronic refrigerant detector, see if refrigerant vapor is present. If any refrigerant is detected, there is a leak. Be careful not to damage the electronic leak detector with water.

USER-LEVEL DIAGNOSTICS

All heat pumps, geothermal and air-source, are becoming more sophisticated. In many cases the microprocessor controls and the user interface are connected and can tell the user what is happening. Some systems are bypassing the user and are sending messages via the Internet to the service company when the system is encountering a problem or is scheduled for some type of maintenance activity. As systems become more sophisticated, technicians and the customer will be notified automatically about a current problem or something that may potentially happen in the near future. In any case, until that happens, it is important to obtain as much information from the owner as possible.

TROUBLESHOOT PROBLEM 1

The customer was complaining about high energy bills. The system was a geothermal open-loop system using a water well for the water supply and a pond to discharge the water—a pump-and-dump system. The customer reported that the system never seemed to shut off. The customer also mentioned that the freeze-protection indicator light had been on for more than a month, but because the emergency heat light never came on, the problem was never reported.

Symptoms: High energy bills; continual operation.

Possible Causes: Lock-out light; low refrigerant charge; low airflow; dirty filters; fouled heat exchanger; low water volume/pressure.

The technician opened the service panel and was confronted with an ice-coated compressor. The compressor had a coating of ice as thick as a half inch at the suction line and the coating covered most of the top of the compressor. The circuit board light-emitting diode (LED) indicated a lock-out condition. After checking, the technician confirmed that no control voltage was being supplied to the coil of the contactor. The contactor was closed and the compressor was running continuously. The technician thought that the contacts must be stuck closed. Shutting off the power, the technician verified that the contactor was stuck closed through visual inspection. The technician discussed the repair and the consequences that might result in the future. Because the contacts were stuck and the compressor continued to run, the compressor could be compromised. The compressor could be internally damaged and could fail sometime in the future. Because the compressor oil could have washed out of the compressor, due to running for long periods under low superheat, failure could occur. Due to the large buildup of ice, the technician recognized that there was the potential for damage to other system components. He checked the condition of all the other components. The technician wanted to make the customer aware that a new compressor might need to be installed at some point in time. The customer agreed to replace the contactor, but not the compressor. Several months after this repair the compressor failed and needed to be replaced.

Thermostat

The thermostat is the primary controller and user interface. Many thermostats are being installed that allow the user to see much more information than ever before (Figure 23-1). A basic thermostat can tell the user that the temperature matches the setting. At the other extreme, the thermostat can tell the user what systems are operating and if there are any problems. This type of thermostat might also be able to communicate via the Internet. A communicating (also called smart) thermostat can supply the technician with information about faults that occur and other system operating conditions. When a service call is being initiated over the phone, the company representative or the technician should ask the user to consult the thermostat and to read what it is displaying. Table 23-1 lists possible indicators displayed on a thermostat.

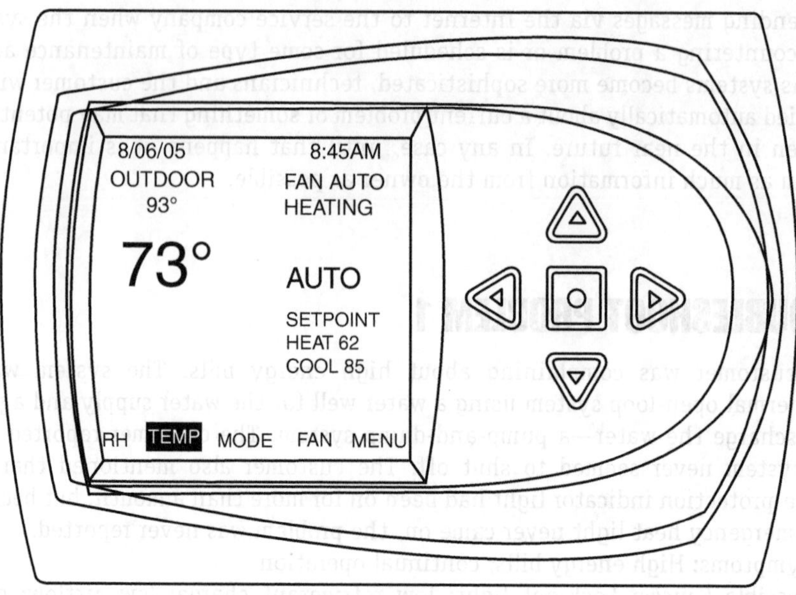

Figure 23–1
Digital thermostats can be touchscreens or have buttons, such as this representation. Most can be programmed for dealer contact information and events per day. Some have the ability to obtain system information and display it for the user. (Courtesy of WaterFurnace International)

Table 23–1 Thermostat Type and Possible Indicators

Thermostat Type	Possible Indicators
System information displays	Lock-out display and fault codes
	a. Compressor fault
	b. High- or low-pressure fault
	c. High- or low-temperature fault
	System component operation
	a. System temperatures
	b. System pressures
	c. Loop pump operation
	Supplemental heat operation
	a. Number of heat strips operating
	b. Outdoor temperature
	Service-needed information
	a. Filter change
	b. System service by date
	c. Call for service (linked to fault code)
	(plus all of the functions below)
Fault information indicator	Flashing fault LED
	Call for service LED
	Filter change LED
	(plus all of the functions below)
No fault or system information	Thermostat setting (heat, cool, or automatic)
	Thermostat on or off condition
	Thermostat setpoint

TROUBLESHOOT PROBLEM 2

The customer was surprised by a call from the installation company that put in the geothermal heat pump. The system was a closed-loop, 4-ton system with electric supplemental heat and a communicating thermostat. It was early summer and the yearly check was not scheduled for another 2 months. The service representative for the installation company reported that it had received a message from the heat pump system of an impending problem and wondered if the company could perform the yearly check early. The customer agreed.

Symptom: The thermostat sent a message through the Internet that system pressures were low.

Possible Causes: Refrigerant leak; low flow rates in the loop; liquid line restriction; low airflow.

The technician arrived and was directed to the heat pump. Checking external pressure and temperature rise across the coaxial heat exchanger, the technician found the temperature rise to be low. Refrigerant gauges were connected to the system at this time. Checking the refrigerant saturation temperatures/pressures indicated that the system seemed to be running with high superheat and low subcooling. Visually looking for refrigerant leaks, the technician found a small puddle of oil below the drier. The brazed filter drier fitting leaked. The remaining charge was recovered; the system was purged; the filter drier was changed; the system was leak checked; the system was evacuated; and then the system was charged to the to the manufacturer's specifications (the system superheat and subcooling were checked again to verify proper system operation).

Owner Maintenance

Every piece of equipment has an owner's manual, typically placed in a holder on the side of the unit by the installation crew. Inside the manual is a list of maintenance tasks that the owner should perform. Table 23–2 is a basic list similar to those found in most user manuals. Most homeowners rely on their service contractor to perform these maintenance procedures. Some user manuals identify the frequency for each of these checks.

Most of these are routine. If the owner is monitoring the system LEDs, they may be able to identify the problem and communicate the problem to the service company.

Monitoring energy bills could be as simple as comparing the bill each month and determining if the amount of energy usage is reasonable. It could also mean that the customer could be monitoring both the usage and the degree days for the structure location. At either extreme, the customer may be aware of how the system has been operating over an extended period of time.

Table 23–2 Owner Maintenance

Owner Maintenance
Check the thermostat setting(s)
Replace the disposable air filter
Clean the electronic air filter
Check the condensate drain/pan
Check for external fluid leaks
Check the thermostat indicators
Check the system LED indicators
Monitor energy bills

TROUBLESHOOT PROBLEM 3

The customer had neglected to have scheduled maintenance performed on the geothermal heat pump, because it continued to work flawlessly over the many years since it had been installed. It was a geothermal, horizontal closed-loop, 3-ton system with electric resistance supplemental heat. With the system emergency heat light coming on intermittently, a call was placed to the service company. The technician arrived and confirmed the complaint with the customer and the symptoms.

Symptoms: System not producing sufficient heat and supplemental heat was supplying the heating needs.

Possible Causes: Mechanical system problems; low water flow or low airflow; poor heat transfer.

The technician conducted some sensory checks and noticed a distinct whistle at the unit. Removing the air filter, the whistle went away and the technician was left holding a solidly packed, dirty filter. Another air filter was sitting beside the unit in the wrapper and the technician installed the new air filter. Because there might be another cause that could be overlooked, the technician took the opportunity to check and record the operation of the system in an effort to uncover any other problems—everything checked out fine. After showing the customer the dirty air filter and explaining that there was no other problem, the technician took time to review the owner's handbook with the customer and pointed out all of the basic maintenance that could be done to save money and increase energy savings.

System Indicators

The user is provided a chart of system indicators and a troubleshooting table with the manufacturer's user manual. The troubleshooting table should provide the owner enough useful information to be able to make basic judgments about system operation. This chart is useful for the homeowner and could be used to determine if a service call is required. Table 23–3 is an example of a troubleshooting chart.

Table 23–3

Symptom	What Can Be Done
System is not running.	Check for power outage. Check LEDs at system.
No heating or cooling or inadequate heating or cooling.	Check air filter and replace if dirty. Check LEDs at system.
Unit hums or buzzes but does not heat or cool.	Disconnect power and call for service.
Electric bills too high.	Check to see if system is indicating that it is running on supplemental heat. Check at thermostat to see if switch is set to continuous fan.
One room is too cold/hot.	Check for closed registers. Check for disconnected ductwork.
Thermostat setpoint is not the room temperature.	Check to see if system is indicating that it is running on supplemental heat.
Whistling noises coming from system.	Check for leaking ductwork. Check filter.
Too much air at diffuser.	Check for closed diffusers in other rooms.

Thermostat settings suddenly changed automatically.	Check thermostat settings for both temperature setpoints (heating and cooling). Check thermostat setting for system switch and set for heating, cooling, or automatic. Check for power loss. Check or replace thermostat battery.
Piping has been damaged and is leaking.	Turn off system at breaker and call for service.
LED is on at thermostat.	Check status/fault lights on the system. Read thermostat instructions for more information.
LED is on or flashing at the unit.	Read the fault condition label and reference the user manual for assistance. If unsure, turn off system at breaker and call for service.
System lock-out LED is on.	Reset the system in one of two ways: 1. Turn off the system at the thermostat by pressing the system switch until the thermostat indicates off. 2. Turn off the system at the breaker. Wait 5 minutes and then turn the system back on. If the lock-out LED comes back on, call for service.
Drain light is on or flashing (if installed).	Check condensate pan and drain for overflow or restriction.

TECHNICIAN TROUBLESHOOTING

Basic customer checks can only go as far as the owner is comfortable with. When the customer is knowledgeable and interested in the system operation, the customer can provide a great deal of information about how the system is currently operating and the history of operation. But if the customer is not knowledgeable, or interested, or is a new owner of an existing system, he or she may not be able to help the technician.

TROUBLESHOOT PROBLEM 4

The customer indicated that the technician was the second technician to conduct a service call on the geothermal system over the past year. The heat output was fine, but the emergency heat light was on often and the energy bills were very high. This was the customer's experience since the system was installed. The customer had compared system operating characteristics with those of other heat pump owners. These comparisons prompted the customer to call for service again. It was early in the heating season; the system was a pond loop (closed loop) and the pond temperature was 60°F. The system was 3 years old and the customer indicated that it had never operated any differently. This was not just the second opinion, as the customer was not satisfied with the previous answers or service.

Symptoms: Freeze switch periodically reset by customer; low heat output; loop temperature into the heat exchanger was were 60°F; system not producing sufficient heat and supplemental heat was supplying the heating needs.

Possible Causes: Loop circulation flow problems; air lock in the loop; low refrigerant charge or mechanical refrigeration issues.

The loop was serviced by one circulating pump. There was no history on the pond loop, but the customer remembered three circuits. The pond was approximately 300 feet from the house. The pump suction pressure was low. The loop was not circulating properly. There was also no information on the type of antifreeze protection. The type of antifreeze or the level was not recorded. Drawing a sample, the technician noticed that it was pink, possibly propylene glycol. Using a hydrometer for this type of antifreeze, the hydrometer indicated that the solution was nearly 100% propylene glycol. It was determined that the system was originally installed this way. This high concentration of propylene glycol created a high viscosity and may have accounted for the reason why the loop pump could not move the fluid effectively to produce the right pressure drop across the heat exchanger.

After rescheduling with the customer and bringing the flush cart, the technician removed the solution, diluted the solution with water, and recharged the loop. The pressure and temperature drop came into the manufacturer's tolerances.

The technician checked the flow rates and verified the system was removing heat. Confident that the problem was solved and the heat could now be exchanged with the pond, he explained the solution to the customer.

Microprocessor Indicators

Many geothermal systems have circuit boards and microprocessors that monitor and control the operation of the system. Many of these have indicators or LEDs that illuminate when system components are operating, not operating, or are in need of service. The legend for the indicators is either close to the LED (such as a label), in the user manual near or on the unit, on the door, or on the wiring diagram/schematic of the unit.

If the system has a microprocessor, all or many of the system functions will be performed through the microprocessor. Typically, the microprocessor will have its own set of testing procedures. The LEDs on the microprocessor are displayed through a cabinet window or are connected to a display panel. These outputs can also be routed to the thermostat, if it is capable of displaying system conditions. An example of system LEDs is shown in Table 23-4.

The microprocessor (if installed) may be linked to another module, called a Comfort Alert™, used by many heat pump manufacturers (Figure 23-2). This module is used to operate and monitor the compressor. Output from this module can be linked to another circuit board, display, or microprocessor, depending on the manufacturer. There are four types of modules:

- Single Stage, Single Phase
- Single Stage, Three Phase
- Dual-Capacity Single Phase
- Dual-Capacity Three Phase.

A Comfort Alert™ module has several flash codes. The three LEDs on the module are: a green "POWER" LED, a red "TRIP" LED, and a yellow "ALERT" LED. Each of the LEDs displays flash codes as noted in Table 23-5.

Table 23–4

LED	LED Description	Possible Cause
Drain	The condensate overflow level has been reached for 30 continuous seconds	1. Drain pan plugged 2. Drain line plugged 3. Drain line does not pitch away from unit 4. Vertical vent is not installed on horizontal drain lines over 6 feet long
Water Flow	The freeze thermistor temperature is at or below the selected freeze protection point (well 30°F or loop 15°F) for 30 continuous seconds	1. Water flow may be restricted or inadequate 2. Solenoid valve may not be opening on well water units 3. Pump(s) may be inoperative in the flow center 4. Entering water temperature too low 5. Low charge or partial restriction 6. Freeze protector thermistor not calibrated
High Pressure (heating mode)	The normally closed safety switch is opened momentarily	1. Inoperative or dirty blower 2. Dirty filters 3. Dirty air coil 4. Restricted supply or return duct 5. Unit may be severely overcharged (this only occurs when the condenser [air coil] is filled with subcooled liquid and there is heat to reject)
High Pressure (cooling mode)	The normally closed safety switch is opened momentarily	1. Water flow may be restricted or inadequate 2. Water-to-refrigerant heat exchanger may be fouled with debris 3. Entering water temperature may be too high 4. Entering air temperature may be too high 5. Unit may be severely overcharged (this only occurs when the condenser [coaxial coil] is filled with subcooled liquid and there is heat to reject) 6. Inoperative pump(s) in the flow center
Low Press/Comp (heating mode)	The normally closed switch is opened for 30 continuous seconds	1. Water flow may be restricted or inadequate 2. Entering water temperature may be too low 3. Return air temperature may be below 50°F 4. The system may be undercharged 5. TXV or filter drier restriction
Low Press/Comp (cooling mode)	The normally closed switch is opened for 30 continuous seconds	1. Airflow inadequate 2. Dirty filters 3. Dirty air coil 4. Return air may be below 60°F 5. The system may be undercharged 6. TXV or filter drier restriction
Airflow	The fan RPM falls below the low-RPM limit for 30 continuous seconds	1. Fan may not be spinning freely 2. Line voltage at terminals 4 and 5 may be below 230 volts 3. The signals from the board may need to be checked

Source: Courtesy of WaterFurnace International

Single Speed Comfort Alert Module

Dual Capacity Comfort Alert Module

Figure 23–2
Two types of Comfort Alert™ modules—one for single-speed compressors and the other for dual-capacity compressors. Both modules monitor and control the compressor. Both modules have system indicator LEDs. (Courtesy of WaterFurnace International)

Table 23–5 Comfort Alert LED Flash Codes

Comfort Alert LED	Description	Cause
Red "TRIP" LED	"Y" thermostat demand signal and compressor is not running	1. Compressor protector is open. 2. Compressor circuit breaker or fuse is open. 3. Broken wire or connector is not making contact 4. Compressor contactor has failed.
Yellow Flash Code 1	Long run time—18 consecutive hours of run time	1. Low refrigerant charge. 2. Evaporator blower is not running. 3. Evaporator coil is frozen. 4. Faulty metering device.
Yellow Flash Code 2	System pressure trip—not used	Not applicable
Yellow Flash Code 3	Short-cycling—compressor run time of less than 3 minutes on four consecutive cycles	1. Thermostat demand signal is intermittent. 2. Conditioning equipment is oversized.
Yellow Flash Code 4	Locked rotor—four consecutive compressor protector trips indicating compressor won't start	1. Run capacitor has failed. 2. Low line voltage. 3. Excessive liquid refrigerant in compressor. 4. Compressor bearings are seized.
Yellow Flash Code 5	Open circuit—"Y" thermostat demand signal with no compressor current	1. Compressor circuit breaker or fuse is open. 2. Compressor contactor has failed. 3. Open circuit in compressor supply wiring or connections. 4. Unusually long compressor protector reset time due to extreme ambient temperature. 5. Compressor windings are damaged.

Yellow Flash Code 6	Open start circuit—"Y" thermostat demand signal with no current in the start circuit	1. Run capacitor has failed. 2. Open circuit in compressor start wiring or connections. 3. Compressor start winding is damaged.
Yellow Flash Code 7	Open run circuit—"Y" thermostat demand signal with no current in the run circuit	1. Open circuit in compressor run wiring or connections. 2. Compressor run winding is damaged.
Yellow Flash Code 8	Welded contactor—current detected with no "Y" thermostat demand signal present	1. Compressor contactor has failed. 2. Thermostat demand signal not connected to module.
Yellow Flash Code 9	Low voltage—less than 17 VAC detected in control circuit	1. Control circuit transformer breaker is tripped. 2. Low line voltage.

*Flash code number corresponds to the number of LED flashes, followed by a pause, then repeated. *TRIP and ALERT LED's flashing at the same time indicates control circuit voltage is too low for operation. *Reset ALERT flash code by removing 24 VAC power from module. *Last ALERT flash code is displayed for 1 minute after the module is powered on.

Source: Courtesy of WaterFurnace International

TROUBLESHOOT PROBLEM 5

The customer complaint was no hot water on a dedicated domestic hot water staged geothermal heat pump. The customer stated that the high pressure (HP) LED light was blinking. The customer stated that after resetting the unit, the domestic hot water would operate, but the next day the same failure would occur. She also stated that there had been no service issues with the unit for some time.

Symptoms: HP LED blinking, indicating a high-pressure condition and safety shutdown.

Possible Causes: Insufficient water flow (pump, filter); fouled coaxial coil; mechanical refrigeration issues such as noncondensables.

The technician confirmed that the pump seemed to be operating and there was water flow by checking pump amp draw and temperature difference of water going through the heat exchanger. Temperature rise was excessively high, indicating possible low flow. After finding a strainer in the water loop at the water inlet side of the loop, the technician shut the unit down, isolated the water valves (incoming water and each side of the water strainer), and removed the strainer, only to find it mostly plugged with sediment. He removed the strainer, cleaned it up with soap and water, and replaced it in the water loop. After opening the isolation valves and putting the unit back in operation, the problem was solved. After verifying a proper pressure drop and temperature difference, the technician explained the problem to the customer and suggested that a scheduled maintenance program be purchased.

TROUBLESHOOT PROBLEM 6

The heat pump was not operating. The customer called to say the emergency heat light was on. There was no noticeable event that caused the light to come on, according to the homeowner.

Symptoms: Lock-out; emergency heat indicator at thermostat is on.

Possible Causes: Compressor mechanical or electrical problem; contactor malfunction; control system malfunction.

The technician reviewed the system indicator lights to try to determine the problem. The system was not running, but the power indicator was on. The low-pressure LED was lit. Checking across the low-pressure control, the technician confirmed that the switch was open (a voltage reading was obtained). Low pressure may mean that not enough heat is being absorbed by the ground coil. Moving to the circulator, the technician measured for power across the pump leads and found none. Something was preventing the circulator from operating. Looking at the electrical schematics, the technician found CR2 and checked for an open switch (refer to complete wiring diagram in Figure 23-3). Voltage across the contacts indicated that it was open. Rechecking that the thermostat was calling for operation by checking voltage on the secondary (control) side of the circuit, the technician assumed that the circuit board was supplying the electrical signal to close CR2 (that signal would have to be measured under the board, which is not recommended). Since the relay was part of the solid state circuit card, the entire circuit card was replaced. The technician confirmed the diagnosis when the circulator started and the low-pressure control closed, causing the system to resume normal operation.

Mechanical

The mechanical system of a geothermal heat pump includes the refrigeration system, geothermal loop, and the air delivery system. Each of these systems is diagnosed differently, but the technician must sort out the system that is causing the problem.

Mechanical troubleshooting starts with using the senses of touch, smell, hearing, and vision. Often, mechanical troubleshooting can be done with these senses before temperature, flow, and other measurements are required. Table 23-6 lists the most common types of mechanical symptoms, causes, and checks that need to be made.

TROUBLESHOOT PROBLEM 7

The customer reported that the geothermal system was not cooling.

Symptoms: Outside air temperature was high, indoor air temperature was above setpoint, thermostat was set correctly, and the refrigeration system was short-cycling.

Possible Causes: Airflow problem; fluid loop problem; overload problem; sensor problems; mechanical refrigeration problems.

The technician found a clean filter, but the LED was indicating a high-pressure control fault. Checking for pressure drop across the coaxial heat exchanger, the technician found no pressure drop. The loop pump was off. Line voltage and control signal were both present, but the pump relay was not pulled in. As the technician tapped the relay, the relay clicked and the pump started. The relay was removed and when operated manually, the technician could feel a distinct mechanical friction point that may have prevented the relay from closing. The relay was replaced.

TROUBLESHOOT PROBLEM 8

The customer reported that the geothermal system was not cooling. Air temperature from the floor diffuser was cold, but the air volume was low.

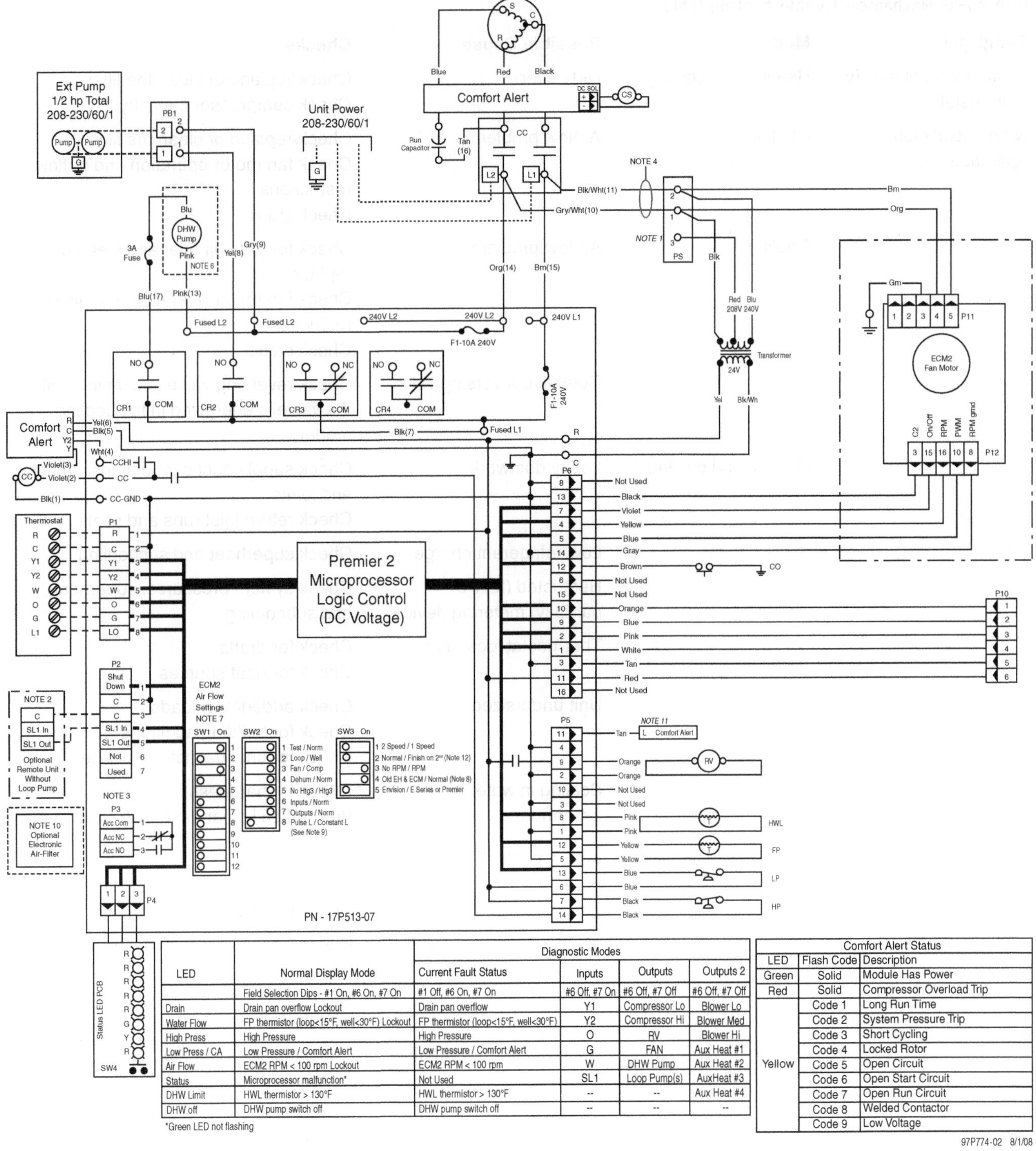

LED	Normal Display Mode	Diagnostic Modes			
		Current Fault Status	Inputs	Outputs	Outputs 2
	Field Selection Dips - #1 On, #6 On, #7 On	#1 Off, #6 On, #7 On	#6 Off, #7 On	#6 Off, #7 Off	#6 Off, #7 Off
Drain	Drain pan overflow Lockout	Drain pan overflow	Y1	Compressor Lo	Blower Lo
Water Flow	FP thermistor (loop<15°F, well<30°F) Lockout	FP thermistor (loop<15°F, well<30°F)	Y2	Compressor Hi	Blower Med
High Press	High Pressure	High Pressure	O	RV	Blower Hi
Low Press / CA	Low Pressure / Comfort Alert	Low Pressure / Comfort Alert	G	FAN	Aux Heat #1
Air Flow	ECM2 RPM < 100 rpm Lockout	ECM2 RPM < 100 rpm	W	DHW Pump	Aux Heat #2
Status	Microprocessor malfunction*	Not Used	SL1	Loop Pump(s)	AuxHeat #3
DHW Limit	HWL thermistor > 130°F	HWL thermistor > 130°F	--	--	Aux Heat #4
DHW off	DHW pump switch off	DHW pump switch off	--	--	--

*Green LED not flashing

Comfort Alert Status		
LED	Flash Code	Description
Green	Solid	Module Has Power
Red	Solid	Compressor Overload Trip
Yellow	Code 1	Long Run Time
	Code 2	System Pressure Trip
	Code 3	Short Cycling
	Code 4	Locked Rotor
	Code 5	Open Circuit
	Code 6	Open Start Circuit
	Code 7	Open Run Circuit
	Code 8	Welded Contactor
	Code 9	Low Voltage

Figure 23–3
Dual-stage ground-loop heat pump diagram. (Courtesy of WaterFurnace International)

Symptoms: Low air volume; insufficient cooling.

Possible Causes: Blower problems; dirty filter; duct connections broken; blower motor problems.

The technician removed the service panel and checked the blower. Everything was clean. The customer had replaced the filter recently. Visual inspection of the ductwork showed nothing wrong. The technician pressed the interlock switch and

Table 23–6 Mechanical Troubleshooting Table

Symptoms	Mode	Possible Cause	Checks
System will not satisfy thermostat	Heating and cooling	Dirty filter	Check/replace or clean the filter Check compressor capacity
No or insufficient operation	Heating	Airflow problem	Check/replace or clean the filter Check fan motor operation and airflow restrictions Check static
	Cooling	Airflow problem	Check for dirty air filter and clean or replace Check fan motor operation and airflow restrictions Check static
		Defective reversing valve	Check reversing valve (assuming that the valve is energized in the cooling mode)
	Heating and cooling	Leaky ductwork	Check supply duct connections, fittings, and joints Check return joist runs and seal
		Low refrigerant charge	Check superheat and subcooling
		Restricted (fully or partially) metering device	Check system pressures, superheat, and subcooling
		Thermostat location	Check for drafts Check for heat sources
		Unit undersized	Check added/new loads Check for building additions Check for unusual activity or use
		Scaling in water heat exchanger	Check for hardness Perform coil clean and flush
		Inlet water too hot/cold	Check load Check loop size Check ground moisture.
High head pressure	Heating	Insufficient airflow over indoor coil	Check/replace or clean the filter Check fan motor Check airflow Check static
	Cooling	Reduced or no water flow	Check pump operation Check valve operation Check water flow
		Air temperature out of specification	Check airflow Check water flow
	Heating	Inlet water too hot	Check load Check loop size Check ground moisture
	Heating and cooling	Scaling in coaxial heat exchanger	Check for hardness Perform coil clean and flush
		Unit overcharged	Check superheat and subcooling

		Noncondensables	Recover refrigerant Evacuate Recharge
Low suction pressure	Heating	Restricted metering device	Check superheat and subcooling
		Reduced water flow	Check pump operation Check water valve operation Check water strainer or filter Check water flow
	Cooling	Water temperature out of range	Check water flow
		Reduced airflow	Check/replace or clean the filter Check fan motor Check airflow Check static
	Heating and cooling	Air temperature out of range	Check fresh air vent Check water flow
Low discharge air temperature	Heating	Insufficient charge	Check for leaks
		Poor performance	Too high of airflow Check fan motor speed Check airflow
High humidity	Cooling	Refer to "Insufficient capacity"	Too high of airflow Check fan motor speed Check airflow
		Unit oversized	Check loads Check for reduced use Check for structural changes (insulation or de-construction)
Water pressure drop is low across heat exchanger	Heating	Too much or too little antifreeze solution Flow valve set too low Pump problems Water pressure too low (open loop)	Check antifreeze level with hydrometer Reset and check pressure drop Check pump operation Check well pump and system
Water pressure drop is high across heat exchanger	Heating and cooling	Fouled heat exchanger	Check pressure drop Flush with acid

This table is an example of a mechanical troubleshooting table.

the motor started, but while watching the blower wheel, it seemed that the wheel slowly came up to speed. The motor was a permanent split capacitor (PSC) and started normally. Releasing the interlock switch, the technician held the motor shaft and attempted to move the blower wheel—it moved on the shaft. Removing the blower and the motor, the technician inspected the shaft where the setscrew had left a deep furrow. The motor was replaced. After tightening the setscrew applied with a thread locker, the technician rechecked operation and confirmed normal air volume.

The refrigerant system trouble could be diagnosed using the symptoms table shown in Table 23-7.

Electrical

GSHP electrical systems are very similar to those for ASHPs. The exception is that they trade the ASHP defrost system with the freeze-protection controls of a GSHP. The freeze-protection control is typically a thermistor that senses the water-based fluid temperature in the refrigerant-to-water heat exchanger. If the temperature goes below setpoint, the heat pump shuts down until the temperature rises as the loop pumps continue to operate.

After conducting a general check using the senses of touch, smell, hearing, and vision, the technician may have made some mechanical observations. When those observations require electrical diagnosis, test meters must be used. Electrical problems account for the majority of the problems associated with geothermal heat pumps. Notice that some symptoms in the electrical table in Table 23-8 are the same symptoms as found in the mechanical table (Table 23-7).

When checking thermistors, the manufacturer of the thermistor or the manufacturer of the heat pump will provide a table for reference (similar to Table 23-9). Measurement is done with an accurate ohmmeter and thermometer.

TROUBLESHOOT PROBLEM 9

The new installation start-up proved to be more difficult than usual. The system was in the heating mode when it should have been in the cooling mode, and the installer left the problem with the service technician.

Symptoms: Thermostat set for cooling and the system operates in heating mode.

Possible Causes: Open-circuited or incorrect control wiring; incorrect thermostat wiring; stuck reversing valve.

The technician noted that this installation used an existing thermostat. It was set correctly, but when operated the system would not cool. The reversing valve was powered. Tracing the control circuit back from the system board to the thermostat, the technician noticed that the wiring was incorrect. After switching the O and B terminals and operating the system, the geothermal unit immediately went into the cooling mode.

Tech Tip

When measuring resistance values, the component must be disconnected on at least one side to isolate that component. Leaving the component in the circuit may provide a parallel path that the ohmmeter will read.

Tech Tip

Most thermostats come with both terminals B and O so that they can be used on both types of systems. Some systems have the reversing valve energized for cooling, while in others it is energized for heating. Typically the thermostat terminal is powered at B for heating and O for cooling. In some thermostats that are programmable, terminal O can be programmed to energize the reversing valve in cooling or heating.

Table 23–7 Mechanical Troubleshooting Table for Water Source Heat Pumps

Symptoms	Amp Draw	Head Pressure	Suction Pressure	Superheat	Subcooling	Air Temp. Diff.	Water Temp. Diff.
Low Airflow Heating	High	High	High	High/Normal	Low	High	Low
Low Airflow Cooling	Low	Low	Low	Low/Normal	Normal/ High	High	Low
Low Water Flow Heating	Low	Low/Normal	Low/Normal	Low	High	Low	High
Low Water Flow Cooling	High	High	High	High	Low	Low	High
High Indoor Air Temperature Heating	High	High	High	Normal/High	Normal/Low	Low	Normal
High Indoor Air Temperature Cooling	High	High	High	High	Low	Low	High
High Airflow Heating	Low	Low	Low	Low	High	Low	Low
High Airflow Cooling	Normal	Low	High	High	Low	Low	Normal
High Water Flow Heating	Normal	Normal	Low	High	Normal	Normal	Low
High Water Flow Cooling	Low	Low	Low	Low	High	Normal	Low
Scaled Coaxial Heat Exchanger Heating	Low	Low	Low	Normal/Low	High	Low	Low
Scaled Coaxial Heat Exchanger Cooling	High	High	High	Normal/Low	Low	Low	Low
Overcharged System	Low/Normal	Low/Normal	High	Low/Normal	High	Normal/Low	Normal
Undercharged System	Low	Low	Low	High	Low	Low	Low
Partial Restriction	Normal/Low	Low/Normal	Low	High	High	Low	Low
Inefficient Compressor	Low	Low	High	High	Normal/High	Low	Low
TXV—Bulb Loss of Charge	Low	Low	Low	High	High	Low	Low
Restricted Filter Drier	Check temperature difference (delta T) across filter drier.						

This is a sample chart that assumes only one problem. There are many variables in mechanical troubleshooting that must be considered.

TROUBLESHOOT PROBLEM 10

The system display was flashing a fault code and the customer called for service. The geothermal open-loop system was operating and keeping the house cool. This type of system has two thermistors on the inlet and outlet side of the indoor air coil to measure the temperature difference.

Symptoms: Fault code light.

Possible Causes: Fault code meant a problem with the air coil temperature.

Checking the air temperature and finding no problems, the technician reset the system and waited for it to come back on. After a few minutes of operation

Table 23–8 Electrical Troubleshooting Table

Symptom	Mode	Possible Cause	Checks
Green Status LED Off	Heating and cooling	Main power problems	Check breaker and disconnect Check for line voltage on the contactor Check for 24 VAC between R and C Check for primary/secondary voltage on transformer
High-Pressure Fault	Cooling	Reduced or no water flow	Check for pump or valve operation
		Water temperature out of specification	Bring water temperature within design parameters
	Heating	Reduced or no airflow	Check fan motor operation
	Heating and cooling	Bad high-pressure control	Check control continuity or voltage drop
Water Coil Low-Temperature Limit Fault	Heating and cooling	Bad thermistor	Check temperature and impedance correlation per chart
Air Temperature Low Limit Fault	Cooling	Reduced or no airflow	Check blower motor
Air Coil Low-Temperature Limit	Heating and cooling	Bad thermistor	Check temperature/resistance table
Condensate Fault	Cooling	Moisture on sensor	Check for moisture shorting
Over/Under Voltage	Heating and cooling	Under voltage	Check line-voltage input Check power-supply wire size Check compressor starting amperage and voltage Check low-voltage output
		Over voltage	Check low-voltage output
Unit Performance Fault	Heating	Heating sensor	Check heat sensor operation Check for overcharge
	Cooling	Cooling sensor	Check cooling sensor operation
No Fault Code Shown	Heating and cooling	No compressor operation	Check for fan operation
	Heating and cooling	Compressor overload	Check overload
		Control board	Reset power and check operation
Unit Short-Cycles	Heating and cooling	Unit in test mode	Reset power or wait 20 minutes for auto exit
		Compressor overload	Check overload
Only Fan Runs	Heating and cooling	Thermostat position	Check thermostat operation
		Locked out	Check for lock-out codes Reset power
		Compressor overload	Check overload
		Thermostat wiring	Check thermostat wiring at heat pump jumpers Y and R
Only Compressor Runs	Heating and cooling	Thermostat wiring	Check G wiring at heat pump Check jumpers G and R for fan operation
		Fan motor relay	Check jumpers G and R for fan operation Check for line voltage across BR contacts Check fan power Check fan relay

		Fan motor	Check for line voltage Check capacitor
		Thermostat wiring	Check thermostat wiring Check Y to R
Unit Doesn't Operate in Cooling	Heating and cooling	Reversing valve	Set for heating mode (on) Set for cooling mode (off)
		Thermostat setup	Check for O and B terminal reversal
		Thermostat wiring	Check O wiring at heat pump
		Thermostat wiring	Check low-voltage O to C in the cooling mode and not W to C

This is one manufacturer's troubleshooting guide and may not be applicable to all systems.

Table 23–9 Example of a Thermistor Troubleshooting Table

Thermistor Temperature (°F)	Resistance
78.5	9,230–10,007 ohms
77.5	9,460–10,032 ohms
76.5	9,690–10,580 ohms
75.5	9,930–10,840 ohms
33.5	30,490–32,080 ohms
32.5	31,370–33,010 ohms
31.5	32,270–33,690 ohms
30.5	33,190–34,940 ohms
1.5	79,110– 83,750 ohms
0.5	81,860–86,460 ohms
0.0	82,960–87,860 ohms

Source: Courtesy of WaterFurnace International

in the cooling mode, the fault light started flashing and indicating the same problem—air temperature. After disconnecting the thermistor and placing a temperature probe in close proximity, the technician used an accurate ohmmeter to compare the resistance of the thermistor to the manufacturer's thermistor table. The air temperature was 75.0°F and the ohm reading was 30,500 ohms, indicating an air temp of 33.5°F, so the thermistor was out of range. See Table 23-9. The thermistor was replaced and the system was placed back into operation.

SUMMARY

In this chapter we have discussed troubleshooting for geothermal heat pumps. We have noted that there are a few differences between air-source and geo-source systems, but that the differences are minor. Chapter 13 and this chapter are related in terms of information. This chapter expands on the information provided in Chapter 13.

User-level diagnostics were discussed, as well as how the customer is becoming or might become more involved with the geothermal system. It was also noted that the thermostat can provide information to both the user and the service company through the Internet. Certain thermostats have the ability to alert the service company that service is needed before the customer is aware of a system problem.

The chapter concluded with troubleshooting where the technician is involved. Troubleshooting was divided into mechanical and electrical. Tables were provided with common mechanical and electrical checks.

REVIEW QUESTIONS

1. Describe user-level troubleshooting.
2. Describe how a thermostat might send a message for service directly to the service company.
3. Relate what observational issues are part of owner maintenance.
4. Describe the relationship between electronic modules and the microprocessor.
5. Explain how electronic modules and microprocessors are used in troubleshooting.
6. Describe the difference between mechanical and electrical troubleshooting.

CHAPTER 24

Energy and Efficiency Calculations

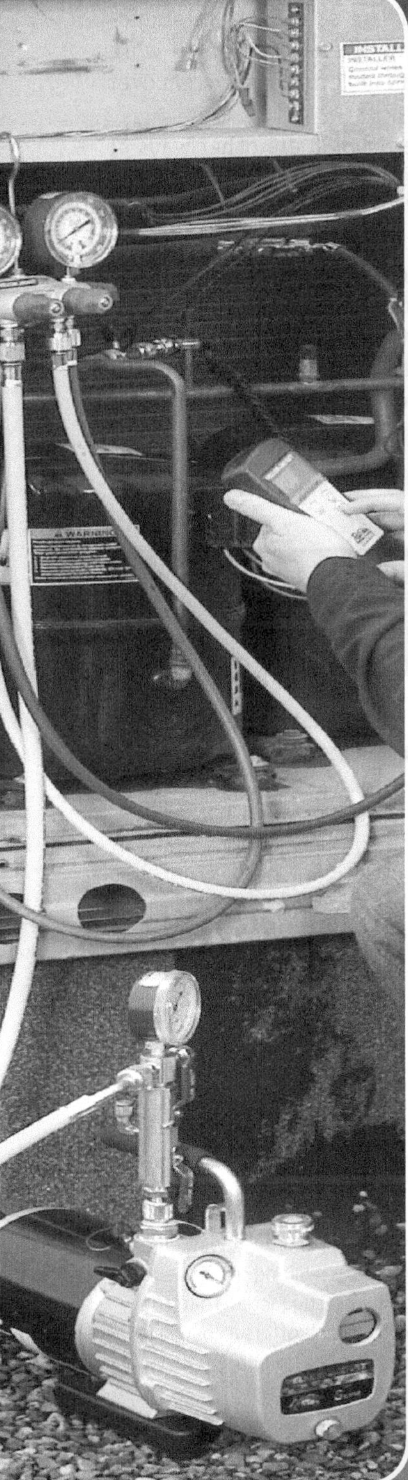

LEARNING OBJECTIVES

The student will:

- Define *coefficient of performance* (COP)
- Describe how COP is used
- Define *energy efficiency ratio* (EER)
- Describe the difference between seasonal energy efficiency ratio (SEER) and heating seasonal performance factor (HSPF)
- Describe where annual fuel utilization efficiency (AFUE) is used
- How does return on investment (ROI) affect purchasing decisions?
- Relate how comparing systems may help with purchasing decisions
- How can incentives help in purchases of systems and how can they be found?

INTRODUCTION

Heat pump systems are compared from system type to system type and from heat pump system to other conventional systems. Making comparisons can be very difficult when every system is different. Some manufacturers offer larger or smaller system components, different types of heat exchangers, and multispeed and two-stage compressors. Ultimately, the amount of work accomplished compared to the amount of energy used seems to be the best comparison point. Even with this basic comparison, how that comparison is made for how long will result in different outcomes.

In an effort to reduce confusion, the industry has come up with several solutions that help with comparing systems. Some of the solutions are single-measurement methods of determining the relationship of energy used to work done. Other solutions take into consideration the effects of running a system over a longer period of time, such as over a season of use.

The costs of systems and how much savings can be reached are other ways of comparing systems. In some cases it is important to know how much lower the energy costs of one system might be over another. The savings can be used to determine how long it will take to recoup the cost of the more efficient system.

Incentives are a way of reducing the initial cost of a more efficient system and/or a way of reducing the operating costs of a new system. Utility, local, state, and federal programs may be available to offset the cost of a more efficient system. Manufacturers may offer incentives to upgrade to higher-efficiency systems. Power suppliers may offer programs to reduce the cost of fuel for new, more efficient heating and cooling systems along with one-time rebates.

Field Problem

After a routine heat pump maintenance call, the homeowner engaged the service technician in a short conversation about his system and energy efficiency. The neighbor had just purchased a new heat pump and was relating all of the expectations of the system. The homeowner, who had a 7-year-old system, was trying to make comparisons with his neighbor. More important, he wanted to know if he should start thinking about a new heat pump system. He wondered if his system was less efficient enough that a new heat pump replacement would pay for itself.

After writing down a few of the efficiency numbers that the homeowner related and checking his log of the homeowner's system, the technician said that he could answer a few of the basic questions, but would like to check in with the office first. Going to the service vehicle, he contacted the office and the sales manager, and related the numbers he had been given, as shown in Table 24–1.

Table 24–1 Comparison of Two Sample Heat Pump Systems

Heat Pump System	Homeowner	Neighbor
Age of system	7	1
Size of system	3 ton	3 ton
Type of system	ASHP	ASHP
Defrost system	Time	Demand
SEER	12	13
HSPF	7.7	8.2
Refrigerant	R-22	R-410A
Energy Star Qualified	Yes	Yes

Returning from the vehicle, the technician related to the customer that the neighbor might be saving a little more money on electrical bills because his system has a higher seasonal energy efficiency ratio (SEER) and heating seasonal performance factor (HSPF) rating, but there may be a greater savings because of the neighbor's "demand" defrost option. The homeowner's system was approximately in the middle of the life expectancy for heat pumps—13 to 15 years. Finishing his conversation, the technician related that his company would be happy to send out a sales representative who would conduct an energy review and do a life-cycle cost analysis (LCCA) of his system and recommend replacement options. In the meantime, he suggested that the homeowner visit the Energy Star website, which offers a comparison calculator and more information about heat pump products, options, and incentives.

COEFFICIENT OF PERFORMANCE (COP)

Coefficient of performance (COP) is a steady-state ratio of the amount of useful work accomplished to the amount of energy needed to perform the work. The amount of work for a heat pump is measured in the amount of heat that is moved or output in BTUH. The amount of energy required to move heat is measured in the amount of electricity needed to operate the heat pump system. One kilowatt hour of electrical energy is equivalent to 3,413 BTUH. The basic equation for COP is:

$$\text{COP} = \text{BTUH}/(\text{kWh} \times 3{,}413\ \text{BTUH})$$

Example:

$$\text{COP} = 36{,}000\ \text{BTUH}/(3.5\ \text{kWh} \times 3{,}413\ \text{BTUH})$$

$$\text{COP} = 36{,}000\ \text{BTUH}/12{,}000\ \text{BTUH}$$

$$\text{COP} = 3$$

The result is a ratio of work done to energy used and is expressed as a single number; the higher (larger) the number the higher the COP. For example, if one heat pump has a COP of 3 and another of the same size has a COP of 4, the unit with the higher number produces more work with less wattage used. Potential owners can use the COP number to make energy-efficiency decisions for heat pumps that are being used in the heating mode.

ENERGY EFFICIENCY RATIO (EER)

The energy efficiency ratio (EER) is typically used in air conditioning to determine efficiency. Comparisons of operating conditions are made by the rating agency, the Air-Conditioning, Heating and Refrigeration Institute (AHRI), with outdoor air conditions of 95°F. The air conditioning system must operate to keep indoor temperatures cool when outdoor temperatures are at 95°F. An EER is generally applied to split air conditioning systems to make comparisons with standard air conditioning systems, when the heat pump is being used to cool. An EER rating is also considered a steady-state condition.

Tech Tip

Steady state is considered to be a condition, output, or input that is measured for a certain set of conditions. In the case of air conditioning, the steady-state outside temperature would be referenced as 95°F. Measurements of an operating air conditioning system at that outside condition only reflect how the system can perform at that condition. Steady-state conditions are typically used to make comparisons to similar systems or conditions.

An EER is calculated by dividing the BTUH absorbed (input BTUs) at steady-state outdoor condition by the energy consumed in watts. The result is a single number that is expressed as relative ratio—what the machine can do at a particular operating condition:

$$\text{EER} = \text{BTUH}/\text{W}$$

Example:

$$\text{EER} = 36{,}000\ \text{BTUH}/3{,}600\ \text{W}$$

$$\text{EER} = 10$$

SEASONAL ENERGY EFFICIENCY RATIO (SEER)

The seasonal energy efficiency ratio (SEER) is applied to cooling equipment. In January of 2006, all residential air conditioners sold in the United States were required to have a SEER of at least 13, and Energy Star–certified central air conditioners need to be SEER 14. It is expected that higher SEERs will be required over time. An SEER is based on an EER, but includes all operating conditions encountered for a year. These conditions include: start-up, shutdown, varying environmental conditions, and varying energy usage. In essence, SEER is the total amount of heat moved divided by the total amount of energy used over an entire season.

SEER is calculated by dividing the total number of BTUs moved over a season by the total amount of electricity used.

If we assume 4 months operating at 10 hours per day, for a total of 1200 hours:

SEER = BTUH × hours of operation/watts used over the same period of time

Example:

SEER = 36,000 BTUH × 1,200 hours/3,323.07 kW (or 3,323,076.923 W)
SEER = 43,200,000 BTU/3,323,076.923 W
SEER = 13.4

Tech Tip

Using the 13 SEER calculation, electrical usage can be determined by dividing SEER into total BTUs, which would result in the number 3,323,076.923 watts (or 3,323.07 kW). Multiplying this number by the cost of electricity per watt or per kilowatt will yield the total cost of operation over the season.

If electrical costs were .10 cents per kWh, then 3,323.07 kWh times .10 kWh will equal $332.30. Divide this amount for the entire season by 4 months and the cost per month is $83.08. If the system were to be upgraded to 15 SEER, the cost per month would be $72. Here is how it is calculated from the SEER example:

New cost per month = ((BTUH × hours of operation)/new SEER)/(W/kilowatt) × cost per kW)/number of months

Example:

New cost per month = (2,880,000 watts/1,000 W/kilowatt)
New cost per month = (2,880 kW × .10 per kW)
New cost per month = $288/4 months
New cost per month = $72

Because SEER compares the annual operating efficiency, similar systems that could be installed in a single location can also be compared. The ability to make one number comparison helps the consumer and the designer make system choices based on return on investment (ROI), or payback.

HEATING SEASONAL PERFORMANCE FACTOR (HSPF)

Heating seasonal performance factor (HSPF) is a way of comparing air-source heat pump (ASHP) systems in the heating mode. The HSPF rating uses the total heating output of a heat pump (and auxiliary electric heat needed), measured in BTUs, and the total amount of electricity used (in watt-hours) during an entire heating season. AHRI 210/240 (formerly ARI Standard 210/240) is the reference

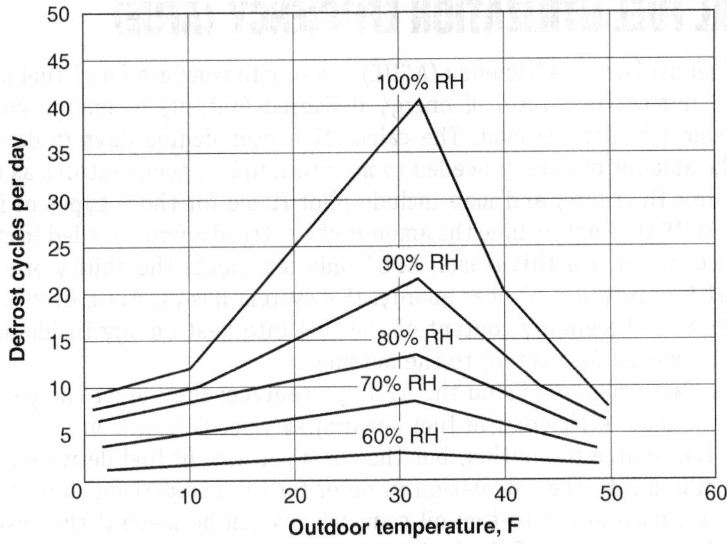

Figure 24-1
Defrost cycles per day due to the outdoor humidity condition at a specific location or micro-climate. (Courtesy of RSES.)

research that helped to provide the basis for HSPF and SEER. HSPF includes a defrost penalty for ASHPs, but local conditions may increase the penalty that has to do with the number of defrost cycles required (Figure 24-1).

HSPF is calculated by dividing the total number of BTUs produced throughout an entire heating season by the number of watt-hours of electricity consumed for the same period:

HSPF = BTUH × hours of operation/watts used over the same period of time

Example:

HSPF = 36,000 BTUH × 1,200 hours/3,323.07 kW (or 3,323,076.923 W)

HSPF = 43,200,000 BTUs/3,323.07 kW (or 3,323,076.923 W)

HSPF = 13

Most heat pumps are above 8 HSPF and operate with seasonal efficiencies above 234%. This means that for every BTU of heat energy (1 watt = 3.413 BTU) used over the entire heating season, ASHP systems will deliver 2.34 BTUs of heat. In comparison, an electric furnace is rated as having 100% efficiency (discounting fuel conversion at the power company). A 90% efficiency gas furnace will deliver .90 BTU from each unit of fuel (not including electrical consumption).

Tech Tip

To convert HSPF to a percentage, divide HSPF by 3.413 watts per BTU. For example, divide 8 by 3.413, which is 2.34, or 234% (Percentage = HSPF/W per BTU).

Table 24-2 compares ASHP system efficiency relationships.

Table 24-2 ASHP System Efficiency Relationships

Type of System	HSPF	SEER	EER
Split System	8.2	14.5	12
Single Packaged System	8.0	14	11

ANNUAL FUEL UTILIZATION EFFICIENCY (AFUE)

Annual fuel utilization efficiency (AFUE) is a way to compare fossil-fuel systems. The AFUE number is a ratio of energy delivered (output) to energy consumed (input) over a heating season. The calculation uses degree days (a determination of the amount of energy needed to maintain indoor temperatures as outdoor temperatures fluctuate) and may include pilot flame for those types of ignition systems. AFUE does not include the amount of electrical energy needed to circulate either air or water. For this reason AFUE only represents the ability of a system to convert fuel into useful heat energy. If a system has an AFUE of 90%, it can convert 90% of the energy content of the fuel into heat energy inside and only 10% will be wasted (or vented) to the outside.

Systems are compared using the AFUE percentage; the higher the percentage number, the more efficient the fuel-burning system. Comparisons can be made from one fuel source to another, but the costs per unit of fuel delivered need to be incorporated into the calculation in order for the comparison to be accurate. For instance, natural-gas-to-fuel-oil comparisons can be made if the cost of the fuel and the BTU content of the fuel can be included in the calculation, as well as the AFUE. Table 24–3 provides an AFUE comparison chart.

RETURN ON INVESTMENT (ROI)

Return on investment (ROI) is another way of saying "payback period." If comparisons are being made between systems with higher efficiencies that have higher costs and lower-efficiency systems with lower costs, it is hard to make the comparison without trying to determine the length of investment time. Simply, should you buy a higher-efficiency unit, which costs more, if you can recoup the extra money spent within a reasonable amount of time? This also assumes that after that time, the extra costs for the higher-efficiency unit will start paying you back or saving you money.

Considerations made when comparing a natural-gas furnace with an ASHP might be those shown in Table 24–4. All of the $ blanks would need to be filled in depending on the local costs of labor, materials, and energy.

For illustration, let's say that the total cost for the gas furnace was $8,575 and the total cost of the ASHP was $10,950. To figure the ROI, the amount of fuel savings would have to be determined. If the gas furnace was determined to cost $175 per month and the ASHP was expected to be $95 per month, the ROI could be determined. The difference in the total cost is $2,375. There are two ways to state the difference: (1) the gas furnace costs $2,375 less, and (2) the ASHP costs

Table 24–3 AFUE Comparison Chart

	New System AFUE			
Existing System AFUE	87%	90%	95%	97%
60%	$31.03	$33.33	$36.84	$38.14
65%	$25.29	$27.78	$31.58	$32.99
70%	$19.54	$22.22	$26.32	$27.84
75%	$13.79	$16.67	$21.05	$22.68
80%	$8.05	$11.11	$15.79	$17.53

Note: This chart is based on $100 worth of energy costs. The savings can be cross-referenced. If a 60% system is upgraded to an 87% unit, the savings will be $31.03 for every $100 spent.

Table 24–4 Comparison of Natural-Gas Furnace and Air-Source Heat Pump

System Type	Consideration	Component/Unit/ Measure	Costs
Natural-Gas System	Gas	Total Therms Used	$
	Fuel Conversion	AFUE	$
	Estimated Electrical	Watts per Hour	$
	Electrical Cost	kWh	$
	Installation		
		Labor	$
		Furnace	$
		Split System A/C	$
		Gas Hook-up	$
		Ductwork	$
		Venting	$
		Drain	$
		Electrical	$
			Total Costs—Gas
System Type	**Consideration**	**Component/ Unit/Measure**	**Costs**
Air-Source Heat Pump	Electricity	Total kWh Used	$
	Fuel Conversion	HSPF	$
	Auxiliary Electricity		Total kWh Used
	Installation		
		Labor	$
		Heat Pump	$
		Auxiliary System	$
		Air Handler	$
		Ductwork	$
		Drain	$
		Electrical	$
			Total Costs—ASHP

$2,375 more. There is also an $80 difference in the monthly cost of each system. Again, the gas furnace costs $80 more to run per month or the ASHP costs $80 less. If the ROI was determined based on the difference in cost to operate per month, the question would be how many months of operation would be required to make up the difference in total costs. The equation would be:

ROI of monthly operation = (ASHP total costs − gas total costs)/(gas operating costs − ASHP operating costs)

Example:

$$\text{ROI} = \text{annual savings} \div \text{invested cost}$$
$$\text{ROI} = 1 \div \text{payback}$$

ROI of monthly operation = ($10,950 − $8,575)/($175 − $95)

ROI of monthly operation = ($2,375)/($80)

ROI of monthly operation = 29 months or 2.4 years

After 29 months, the extra cost of the system would be recouped and at the end of the 30th month; $80 would be saved from this month forward by the ASHP system. Another, more complicated, ROI involves the life cycle of the equipment. In this type of ROI, the expected life of the equipment and the cost of replacement and repair would be figured into any cost savings that might be expected. Commonly called a life-cycle cost analysis (LCCA), this type of ROI requires a more in-depth calculation that includes: total system cost, total operational costs, life span of the equipment (for heat pumps, generally around 15 years), and total maintenance costs.

HEAT PUMP COMPARISONS

Comparing other systems with different fuel sources can be fairly complicated. It becomes less complicated when the same types of systems are compared. This doesn't mean that it is simple. Each system in the comparison should include all of the same attributes as the others being compared. Depending on where the heat pump is installed, the comparison for an ASHP may be between two of the factors previously discussed. If the climate conditions require more cooling days, the heat pump comparison may focus on the SEER. If the ASHP is in a colder region where there are more heating days, the focus of comparison may be on the HSPF. In this section various heat pumps will be compared. The comparison will be between three types of heat pumps: air-source heat pump (ASHP), water-source heat pump (WSHP), and ground-source heat pump (GSHP).

It has been noted already that there is a heat source advantage for either a WSHP or GSHP over an ASHP. Air temperatures fluctuate greater than earth temperatures (Figure 24-2).

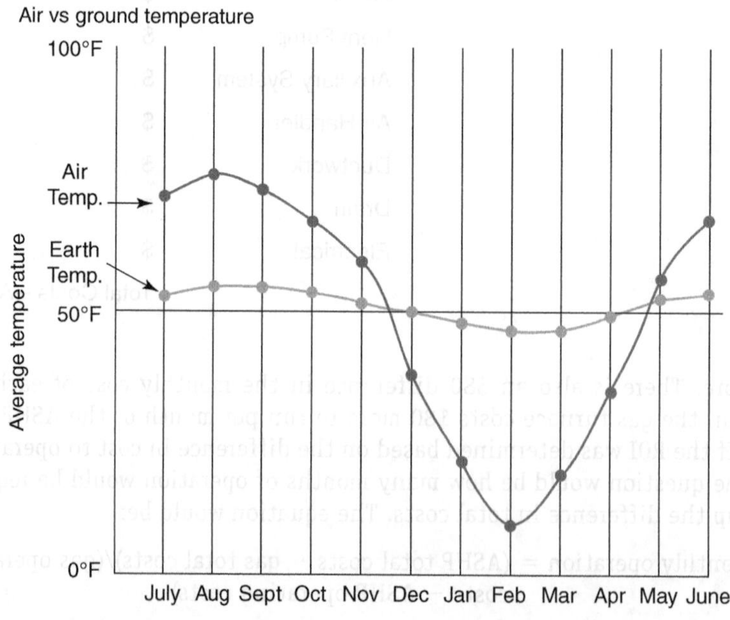

Figure 24–2
Air temperatures will vary greatly depending on the part of the country. Ground temperatures from 4 to 6 feet vary only slightly. Earth temperature tends to stabilize around 50°F at 6 feet. (Courtesy of Delmar/Cengage Learning)

For this comparison, it will be assumed that a heat pump is being considered for installation. A heat gain/loss calculation has been done and the result shows that this example home has a cooling gain of 48,000 BTUH and a heating loss of 38,000 BTUH. The home is to be ducted to deliver approximately 450 cubic feet per minute (CFM) of air per ton of system capacity. Air filtration will be the same with primary and secondary filter systems. The duct system has been designed for a heat pump, and the cost of this delivery system is approximately the same for all comparisons. Each of the systems will heat domestic hot water during the cooling season using a desuperheater. Also, each system will be controlled with the same programmable thermostat. Table 24–5 shows the comparison considerations.

All of the heat pumps are single-speed, 4-ton units. The air-source system is similar to a split air conditioning system. It has an outside unit with an inside coil mounted on an electric furnace with blower. The water-source system takes water from upgrading the water well, which requires a more expensive outdoor coil of copper-nickel for raw water use. The standard well was to be a 5″ well with a ½-horsepwer submersible pump and a 40-gallon expansion tank. The well was upgraded to a 6″ well, ¾-horsepower submersible pump, and an 80-gallon expansion tank. These upgrades were attributed to the heat pump system, not the entire cost of the well. The WSHP will discharge to an existing pond near the house. The GSHP will require 400′ (100′ per ton) of trenching for a concentric horizontal loop (also called a slinky). This system will also include the cost of the loop, pump, and fittings to the heat pump. Because this heat pump configuration will not use raw water, the outdoor heat exchanger does not need to be copper-nickel. Figure 24–3 shows a comparison of these systems.

Table 24–5 Heat Pump Comparison Considerations

	ASHP	WSHP	GSHP
System	• Air-source heat pump with single or multiple speeds • Supplemental heat • Back-up/ emergency heat	• Water-source heat pump • Water source • Water pump	• Ground-source heat pump • Loop configuration • Circulator
Style	Split system	Open loop	Horizontal loop
Efficiency	8.2 HSPF and 13 SEER	2.8 COP and EER	COP and EER
Defrost	Demand or timed defrost	N/A	N/A
Heat Source	Air	Water	Ground
Heat Extraction	Air coil	Water well or source	Earth
Domestic Hot Water	Desuperheater—summer	Desuperheater—summer/winter	Desuperheater—summer/winter
Delivery System	Ductwork	Ductwork	Ductwork
Water Well	N/A	Price per foot of depth	N/A
Area of Land	N/A	N/A	Sq. ft. of land area is in relationship to size of GSHP
Air Filtration	Primary and secondary filters	Primary and secondary filters	Primary and secondary filters

	ASHP	WSHP	GSHP
Heat pump—4 ton	$3,300	$2,899	$2,899
Back-up system—electric strip	$550	$550	$550
Delivery system—duct & filters	$2,755	$2,755	$2,755
Thermostat—programmable	$125	$125	$125
Vibration eliminators	$30	$30	$30
Outside unit pad	$20	N/A	N/A
Water well—6 inch @200 ft depth (upgrade cost only) with discharge to pond.	N/A	$7,200	N/A
Water well pump	$0	$240	$0
Trenching—400 ft @6 ft depth	N/A	N/A	$2,400
Closed loop slinky, pump & connectors	N/A	N/A	$9,600
Total system cost	$6,780	$13,799	$20,359
Adjusted cost with federal 30% tax credit (until 2016)	$0	$9,659	$14,251
Cost difference	$0	$2,879	$7,471
Operational costs per Year—.10 cents per kWh	$1,200	$780	$675
ROI in Years	0	6.8	8
% ROI	0	14.7	12.5

Note: The costs offered in this example are only for comparison purposes and may not reflect the true costs of any single system or configuration. Always check and compare prices of equipment to be specified for a particular application. Do not rely on this example as a true estimate for any installation. Labor costs were intentionally not included.

Figure 24–3
When comparing system costs, ASHPs are lowest for installed system costs. When comparing energy operating costs, ASHPs cost more to run. (Courtesy of Delmar/Cengage Learning)

INCENTIVES

Heat pump manufacturers generally subscribe to the EPA Energy Star requirements to qualify equipment and related products. Higher energy ratings for equipment may result in qualification for energy incentives that could have the potential to reduce overall costs. Incentives are offered by power companies and by governmental agencies. Some utility companies adjust prices if homeowners and businesses switch from fossil-fuel systems to heat pumps. The Internal Revenue Service (IRS) may also provide incentives, such as tax credits, to compel owners to switch to higher-efficiency equipment.

Before equipment is purchased or specified, it is suggested that all local, state, and federal websites be checked for energy incentives and qualifying language. Also, check with energy providers and the manufacturer of similar equipment to see if there is a rebate, upgrade, or changeover program. Manufacturers may offer incentives to switch from one fuel to another or from one type of system to another. There may also be an upgrade program where basic systems replaced with energy-efficient systems may qualify for a discount. In cases where older systems have lifetime warranties, companies may also offer discounts on new systems in place of warranty replacement parts.

SUMMARY

In this chapter the popular and most used energy relationships have been discussed. COP, EER, SEER, and HSPF are all used to compare one system to another. Some of these are steady-state or single-measurement relationships. SEER and HSPF are seasonal energy relationships and attempt to estimate the ratio of the amount of work to energy used over a cooling or heating season.

AFUE is reserved for use with systems that convert fossil fuel to usable heat energy. This energy conversion rating provides a number that is a percentage of heat used for heating. Subtracting this number from 100% represents that part of the unit of fuel that is wasted.

ROI is a way to compare systems and determine how long it will take to pay back the amount of money used to purchase higher-efficiency equipment. Older or existing systems are used for comparison with newer and higher-efficiency equipment. The lower cost of operation and the money saved is used to determine the number of months or years it will take to recoup the money spent for the higher-efficiency system.

Incentives to reduce the cost of the higher-efficiency system or to reduce the cost of energy used may be available. Local, state, and federal incentives may be available as a way of reducing the purchase cost of the new system. Manufacturers may offer incentives to upgrade to newer systems. Utility companies may offer incentives to switch to higher-efficiency equipment or to change from one fuel source to another.

REVIEW QUESTIONS

1. Define *coefficient of performance* (COP).
2. Describe how COP is used.
3. Define *energy efficiency ratio* (EER).
4. Describe the difference between seasonal energy efficiency ratio (SEER) and heating seasonal performance factor (HSPF).
5. Describe where annual fuel utilization efficiency (AFUE) is used.
6. How does return on investment (ROI) affect purchasing decisions?
7. Relate how comparing systems may help with purchasing decisions.
8. How can incentives help in purchases of systems and how can they be found?

Answer Key

Chapter 1

1. Describe how the basic refrigeration process functions. *The basic refrigeration process can start with any of the four components. Starting with the compressor, the refrigerant is compressed and sent to the condenser. In the condenser, the refrigerant loses its heat and changes state to a liquid. The liquid is sent to the metering device, where the pressure and temperature of the liquid refrigerant are dropped to the evaporator saturation temperature and pressure. In the evaporator, the liquid changes state while absorbing heat. Vapor leaving the evaporator is sent to the compressor, where the refrigeration process repeats.*
2. Looking at a real refrigeration system, a picture, or a diagram, identify and name the four basic components of all compression refrigeration systems in the order refrigerant flows through them. *Compressor, condenser, metering device, evaporator.*
3. Describe the function of the compressor. *The compressor is a point of pressure change in the refrigeration process. It compresses the refrigerant, creating a high-pressure, high-temperature vaporous refrigerant. It pulls vaporous refrigerant from the evaporator.*
4. Describe the function of the condenser. *The condenser receives the high-pressure, high-temperature vaporous refrigerant from the compressor and provides the surface area for the refrigerant to exchange its heat with the outside ambient environment. As the refrigerant cools, it condenses to a liquid.*
5. Describe the function of the metering device. *The metering device regulates the amount of refrigerant entering the evaporator. It forms the second point of pressure change in the system, along with the compressor. The metering device allows the liquid refrigerant to drop in pressure to the operating saturation pressure of the evaporator.*
6. Describe the function of the evaporator. *The evaporator allows expansion of the refrigerant to occur as the refrigerant absorbs heat. Both pressure and temperature lower to the working saturation temperature and pressure. As heat is absorbed, the refrigerant changes state from a liquid to a vapor. After all of the refrigerant has changed to a vapor, the vapor is sent to the compressor.*
7. Describe saturation as it applies to refrigerants. *Saturation refers to when the liquid and vapor have reached equilibrium in a closed container. There is no more liquid or vapor being produced. In an operating refrigeration system, it is the point where liquid and vapor are changing state. Saturation is also the basis for the pressure/temperature charts for refrigerants.*
8. Describe superheat as it applies to refrigerants. *Superheat describes refrigerant temperatures above saturation when all of the liquid has turned to vapor. Vapor that is heated above saturation is considered superheated.*
9. Describe subcooling as it applies to refrigerants. *Subcooling describes refrigerant temperatures below saturation when all of the vapor has turned to liquid. Liquid that is cooled below saturation is considered to be subcooled.*

Chapter 2

1. Describe the function of the indoor coil during the heating mode. *The indoor coil functions as the condenser in the heating mode. The condenser adds heat to indoor air to heat occupied spaces.*
2. Describe the function of the outdoor coil during the heating mode. *The outdoor coil functions as the evaporator in the heating mode. The evaporator extracts heat from outdoor air.*
3. Relate how the indoor and outdoor coils change function. *Indoor and outdoor coils change function because they are connected to the compressor in reverse. During cooling, the suction of the compressor is connected to the indoor coil. During heating, the suction of the compressor is connected to the outdoor coil. Connecting the suction to the outdoor coil makes the outdoor coil function as the evaporator.*
4. Name the device that is responsible for changing the function of the indoor and outdoor coils and describe its operation. *The device that reverses the function of the indoor and outdoor coils is called the reversing valve. The reversing valve changes the connections of the indoor and outdoor coils to the compressor. In one mode, the compressor is connected so that hot vapor flows to the outdoor coil; in the reverse mode, the hot vapor flows to the indoor coil.*
5. Describe the function of the indoor coil during the heating mode. *The indoor coil functions as the condenser in the heating mode. The condenser adds heat to indoor air to heat occupied spaces.*
6. Explain how heat is pulled from cold air. *Heat is pulled from outdoor air anytime the outdoor coil is colder than the outdoor air. Heat will move from the*

relatively warmer outdoor air to the colder outdoor coil. As outdoor temperatures go down, the amount of heat extracted will be reduced.

7. Describe what a thermal balance point is and how it relates to the operation of a heat pump. *The thermal balance point is an outdoor temperature where the heat pump capacity is exactly the same as the heat loss of the home or building. At this point the heat pump will run 100% of the time and be able to maintain indoor temperature conditions. Below this point, supplemental heat will be required.*

Chapter 3

1. Describe how the sun influences the amount of heat available in outdoor air. *The sun adds heat to the outdoor air by heating objects and the ground by radiation. Air picks up the heat by conduction as well as radiation and convection as it scrubs the surface of the objects and the ground.*
2. Explain how heat is measured in a pound of outdoor air (or in CuFt). *Heat is measured in BTU/CuFt, and the measurement can be obtained using the psychrometric chart and any two conditions of an air sample. Heat in a cubic foot is another condition of an air sample and can also be read from the psychrometric chart.*
3. Relate what "absolute zero" means in relation to the amount of heat available in outdoor air. *Absolute zero is the absence of heat and no molecular motion. It is measured as a –460°F or –273°C (or zero on the related absolute scales Rankin and Kelvin).*
4. Describe the function of and need for supplemental heat. *Supplemental heat usually is in the form of resistance strip heat. It is used whenever the system goes into defrost mode to ensure that the occupied space does not see temperature drop during defrost or drop below the thermal balance point.*
5. Explain why a defrost mode is needed for air-source heat pumps. *Defrost is necessary because the outside coil, functioning as the evaporator, will condense moisture. This moisture forms frost and acts as an insulator and must be removed.*
6. Describe the similarity in function of the indoor coil in the cooling mode to the outdoor coil during the heating mode. *During cooling, the indoor coil functions as the evaporator. During the heating mode, the outdoor coil functions as the evaporator. In both cases the evaporator is used to absorb heat from the air (indoor or outdoor).*

Chapter 4

1. Describe how the indoor coil acts as a heat rejector. *Discharge vapor from the compressor gives up heat as the indoor coil exchanges the heat from the refrigerant to the indoor air, rejecting the heat from the refrigerant in the indoor coil to the indoor air heating the structure.*
2. Describe the process of desuperheating. *Superheat is the amount of heat above saturation. In order to drop to saturation temperature, the superheat must be removed from the vapor leaving the compressor to the saturation point in the condenser.*
3. Explain what happens to the liquid/vapor of a refrigerant at saturation. *When refrigerant is at saturation and heat is removed, vapor changes state to a liquid. If the saturated refrigerant is heated, liquid refrigerant changes to a vapor.*
4. Explain how a refrigerant can be subcooled. *Refrigerant can only be subcooled when all of the refrigerant vapor has turned to a liquid. At that point, more heat can be removed from the liquid, subcooling the liquid below saturation.*
5. Describe what happens to the quantity of heat in the indoor coil as the outside temperature changes. *The BTUH of heat going to the indoor coil is the result of the amount of heat that can be obtained by the outdoor coil. If the outdoor coil temperature drops, the total heat sent to the indoor coil will drop.*

Chapter 5

1. Describe what a single-stage room thermostat can control. *The single-stage room thermostat is only designed to control one heating and one cooling source. It is the standard thermostat for conventional heating and cooling applications.*
2. Explain the difference between single-stage and multistage thermostats. *A single-stage thermostat controls one heating or cooling source, while a multistage thermostat can control two or more heating and two or more cooling sources/stages with one thermostat.*
3. Describe what a multistage thermostat controls in an air source heat pump application. *During the heating mode of a typical heat pump, the heat pump operation is controlled by the first stage of heat. The second stage of heat is supplemental electric resistance strip heat that is operated when the heat pump operates below the thermal balance point.*
4. Explain the application of single-stage and multistage thermostats. *Single-stage (four-wire) thermostats are used with single-stage thermostats (R, G, W, Y) for additional stages. Multistage thermostats have multiple Y_1 and Y_2 connections (W_1 and W_2, R, G, and O).*
5. Describe what a programmable thermostat can do as compared to a nonprogrammable thermostat. *A programmable thermostat is able to reset the thermostat to several new setpoints during a day. The nonprogrammable thermostat has only one setting.*
6. Explain what a touchscreen thermostat can provide for a user. *Touchscreen thermostats offer the user specific information and user interface control with each feature of the thermostat. They can reduce the number of external buttons and increase user information.*

7. Describe night set-back and the optimum temperature and time settings for energy savings. *It has been shown that a setback of 7°F for a period of 8 hours or longer has the ability to reduce energy costs for heat pumps in the heating mode.*
8. Describe the difficulty of dehumidification using a thermostat and a whole-structure cooling system. *Even though some thermostats incorporate humidity control, dehumidification using a whole-structure cooling system may not work as well as expected. The cooling system has the problem of overcooling the structure when controlled by the humidity feature of the thermostat.*

Chapter 6

1. Describe the operation of the reversing valve. *The reversing valve is designed to connect the suction and discharge of the compressor to either the indoor or outdoor coils, depending on the mode—heating or cooling.*
2. Describe the function of the pilot operator of a reversing valve. *The pilot operator of a reversing valve uses the pressure generated on the discharge and suction sides of the system to move a piston from the heating (default) position to the cooling position. When the piston moves to the new position, the indoor and outdoor coils change function—evaporator or condenser.*
3. Explain how a "default" mode or position for a reversing valve relates to the mode (heating or cooling) of the heat pump. *The default mode is the de-energized position of the reversing valve. If no power is applied to the operator, the valve is in the default position. Typically, the default position of the reversing valve is the heating mode.*
4. Describe the use of two metering devices in one heat pump. *In most heat pumps, two metering devices are used—one for cooling and the other for heating.*
5. Explain how single-direction filter driers can be used in a reverse-flow heat pump system. *A single-direction filter is a standard filter drier. When standard filter driers are used, typically two are needed in conjunction with check valves. Each check valve only allows the flow of liquid refrigerant in one direction through the filter drier.*
6. Describe the construction of a biflow filter drier. *A biflow filter looks, from the outside, like a single filter drier. Inside, there are two driers and each has a check valve. Each check valve operates to divert the flow of liquid refrigerant through a single filter drier.*
7. Describe the purpose and operation of an accumulator. *The purpose of an accumulator is to protect the compressor from liquid returning in the suction line.*
   It operates as a tank to collect liquid refrigerant by allowing the velocity of the refrigerant to slow and the heavy liquid to drop to the bottom of the tank. Only vaporous refrigerant at the top of the tank is allowed to return to the compressor.

Chapter 7

1. Describe one type of high-start-torque motor. *Capacitor start, capacitor run (CSCR) motors are considered high-start-torque motors and use a start capacitor to help the motor start against a force or pressure.*
2. Describe one type of low-start-torque motor. *A permanent split capacitor (PSC) is considered a low-start-torque motor and can only start against a small force. The run capacitor is used to help start the motor in the correct direction and to improve the run performance.*
3. List 3 types of motor-starting devices. *Potential relay, current relay, and PTC starters are the three types of motor-starting devices.*
4. Explain why blowers and fans use a certain type of motor, and name one motor type. *Blowers and fans are not required to start against a pressure. These devices only need to start and rotate in the same direction. Motors used on fans and blowers are low-start-torque motors and could be PSC motors.*
5. Describe how to identify common, start, and run motor terminals. *Common, start, and run are terminals of a split-phase motor. To find common, test all combinations of two terminals to find the highest resistance. The terminal not being tested is common. Next, test from common to the other two terminals. The lowest resistance is between common and run. The terminal left over is the start terminal.*
6. Describe how to electrically check motor windings. *Four tests need to be conducted: amperage, short to ground, continuity, and motor protection. Both FLA and LRA amperages should be taken and compared to the motor nameplate. Short to ground is testing from the motor terminals to the motor housing. Any reading is considered a short. Continuity tests for open windings. If there is no continuity between common and run, the motor may be internally protected. If the motor is hot, letting it sit and cool may reset the motor protector.*
7. Explain what "ECM" stands for and how these motors work. *ECM stands for electronically commutated motor. Electronic components are used to determine the position of the rotor. This type of motor can start slowly and ramp up speed to meet the load. ECMs have more efficient electrical usage and are found on high-efficiency equipment.*

Chapter 8

1. Explain the differences between reciprocating, rotary, and scroll compressors. *Reciprocating compressors use a crank and piston. Rotary compressors use fewer moving parts and one continuous movement to compress vapor. Scroll compressors have the fewest moving parts and operate at higher efficiencies.*

2. Describe how compressors can be staged. *Two or more compressors can be used (in parallel) on one system. Each compressor is brought on as the need for extra capacity is required. The staging process includes: cylinder unloading; varying the speed of the compressor; and/or using multiple compressors.*
3. Explain how compressors are "unloaded." *Unloading is a way to reduce compressor capacity. It may mean holding open a valve or it may be opening a set of bypass ports in a scroll compressor. Compressors start unloaded and operate at reduced capacity during times of low need to reduce energy consumption and avoid short-cycling periods.*
4. Describe how a two-speed compressor operates electrically. *Two-speed compressors are built to operate as either four-pole or two-pole motors. In four-pole operation they run at lower RPMs.*
5. Describe how a variable-speed compressor operates. *Variable compressors change the speed of the electric motor. There are several ways that speed control is accomplished. As the speed is reduced the energy consumption is reduced, and the capacity of the heat pump is reduced to match reduced need.*
6. Describe how to conduct a compressor efficiency check. *Compressor efficiency checks are done in relationship to manufacturer specifications. The heat pump manufacturer provides operating information under various conditions. The technician compares the manufacturer's data with current operating measurements to determine if the system is operating to efficiency standards.*

Chapter 9

1. Describe the reason why air-source heat pumps (ASHPs) require defrost systems. *An ASHP removes heat from outdoor air that may contain high levels of moisture. Moisture condenses on the outdoor coil in the form of frost. Frost accumulation acts as an insulator to heat movement and must be removed by defrosting the coil.*
2. Describe how the outdoor coil is warmed to melt frost. *The method used by manufacturers is hot gas defrost. This is accomplished by placing the heat pump in the cooling mode, sending hot compressor discharge vapor to the outdoor coil.*
3. Describe the reason why certain operations need to occur to have a good defrost cycle. *Six operations need to occur to have good defrost. One, defrost needs to be initiated. Two, the outdoor fan is shut down. Three, the reversing valve is placed in the cooling mode. Four, the indoor blower needs to remain on. Five, the air must be heated if necessary to reduce cold air. Six, the defrost cycle is terminated.*
4. Describe the operation of a pressure-initiated/temperature-terminated defrost system. *Air pressure difference is measured across the outdoor coil. A pressure difference of .4 inches of water column or more means that the coil is becoming blocked, presumably with frost. The pressure difference will initiate the defrost cycle. As the temperature of the outdoor coil rises, the temperature terminator will terminate the defrost cycle when temperatures rise above 40°F.*
5. Explain how electronic controls have helped improve heat pump defrost. *Electronic circuit boards and sensors have taken the place of mechanical controls and are more accurate and less troublesome than the old controls. Electronic controls have improved the reliability and efficiency of heat pump systems by reducing defrost time.*
6. Describe the benefits of demand defrost controls. *Demand defrost systems can measure and store conditions that can be used to modify the initiation, cycle, and termination of heat pump defrost. By being able to modify the conditions of defrost, the demand defrost system is able to meet the needs of defrost as the outdoor conditions change. This ability improves the heat pump's efficiency and reliability.*

Chapter 10

1. Describe what is meant by the term "electromechanical" and identify those types of components. *Electromechanical means that there are open contacts being closed by an electromagnetic coil. Examples of electromechanical devices are: contactor, compressor motor, solenoid, and relay.*
2. Describe the use of a wiring diagram legend. *The wiring diagram legend provides the full name or a description of devices that are labeled with code letters or numbers.*
3. Name the five similar electrical circuits that heat pumps share with split air conditioning systems. *(1) Compressor, (2) Outdoor fan, (3) Indoor blower, (4) Compressor starting controls, (5) Temperature controls*
4. Explain the function of the crankcase heater. *The crankcase heater warms the crankcase, preventing migration and dilution of crankcase oil with refrigerant.*
5. List the four modes of operation for air-source heat pumps. *The four modes of operation are heating, cooling, defrost, and emergency heat.*
6. Relate why a hard-start kit might be used and describe its operation. *If a compressor is required to start while under pressure (short cycle), a hard-start kit is used. The hard-start kit contains a start relay and start capacitor to help the compressor start under difficult circumstances.*

7. Describe three benefits that electronic circuits provide for heat pumps. *(1) A microprocessor may have visual indicators that show the condition of operation and often show faults. (2) The microprocessor may have a self-diagnostic feature that can tell if the test circuit is still shorted. (3) Electronic circuits do not have open contacts that can wear out. (4) The electronic circuit can have built-in timing capabilities that can be used to prevent the compressor from starting too soon.*

Chapter 11

1. Describe the reason for pre-planning the heat pump installation. *Pre-planning involves customer-specific requirements, building load calculations, system design, and equipment selection.*
2. Describe the importance of the manufacturer's clearance recommendations. *Clearance recommendations are a requirement for equipment operation and are necessary to perform installation and service work. In some cases distances need to be increased to allow for a comfortable service area.*
3. Relate how duct leakage can affect heat pump performance. *Outside air leaking into the ducting system on the return side or leaking out of the supply ducting will reduce the heat pump's performance.*
4. Describe how sound is transmitted through metal ductwork. *Metal ductwork can transmit sound by conduction through the entire length of the duct run. It can also reflect (attenuate) sound waves throughout the ducting system if it is not lined.*
5. Explain how to measure airflow from a diffuser. *Airflow is measured in fpm as an average across the entire surface of the diffuser. The average fpm is multiplied by the Ak (free area in square feet) number provided by the manufacturer to determine the volume of air being supplied in CFM.*
6. Describe how the ACCA Quality Installation (QI) standards can be used. *The Quality Installation standard is a formal guide for contractors and installers to use. It provides procedures that help to improve the quality of a heat pump installation.*

Chapter 12

1. Describe the need for standards. *In the past, there was no guide to maintenance activities for heat pumps. ACCA has provided a standardized list of inspection tasks that helps to guide service technicians through all of the important heat pump maintenance procedures.*
2. Explain how standards relate to heat pump maintenance. *Though many of the inspection tasks are general in nature, they also apply to heat pumps. Additionally, where specific heat pump checks are required, they are listed separately.*
3. Discuss how the general inspection tasks relate to heat pump outdoor units. *Outdoor units house the outdoor coil of a heat pump. The general inspection tasks relate to the heat pump outdoor unit through cleaning of the coil and maintenance of the outside electrical and mechanical equipment.*
4. Describe the difference between Quality Maintenance (QM) and Quality Installation (QI) standards. *QM standards are to be used to guide maintenance operations. QI standards have to do with installation and are required if QM standards are to be applied.*
5. Describe what might happen if system controls were neglected in the QI standards. *If the system controls portion of the QI standards are neglected, it is possible that thermostats, for instance, may be misapplied. Having the wrong thermostat or thermostat features would hamper the efficient operation of the heat pump and could affect customer comfort.*

Chapter 13

1. Describe the principles of troubleshooting. *The principles of troubleshooting are: Listen to the customer and obtain information; determine symptoms; list possible causes; rule out possible causes; identify the problem; and verify the problem.*
2. Describe how symptoms lead to possible causes. *For all symptoms there are one or more possible causes. Possible causes can be either mechanical or electrical.*
3. Relate what is to be done with each cause. *Each cause must be either ruled out or identified as the cause.*
4. Describe the "ruling-out" process. *Ruling out is the process of checking for the cause and determining that there is no reason to identify it as the problem. After each cause is ruled out, the next cause is investigated until the problem is identified. The "ruling-out" process is where instrumentation is commonly used.*
5. Explain how verification of a cause is performed. *Once a cause is identified, it must be verified using another form of testing. Verification could be simple visual identification of the cause, or it may require a different test instrument to determine that a particular cause is the actual problem and lead to finding the cause of the problem.*

Chapter 14

1. Describe the importance of conducting a thorough heat gain and loss calculation. *Heat gain and loss take into consideration all of the structural components of a building. If the wrong value or an incorrect size is used, the calculation could be wrong. The correct value is necessary to be able to properly select the heat pump for the cooling load.*

2. Relate how air-source heat pumps (ASHPs) are selected after the heat gain is known. *ASHPs are chosen using the manufacturer's detailed performance data. This data is presented in a table format.*
3. Describe the function of a thermal balance point graph. *The thermal balance point graph is used to determine where system capacity crosses the structure load line. At that intersection, the heat pump can maintain indoor conditions. Below that point, supplemental heat will need to be used to augment heat pump capacity.*
4. Explain why multi-speed or variable speed blowers are beneficial in the removal of moisture for ASHPs. *When heat pumps are sized for the cooling load, the system fan will operate on high speed to maximize system efficiency. More moisture is removed when fan speeds are reduced and the coil becomes colder. This can also drop system efficiency.*
5. Explain how heat is pulled from cold air. *Heat is pulled from outdoor air anytime the outdoor coil is colder than the outdoor air. Heat will move from the relatively warmer outdoor air to the colder outdoor coil. As outdoor temperatures go down, the amount of heat extracted will be reduced.*
6. Describe how to determine and adjust cubic feet per minute (CFM) for an individual room. *Airflow to an individual room is part of the total airflow to the structure. Individual room airflow can be adjusted by measuring the face velocity of the diffuser and using diffuser manufacturer data to determine CFM. Adjustment of a single room may affect the entire system or one other room. It is recommended that all room airflow be checked and balanced at the same time.*

Chapter 15

1. Describe how a split-system heat pump can be installed. At least two different configurations should be compared. *A split-system heat pump can be installed so that the indoor coil is in the basement or attic. Both are connected to a blower or fan to move the air. Attic installations may need additional drain pans to protect against condensate overflow.*
2. Describe where ductless (mini-split) split systems can be installed and their purpose. *A ductless split system is designed to heat/cool a space. The space could be a single room or a small apartment. The indoor unit can be mounted on the floor, wall, or ceiling.*
3. Describe a packaged system and the two different places where it may be installed. *A packaged unit contains both the indoor and outdoor coil, as well as all of the other components. This type of unit can be installed, in one piece, on the ground, through the wall, or on a roof.*
4. Describe a dual-fuel installation and why a fossil fuel might be used. *A dual-fuel system is essentially a heat pump and furnace combination. A fossil-fuel gas furnace might be used for supplemental heat because it may be more economical than electric resistance heat.*
5. Explain what "geothermal" means. *"Geothermal" means heat from the ground. It could take the form of direct geothermal heat, such as a hot spring, or the use of a heat pump to move heat from the ground.*
6. List and describe two methods to transfer heat from the ground or water using a geothermal heat pump. *Two methods of transferring heat from the ground are an open-loop water system and a DX (direct-expansion) system. A heat pump could be used to transfer heat from water pumped from the ground and then discharged to the ground. A heat pump could also be used to pump heat directly from the ground, as with a DX system.*

Chapter 16

1. Describe the difference between geo-power and geo-source systems. *Geo-power systems use high temperature locations in the Earth to generate steam for the production of electrical power. Geo-source describes areas of the ground where lower levels of temperature are used as a heat source.*
2. Describe how a geothermal heat pump transfers heat from the ground to the occupied space. *Heat in the ground is absorbed by a loop that is buried in the ground and filled with a water-based solution, or refrigeration tubing buried in the ground. Heat in the water filled loop is absorbed by refrigerant that transfers it to indoor air with the heat pump.*
3. Relate what ground sources can be used for geothermal heat pumps. *There is only one source of heat, the ground. The soil or ground water can be connected to the heat pump to move source heat from the ground to the living space.*
4. Describe the difference between ground-source heat pumps (GSHPs) and water-source heat pumps (WSHPs). *GSHPs are connected to the ground directly or by using a secondary loop. WSHPs move heat from ground water or a piped building loop to a refrigerant heat exchanger.*
5. Explain the difference between open and closed loops. *An open loop is open to the atmosphere at some point in the system. A closed loop is separated from the atmosphere.*

Chapter 17

1. Describe the difference between an open- and closed-loop system. *A closed-loop system uses the same water solution sealed in a closed loop that is recirculated. An open-loop system uses fresh water from the ground for a "pump and dump" system.*

2. Describe the function and use of geothermal plastic pipe material. *The plastic pipe material is called high-density polyethylene (HDPE) and is used for the outside loop carrying water-based heat transfer solution. It is placed about 6′ deep in the ground or installed vertically in bores.*
3. Relate why the heat transfer solution must be protected from freezing in some installations. *Temperatures of the ground are around 50°F, but can drop to around 40°F when the system is working. The temperature difference of the coil could be another 10°F colder, making the solution cool to the freezing point. The solution needs to be protected from freezing.*
4. Describe why "slinky" systems use less ground area. *The slinky design concentrates the amount of tubing connected to the ground. With this loop arrangement, it is possible to shorten the trench required to bury the coil.*
5. Explain why vertical loops might be used rather than horizontal loops. *Vertical loops do not take up as much room horizontally as horizontal loops. If there is not enough ground surface area, vertical loops may be an option.*

Chapter 18

1. Describe how heat is absorbed and moved by a water-source heat pump (WSHP) in a ground-source application. *Ground water is circulated through the heat exchanger and heat is transferred from the water to the refrigerant. Refrigerant is circulated to the indoor coil, where heat is transferred from the refrigerant to heat air for the occupied space.*
2. Describe why "pump and dump" WSHPs are also referred to as open-loop systems. *WSHPs pull water from wells or other sources that are at atmospheric pressure. Interacting with or being under the influence of atmospheric pressure causes them to be classified as open-loop WSHPs.*
3. Relate how water is used only once in a single-use system. *Water is typically used once, where heat is absorbed, and then the water is discharged. New water is continually pulled into and used in a single-use system.*
4. Describe the relationship between ground-water temperature and ground temperature. *The temperature of the ground is generally the temperature of the ground water. Ground-water temperature and the temperature of the ground have to do with the location from north to south throughout the country.*
5. Explain the difference between water-to-air and water-to-water systems. *These two types of WSHPs use the same outdoor coils, but differ when indoor coils are compared. The water-to-water system uses the same type of coaxial coil on both sides of the system, cooling the water solution rather than air.*
6. Describe how cooling towers can be used with a WSHP. *A cooling tower rejects heat to the outdoor air when the loop temperature gets too high.*

Chapter 19

1. Describe the relationship between feet of head and pressure per square inch gauge (psig). *Feet of head is a measurement of pressure, just like psig. The relationship is for every 1 foot of head, psig will measure .433, or .31 feet of head will equal 1 psig.*
2. Describe the reason for conducting a pressure-drop measurement across the coaxial heat exchanger. *Pressure-drop measurement and the manufacturer's specification sheet will provide the answer to how many gallons per minute are flowing through the heat exchanger.*
3. Relate how temperature difference across the coaxial coil, pressure difference, and flow are related. *Temperature difference (rise or drop) relates to the BTUH being extracted from the loop water, assuming that the flow rate is correct, and flow rate is a function of pressure drop.*
4. Describe what causes frictional loss in a closed loop. *Frictional loss is caused by the length and size of the tubing, and the number and type of fittings that are used. All of these add up to a total frictional loss.*
5. Explain why vertical feet of head is not important for closed loops, but is for open-loop systems. *Vertical rise and fall moves the same weight of water in a closed loop, canceling out vertical feet of head. In an open loop, vertical rise and fall (feet of head) is calculated because the loop is open to the atmosphere.*
6. Describe what components are in a flow center and the function of a flow center. *The flow center is made up of pumps, valves, connectors, and electrical components. Flow centers are responsible for moving the water in the loop, from the ground to the refrigerant-to-water heat exchanger and back to the ground.*

Chapter 20

1. Describe why a defrost system is not necessary for a geothermal heat pump. *Defrost is not necessary because the outdoor coil is in the ground, where year-round temperatures are nearly constant. There is no frost to get rid of.*
2. Describe the construction of a coaxial heat exchanger. *The coaxial heat exchanger is described as a tube within a tube. Refrigerant is contained in the outer tube and surrounds the inner tube. The inner tube contains water.*
3. Describe the construction of a shell-and-coil heat exchanger. *The shell-and-coil heat exchanger is similar to the coaxial exchanger in operation, but the outside shell surrounds the entire coil. Refrigerant in the tank is circulated over the coil and water is circulated inside the coil.*
4. Explain the concept of counter-flow heat exchange. *Counter-flow describes the movement of two fluids in opposition—one flowing in the opposite direction of the other. It is a method used to maintain an equal*

temperature difference from one end of the heat exchanger to the other and generally is done with a coaxial-type exchanger.

5. Describe the difference between open- and closed-loop water-source heat pumps (WSHPs). *An open loop receives water directly from the ground. An expansion tank and water well pump are used to supply the water. Closed-loop systems use water that is continually circulated in a ground loop.*
6. Explain the concept of "approach" temperature. *Approach describes the amount of temperature difference between the refrigerant temperature and the cooled or heated medium (air or water). It generally describes how well heat is exchanged between the refrigerant and the medium.*
7. Describe what happens to resistance heat on a third-stage call for heat (use the wiring diagram in this chapter). *On the third stage of heating, the electric resistance heaters are cycled on when signaled by the thermostat closing from R to W and through the microprocessor to the PCB. Each heater is cycled on with a delay period preceding each heater being powered.*

Chapter 21

1. Describe what surface considerations affect geothermal loop installation. *Surface conditions include all of the obstructions that may affect the installation of a ground coil. Things such as trees, pools, and buried services will affect the placement and the amount of space available for a ground-loop installation.*
2. Describe the difference between closed-loop and open-loop installation. *Closed-loop installations require many square feet of ground surface in order to bury the coil. Open-loop systems may use a well, river, or lake which requires no more space than normal construction.*
3. Relate the importance of flushing and pressurizing the ground loop. *Flushing is necessary to remove the dirt and debris that could plug the loop and the heat exchanger. Pressurizing is necessary to check for leaks.*
4. Describe the fluid-charging operation for a ground loop. *Charging involves exchanging the flush water with an antifreeze solution. After replacing the water, a static pressure is applied to the system that will act as a working pressure.*
5. Describe the reason why thermal balance point is of lesser concern with geothermal heat pumps. *Geothermal heating systems are less concerned with thermal balance points than air-source systems. Many geothermal systems are installed in northern climates where the ability to heat is more important than cooling. In this situation, supplemental heat is controlled with the indoor thermostat and is brought on as the last stage of heat.*

Chapter 22

1. Describe what makes a geo-system different to maintain. *Instead of an air coil located outside, the geo-system has a water coil or draws water from the ground.*
2. Describe the checks performed on the contactor. *The contactor has to be checked for contact wear, operation, and operator amp draw.*
3. Relate why voltages and amperages are recorded. *Voltage and amperage records of the motors and compressors are a record of operation. Reading them over a long period of time can indicate trends in operation. Increases in amperage are a sign of trouble and would need to be checked out.*
4. Describe what to listen for when checking a pump. *The pump should sound normal—a typical motor sound. Any other sounds are a sign of trouble and should be checked.*
5. Describe the function of the indoor coil during the heating mode. *The indoor coil functions as the condenser in the heating mode. The condenser rejects heat to the indoor air to heat occupied spaces.*
6. Explain why the outdoor coil may have to be checked more often for an open-loop system. *The outdoor coil (source coil) in an open-loop system is subjected to continually new, mineral-laden water. These minerals can deposit inside the coil and cause problems with heat exchange.*
7. Describe what a desuperheater is and what needs to be maintained. *A desuperheater is a coil that heats domestic hot water (DHW). It looks like and works in the same way as a source coil on an open-loop system. Therefore it requires the same type of maintenance.*

Chapter 23

1. Describe user-level troubleshooting. *User-level troubleshooting starts with experiencing the system's operation and using the control system. Most users interface with the thermostat, which, depending on type, may provide the user with basic or more sophisticated information.*
2. Describe how a thermostat might send a message for service directly to the service company. *Some electronic thermostats have the ability to send messages over the Internet to alert the service company of pending problems with the system. This can be done to eliminate system downtime due to a mechanical or electrical problem.*
3. Relate what observational issues are part of owner maintenance. *The owner is the first source of diagnostic information and may be directed to perform basic troubleshooting when the system exhibits certain problems. A lock-out, for instance, may require the system to be reset—a job that the user can perform.*

4. Describe the relationship between electronic modules and the microprocessor. *Most electronic modules are wired to the microprocessor and provide information to the microprocessor. Signals sent from the module can be directed through the microprocessor to provide information on the system or thermostat display.*
5. Explain how electronic modules and microprocessors are used in troubleshooting. *Electronic modules and microprocessors generate LED signals in the form of flash codes that provide troubleshooting information.*
6. Describe the difference between mechanical and electrical troubleshooting. *The difference between mechanical and electrical troubleshooting is that most electrical troubleshooting is done with volt/ohm/amp meter instrumentation.*

Chapter 24

1. Define *coefficient of performance* (COP). *COP is a ratio of heat energy moved to the amount of electrical energy used. The result is a single number that can be used to compare one system to another.*
2. Describe how COP is used. *The number that results from a COP calculation can be used to compare system operating costs.*
3. Define *energy efficiency ratio* (EER). *The EER is a calculation that divides the amount of heat energy moved in BTUH by watts. The number is used to compare system energy efficiency.*
4. Describe the difference between seasonal energy efficiency ratio (SEER) and heating seasonal performance factor (HSPF). *The difference between SEER and HSPF is that SEER is used for cooling systems and HSPF is used for heating systems.*
5. Describe where annual fuel utilization efficiency (AFUE) is used. *AFUE is used to relate energy efficiency for devices that convert fossil fuel to heat energy. It represents the amount of heat energy that can be used for useful heating.*
6. How does return on investment (ROI) affect purchasing decisions? *ROI answers the question: How long will it take for the money used for improvements to be paid back in energy savings?*
7. Relate how comparing systems may help with purchasing decisions. *Comparing systems results in purchasing energy-efficient products that can fit any budget.*
8. How can incentives help in purchases of systems and how can they be found? *Incentives can help to reduce the initial cost or offset the operation costs of purchased systems. Incentives can be found on local, state, and federal websites. They can also be found through manufacturers and energy suppliers.*

Index

Italic page numbers indicate material in tables or figures.